TOOLS AND THE ORGANISM

TOOLS AND THE ORGANISM

Technology and the Body in Ancient Greek and Roman Medicine

Colin Webster

THE UNIVERSITY OF CHICAGO PRESS
CHICAGO AND LONDON

The University of Chicago Press, Chicago 60637
The University of Chicago Press, Ltd., London

Published 2023
Printed in the United States of America

32 31 30 29 28 27 26 25 24 23 1 2 3 4 5

ISBN-13: 978-0-226-82877-0 (cloth)
ISBN-13: 978-0-226-82878-7 (e-book)
DOI: https://doi.org/10.7208/chicago/9780226828787.001.0001

Library of Congress Cataloging-in-Publication Data

Names: Webster, Colin (Classicist), author.
Title: Tools and the organism : technology and the body in ancient Greek and Roman medicine / Colin Webster.
Description: Chicago : The University of Chicago Press, 2023. | Includes bibliographical references and index.
Identifiers: LCCN 2023008555 | ISBN 9780226828770 (cloth) | ISBN 9780226828787 (ebook)
Subjects: LCSH: Medicine, Greek and Roman. | Human body (Philosophy)—History. | Philosophy, Ancient.
Classification: LCC R138 .W43 2023 | DDC 610.938—dc23/eng/20230419
LC record available at https://lccn.loc.gov/2023008555

♾ This paper meets the requirements of ANSI/NISO Z39.48-1992 (Permanence of Paper).

Contents

Abbreviations

Aëtius (Aët.)
Aelian (Ael.)
Nat. anim. ▸ *De natura animalium/Nature of Animals*
Anonymus Londinensis (*Anon. Lond.*)
Anonymus (Anon.)
Morb. Ac. et Chr. ▸ *De morbis acutis et chroniis/On Acute and Chronic Diseases*
Aristocritus (Aristocr.)
Theos. ▸ *Theosophy*
Aristophanes (Ar.)
Eccl. ▸ *Ecclesiazusae/Assembly Women*
Fr. ▸ *Fragments*
Plut. ▸ *Plutus/Wealth*
Ran. ▸ *Ranae/Frogs*
Aristotle (Arist.)
Cael. ▸ *De caelo/On the Heavens*
De an. ▸ *De anima/On the Soul*
De motu an. ▸ *De motu animalium/On the Movement of Animals*
Gen. an. ▸ *De generatione animalium/On the Generation of Animals*
Hist. an. ▸ *Historia animalium*
Juv. ▸ *De juventute et senectute/On Youth and Old Age*
Metaph. ▸ *Metaphysica/Metaphysics*
Mete. ▸ *Meteorologica/Meteorology*
Part. an. ▸ *De partibus animalium/Parts of Animals*
Ph. ▸ *Physica/Physics*
Poet. ▸ *Poetica/Poetics*
Pol. ▸ *Politica/Politics*
Pr. ▸ *Problemata/Problems*
Resp. ▸ *De respiratione/On Breathing*
Rhet. ▸ *Ars Rhetorica/Rhetoric*

Sens. ▸ *De sensu et sensibilibus/On Sense and Sensible Things*
Somn. ▸ *De somno et vigilia/On Sleep and Wakefulness*
Athenaeus (Ath.)
Deipnosophistae
Athenaeus Mechanicus (Athen. Mech.)
Caelius Aurelianus (Cael. Aur.)
Tard. Pass. ▸ *Tardae passiones/De morbis chronicis/Chronic Affections*
Acut. Pass. ▸ *Celeres passiones/De morbis acutis/Acute Diseases*
Calcidius (Calcid.)
In Tim. ▸ *In Platonis Timaeum commentarius/Commentary on Plato's* Timaeus
Cassius Iatrosophista (Cass. Iatr.)
Pr. ▸ *Problemata/Problems*
Celsus
Med. ▸ *De medicina/On Medicine*
Censorinus (Cens.)
Die nat. ▸ *Dies natalis/On Birthdays*
Cicero (Cic.)
Att. ▸ *Epistulae ad Atticum/Letters to Atticus*
De or. ▸ *De oratore/On Oratory*
Clement of Alexandria (Clem. Al.)
Protr. ▸ *Protrepticus*
Strom. ▸ *Stromata*
CMG ▸ *Corpus Medicorum Graecorum*
Dio Chrysostom (Dio Chrys.)
Or. ▸ *Orationes/Discourses*
Diodorus Siculus (Diod. Sic.)
Diogenes of Apollonia (Diog. Ap.)
Diogenes Laertius (Diog. Laert.)
Dionysius of Halicarnassus (Dion. Hal.)
Ant. Rom. ▸ *Roman Antiquities*
Dioscorides (Dioscor.)
MM ▸ *De materia medica*
DK ▸ Hermann Diels and Walther Kranz, eds. (1903) 2018. *Die Fragmente der Vorsokratiker*. New York: Cambridge University Press.
Empedocles (Emped.)
Frontinus (Frontin.)
Aq. ▸ *De aquaeductu urbis romae/On the Water Supply of the City of Rome*
Galen (Gal.)
AA ▸ *De anatomicis administrationibus/Anatomical Procedures*
Ad. Lyc. ▸ *Adversus Lycum/Against Lycus*

Aff. Dig. ▸ *De propriorum animi cuiusque affectum dignotione et curatione; De animi cuiuslibet peccatorum dignotione et curatione/The Diagnoses and Treatments of Affections and Errors*

Art. Sang. ▸ *An in arteriis natura sanguis contineatur/On Whether Blood is Naturally Contained in the Arteries*

At. Bil. ▸ *De atra bile/Black Bile*

Comp. Med. Gen. ▸ *De compositione medicamentorum per genera/The Composition of Drugs According to Kind*

Comp. Med. Loc. ▸ *De compositione medicamentorum secundum locos/The Composition of Drugs According to Places*

Const. Art. Med. ▸ *De constitutione artis medicae/The Composition of the Art of Medicine*

[*Def. Med.*] ▸ *Definitiones medicae/Medical Definitions*

Di. Dec. ▸ *De diebus decretoriis/Critical Days*

Diff. Puls. ▸ *De differentiis pulsum/The Distinct Types of Pulse*

Dig. Puls. ▸ *De dignoscendis pulsibus/Diagnosing the Pulses*

Elem. ▸ *De elementis ex Hippocrate/Elements According to Hippocrates*

Foet. Form. ▸ *De foetuum formatione/On the Formation of the Foetus*

Hipp. Aph. ▸ *In Hippocratis Aphorismos/Commentary on Hippocrates's* Aphorisms

Hipp. Art. ▸ *In Hippocratis Articulos/Commentary on Hippocrates's* Joints

Hipp. Epid. VI ▸ *In Hippocratis Epidemiarum/Commentary on Hippocrates's Epidemics*

[*Hist. phil.*] ▸ *Historia philosophica/History of Philosophy*

In Hipp. Nat. Hom. ▸ *In Hippocratis Naturam Hominis/Commentary on Hippocrates's On the Nature of the Human*

Ind. ▸ *De indolentia/Περὶ Ἀλυπίας/Avoiding Distress*

Inst. Od. ▸ *De instrumento odoratus/The Organ of Smell*

[*Int.*] ▸ *Introductio seu medicus/Introduction*

Lib. Prop. ▸ *De libris propriis/My Own Books*

Loc. Aff. ▸ *De locis affectis/Affected Places*

Med. Exp. ▸ *De experientia medica/Medical Experience*

MM ▸ *De methodo medendi/Method of Healing*

Nat. Fac. ▸ *De facultatibus naturalibus/Natural Faculties*

Opt. Med. Cogn. ▸ *De optimo medico cognoscendo/Discovering the Best Physician*

Ord. Lib. Prop. ▸ *De ordine librorum propriorum/The Order of My Own Books*

PHP ▸ *De placitis Hippocratis et Platonis/The Opinions of Hippocrates and Plato*

Praec. ▸ *De praecognitione ad Epigenem/Prognosis, for Epigenes*

Prop. Plac. ▸ *De propriis placitis/My Own Opinions*

Ptis. ▸ *De ptisana/Barley Gruel*

San. Tu. ▸ *De Sanitate Tuenda/On the Preservation of Health*

Sect. ▸ *De sectis ad eos qui introducuntur/On Sects for Beginners*)

Sem. ▸ *De semine/Semen*

SMT ▸ *De simplicium medicamentorum temperamentis ac facultatibus/The Properties of Simple Drugs*

Syn. Puls. ▸ *Synopsis de pulsibus/A Synopsis of the Pulse*

Temp. ▸ *De temperamentis/Mixtures*

[*Ther. Pis.*] ▸ *De theriaca ad Pisonem/Theriac, for Piso*

Thras. ▸ *Thrasybulus sive Utrum medicinae sit aut gymnasticae hygiene/Thrasybulus, whether Hygiene Belongs to Medicine or Physical Training*

Trem. Palp. ▸ *De tremore, palpitatione, convulsione, et rigore/Tremor, Spasm, Convulsion, and Shivering*

UP ▸ *De usu partium/On the Function of the Parts*

Us. Puls. ▸ *De usu pulsuum/On the Function of the Pulse*

Us. Resp. ▸ *De usu respirationis/On the Function of Breathing*

Ut. Diss. ▸ *De uteri dissectione/On the Anatomy of the Womb*

Ven. Sect. Er. ▸ *De venae sectione adversus Erasistratum/Bloodletting, Against Erasistratus*

Herodotus

Hist. ▸ *Histories*

Herophilus (Heroph.)

Heron

Aut. ▸ *De automatis/Automata*

Def. ▸ *Definitiones/Definitions*

Pneum. ▸ *Pneumatica/Pneumatics*

Hippocrates (Hipp.)

Acut. ▸ *De victu acutorum/On Regimen in Acute Diseases*

Aer. ▸ *De aere, aquis, locis/Airs, Waters, Places*

Aff. ▸ *De affectionibus/Affections*

Art. ▸ *De arte/On the Art*

Artic. ▸ *De articulis/On Joints*

Alim. ▸ *De alimento/Nutriment*

Anat. ▸ *De anatome/Anatomy*

Aph. ▸ *Aphorismi/Aphorisms*

Carn. ▸ *De carnibus/Flesh*

Cord. ▸ *De corde/On the Heart*

Decent. ▸ *De decenti habitu/Decorum*

Ep. ▸ *Epistulae/Letters*

Epid. ▸ *De morbis popularibus/Epidemiae/Epidemics*

Flat. ▸ *De flatibus/Winds*

Fract. ▸ *De fracturis/Fractures*

Genit. ▸ *De genitura/Generation*

Gland. ▸ *De glandulis/Glands*

Haem. ▸ *De haemorrhoidibus/Haemorrhoids*

Int. ▸ *De internis affectionibus/Internal Affections*

Iusi. ▸ *Iusiurandum/The Oath*

Loc. Hom. ▸ *De locis in homine/Places in the Human*

Medic. ▸ *De medico/Physician*

Morb. 1 ▸ *De morbis* 1/*Diseases* 1

Morb. 2 ▸ *De morbis* 2/*Diseases* 2

Morb. 3 ▸ *De morbis* 3/*Diseases* 3

Morb. 4 ▸ *De morbis* 4/*Diseases* 4

Morb. Sacr. ▸ *De morbo sacro/On the Sacred Disease*

Mul. 1, 2, and 3 ▸ *De morbis mulierum/De mulierum affectibus/Diseases of Women* 1, 2, and 3

Nat. Hom. ▸ *De natura hominis/Nature of the Human*

Nat. Mul. ▸ *De natura muliebri/Nature of Woman*

Nat. Pue. ▸ *De natura pueri/Nature of the Child*

Oss. ▸ *De ossium natura/Nature of Bones*

Pharm. ▸ *Peri pharmakon/On Drugs*

Prog. ▸ *Prognosticum/Prognostic*

Salubr. ▸ *De salubri diaeta/Regimen in Health*

Ulc. ▸ *De ulceribus/Wounds*

Vict. ▸ *De victu/De diaeta/Regimen*

VM ▸ *De vetere medicina/On Ancient Medicine*

Hippolytus (Hippol.)

Haer. ▸ *Refutatio omnium haeresium/Refutation of All Heresy*

Homer (Hom.)

Il. ▸ *Iliad*

Od. ▸ *Odyssey*

Hom. Hymn Dem. ▸ *Homeric Hymn to Demeter*

Iamblichus (Iambl.)

Myst. ▸ *De mysteriis/On the Mysteries*

LM ▸ Laks, André, and Glenn W. Most, eds. (2016). *Early Greek Philosophy*, vols. I–IX. Loeb Classical Library. Cambridge, MA: Harvard University Press.

LSJ ▸ Liddell, George Henry, Robert Scott, and Henry Stuart Jones, eds. (1940). *Greek-English Lexicon*, with a revised supplement, 9th ed. Cambridge: Cambridge University Press.

Ibn Abī Uṣaybiʿah

Uyūn al-anbāʾ ▸ *Uyūn al-anbāʾ fī ṭabaqāt al-aṭibbā/Literary History of Medicine*

Lucretius (*Lucr.*)

DRN ▸ *De rerum natura/On the Nature of Things*

Macrobius (Macrob.)

In Somn. ▸ *Commentarii in somnium Scipionis/Commentary on the Dream of Scipio*

Musonius Rufus (Mus. Ruf.)

Nicophon

Fr. ▸ *Fragments*

Oribasius

Coll. Med. ▸ *Collectiones medicae/Medical Collections*

Philo

Pneum. ▸ *Pneumatica/Pneumatics*

Plato (Pl.)

Phdr. ▸ *Phaedrus*

Plt. ▸ *Politicus*

Prt. ▸ *Protagoras*

Rep. ▸ *Republic*

Tht. ▸ *Thaeatetus*

Ti. ▸ *Timaeus*

Pliny (Plin.)

HN ▸ *Historia naturalis/Natural History*

Plutarch (Plut.)

Adv. Col. ▸ *Adversus Colotem/Against Colotes*

De Amic. ▸ *De amicorum multitudine/Having Many Friends*

Per. ▸ *Pericles*

Prim. frig. ▸ *De primo frigido*

Quaest. conv. ▸ *Quaestiones convivales*

Polybius (Polyb.)

Hist. ▸ *Historiae/Histories*

Porphyry (Porph.)

Iliad ▸ *Quaestiones Homericarum ad Iliadum pertinentium reliquiae/Homeric Questions on the Iliad*

Praxagoras (Praxag.)

Pseudo-Aristotle ([Arist.])

Sud. ▸ *De sudore/On Sweat*

Pseudo-Iamblichus (Ps.-Iambl.)

Theol. ▸ *Theologoumena arithmeticae*

Pseudo-Plutarch (Ps.-Plut.)

Plac. ▸ *Placita philosophorum*

Strom. ▸ *Stromateis/Miscellanies*

Pseudo-Socrates ([Soc.])

Ep. ▸ *Epistulae/Letters* (*Epistolographi Graeci*)

Rufus of Ephesus (Ruf. Eph.)

Anat. ▸ *De partibus corporis humani/On the Anatomy of the Parts of the Body*

Onom. ▸ *De corporis humani appellationibus/On the Names of the Parts of the Body*

Scholia in Euripidem (*Schol. in Eur.*)

Seneca

Ep. ▸ *Epistulae/Letters*

Sextus Empiricus (Sext. Emp.)

Math. ▸ *Adversus mathematicos/Against the Logicians*

Pyr. ▸ *Πυρρώνειοι ὑποτυπώσεις/Outlines of Pyrrhonism*

Simpl.

In Cael. ▸ *In Aristotelis de caelo commentaria/Commentary on Aristotle's* On the Heavens

In Phys. ▸ *In Aristotelis Physicorum libros quattuor posteriors commentaria/Commentary on Aristotle's* Physics

Soranus (Sor.)

Gyn. ▸ *Gynaeciorum libri iv/Gynecology*

[*Quaest. med.*] ▸ *Quaestiones medicinales/Medical Questions*

Vita Hippocr. ▸ *Vita Hippocratis/Life of Hippocrates*

Sophocles (Soph.)

Fr. ▸ *Fragments*

Stobaeus (Stob.)

Anth. ▸ *Anthologium/Anthology*

Tatian (Tat.)

Or. ▸ *Oratio ad Graecos/Oration to the Greeks*

Tertullian (Tert.)

De an. ▸ *De testimonio animae*

Theophrastus (Theophr.)

Hist. pl. ▸ *Historia plantarum/Inquiry into Plants*

Sens. ▸ *De sensibus*

Themistius (Them.)

In Phys. ▸ *In Aristotelis Physica/On Aristotle's* Physics 1–3

Vitruvius (Vitr.)

De arch. ▸ *De architectura/On Architecture*

Notes on Translations, Names, Citations, and Editions

All translations are my own, except where otherwise indicated.

After much consultation and internal debate, I have used the more common Latinized versions of Greek names. Despite my own ideological objections to the compression of the ancient Greek world into its Roman reception, I have done this so as not to alienate nonspecialist audiences, who might be unnecessary perplexed by references to Hippokrates and Klearkhos rather than Hippocrates and Clearchus. As for the titles of texts themselves, I generally use an English translation, except where doing so would add confusion rather than reduce it; and when a Latinized version is in wide and common usage, I have kept it.

For all *physikoi*, I cite the Diels-Kranz passages, followed by the Laks and Most Harvard Loeb edition fragment numbers and the original author and work from which the passage was excerpted. While expanding these citations considerably, this allows for both specialists and nonspecialists to find these references without consulting multiple editions.

When citing Hippocratic texts, I have consulted modern scholarly editions wherever possible, although I have cited the divisions presented within the Loeb Harvard editions, since these are the most readily accessible editions for most English readers. Each Hippocratic citation is also cross-referenced with the Littré editions to ease finding passages within critical editions in various languages. My citations also employ only arabic numerals. What each of these numerals refers to depends on the individual text, but citations always operate from largest to smallest divisions, i.e., book, chapter/section, paragraph/entry. For example, the third entry in the first section of *Epidemics* 6 would be cited as *Epid.* 6.1.3, whereas the second chapter of *Places in the Human*, which does not have book divisions, would be cited as *Loc. Hom.* 2.

For Galen, I have again used Loeb citations where possible, or used section divisions seen in the most common English translations, but since these translations are not always obvious or readily available, I have generally indicated which edition I have relied upon. A list of current critical editions and translations can be found in Nutton 2020. All Galenic citations are cross-referenced with the Kühn volumes where possible for clarification, except where no Kühn volume is available. Citations for Galen likewise use only arabic numerals according to the same descending divisions (book, chapter, paragraph).

For other medical authors, I cite the collection or edition I have used when relevant, and these editions can be found in the general bibliography.

Introduction

0.1 TECHNOLOGIES AND THE CONSOLIDATION OF THE BODY

No narrative of the future would be complete without bodies augmented or integrated with technologies. Synthetic amalgams of machine and animal, robot and human, and computer and consciousness form key signs in the semiotics of our imagined futures. Our literary and visual imaginations remain fascinated with technologies infiltrating corporeal structures, mimicking creatures, and synthesizing life. Of course, this techno-future does not belong to some upcoming era but has already exploded our present. Modern lives teem with interfaces, wearables, and websites, such that knowledge of our own bodies arrives heavily mediated by tools, devices, and apparatuses. Even the standard medical encounter starts with a blood-pressure monitor and pulse oximeter, and any ailments that require specialized diagnosis involve higher echelons of technologies and techniques. Even as I write this, my watch tells me that I have failed to exercise sufficiently and that I have not stood frequently enough to satisfy its demands. Tools mediate our understanding of corporeality as well as forming integral parts of our own embodied experiences.

The aura of futurity emanating from such notions can provoke questions about the relationship between technologies and medicine in the deep past. How did tools shape notions of the body before X-rays, MRIs and the marriage of diagnostics and machines? How did technologies interface with humans at both physical and conceptual levels in a world without centrifuges, robot surgeries, or sonograms? This book aims to show that despite a medical environment populated with fewer machines than today, technologies were still crucial to ancient Greek and Roman ideas about corporeality from the very moment the body emerged as the primary object of medical expertise in the fifth century BCE. They continued to be a central part of medical discourse across classical antiquity into the second century CE and beyond, structuring questions about what type of object the body is, what types of substances it contains, and how its parts function. In fact, tools and technologies were essential to the development of the idea that the body functioned in the first place.

It requires some effort to recognize how some of our most basic assumptions about corporeality are indebted to tools. For example, in her *Cyborg Manifesto*, Donna Haraway boldly declared that the microelectronic communications and synthetic hormones of the late twentieth century (and now twenty-first) have produced new corporeal configurations and fresh embodied extensions. She claims that as a result "we are all chimeras, theorized and fabricated hybrids of machine and organism; in short, we are cyborgs."[1] Haraway's cyborg presents a useful image in two ways. On the one hand, even if the opportunities for transformative technological interventions have multiplied, the following chapters will demonstrate that the interpenetration of technologies and bodies occurred in various ways throughout ancient medical contexts, sometimes creating cyborg-like arrangements. On the other hand, Haraway's juxtaposition of "machines and organisms" captures a linguistic oddity that reveals even more clearly how conceptions of tools and bodies did not only start merging in the recent past.[2] In current English usage, "organisms" are living objects, while "organs" are the major parts that compose such beings and perform hierarchically arranged somatic tasks. Despite the associations of the English term with life, growth, and biotic purity, the words "organic" and "organism" both derive from the Greek word *organon*, which means "tool." Calling a living thing an "organism" is already to state that it is tool-like. This book traces the origins of this linguistic flip and the other ways that technologies impacted ancient theories of the body in biomedical discourses.

Although there are many ways to examine the relationship of tools and bodies, this book argues that understanding how technologies influenced medical theories requires traveling along at least three interrelated axes, which we might call the structural, investigative, and analogical. The first follows how the concept of a "tool" began to structure ancient ideas about the corporeal whole and how its component parts relate to one another. This was not always the case. Hippocratic medical authors of the fifth

1. Haraway 1991: 150.

2. We might also consider Canguilhem's 1952 lecture "Machine and Organism," which provides another classic example of treating these categories as ontological opposites, even as he attempts to break down barriers between them; see Canguilhem 2008: 75–97.

century BCE never use the term "tools" [*organa*] to describe inner parts of the body. When they mention the heart, lungs, liver, gallbladder, kidneys, bladder, spleen, and other internal structures that we call organs, they use terms like "viscera" [σπλάγχνα], "places" [τόποι], "spaces" [χωρία], "parts" [μέρεα], or "structures" [σχήματα].[3] Rather than being a simple variance in nomenclature, this terminology reflects substantially different notions about what internal parts do and how they relate to the human whole. Hippocratic bodies generally did not "work," "operate," or "function" as hierarchical physical systems that sustained life. Instead, Hippocratic authors consolidated corporeality in other ways, notably around the concept of an intrinsic "nature" [ἡ φύσις] and the maintenance of a homeostatic balance of humors, all without employing the model of a multipart machine. Nevertheless, the seeds of tool-like functional parts appear in their texts and even predated the Hippocratics, appearing among pre-Socratic natural philosophers such as Empedocles. It was Plato, however, who first explicitly referred to corporeal parts as tools, and Aristotle who expanded this idea by characterizing the body as a tool-like functional object whose parts contributed to the maintenance of life. This view of the body flourished in the Hellenistic period, after Aristotle's death, and in the second century CE, Galen cemented its primacy as the core way to think about corporeality. We still live with this conceptual legacy today.

Recognizing the conceptual impact of organa on ideas about corporeal structures contextualizes a second axis of inquiry, one that emphasizes how material tools, implements, and technological procedures supported and influenced competing medical theories. Notions about the body changed quite dramatically across classical antiquity, and different ancient authors proposed divergent ideas about somatic components, capacities, and behaviors. This book emphasizes that these theories did not develop in an intellectual vacuum but were supported by varying sets of tools and techniques. In fact, as Peter Galison has stressed, certain technologies can embody and imply larger theoretical and methodological commitments.[4] The corollary of this observation is that new theoretical commitments can also imply new tools. Indeed, the rise of the organized body in the Hellenistic period, with its emphasis on functional parts, led to the rise of dissection as an investigative mode. These epistemic practices structured corporeality in certain ways, consolidating the body

3. See Byl 1971; Ioannidi 1981.

4. Galison 1997: 2. For this reason, he states that laboratory machines are "dense with meaning."

around solid components rather than liquids or humors. Tools of dissection therefore encoded a theory about what type of object the body was.

It is also crucial to recognize that dissection and other modes of investigation did not simply reveal previously hidden corporeal features and activities. To be sure, many corporeal behaviors that we now take for granted did not seem obvious to ancient authors, including the assertion that the lungs expand and contract, that the heart beats, or that the blood circulates. Rather than attribute these discrepancies to the fumbles of crude thinkers, we should acknowledge that these features needed to be made visible and, more important, made relevant to models of the body. Yet, as Ian Hacking notes, investigations can structure and stabilize the world to the point that they can create new phenomena.[5] This reciprocal relationship can go even deeper. Scholars from fields as diverse as the history of science, performance studies, and anthropology have made use of Bruno Latour's idea of the body as an "interface." For Latour, the body does not have some essential, unchanging nature but is constructed in an ongoing affective relationship with the objects it encounters. Tools therefore "articulate" it by teaching it to be affected in certain ways.[6] Latour uses the example of a fragrance kit that an aspiring perfumer uses to gain the ability to discern different, hitherto unregistered scents and chemicals. By teaching an individual how to discern these smells, this kit articulates the nose and its capacities in ways that would not have existed without its intervention. Ancient authors, too, use various techniques to "articulate" the substances, properties, and behaviors of the body, especially insofar as many corporeal features rely on some technological interface to be seen. For example, numerous commentators have attempted to identify the corporeal substance that Galen or the Hippocratic author of *Nature of the Human* are describing when they discuss "black bile," since no obvious correlate can be distinguished in bodies as we now experience them. Yet both authors insist that they draw this humor out through the application of certain emetic drugs, especially at certain times of year. Rather than reject their assertions as incorrect observations or sloppy inferences, it is more useful to acknowledge that "black bile" was not a simple corporeal fluid that we can locate and identify. It was a substance that was articulated through a body's interaction with a certain set of therapeutic substances. Of course, the notion that tools articulate, rather than simply reveal, different corporeal entities and behaviors is easier to accept when the substances appearing in these interactions are no longer held to be

5. Hacking 1983: 230.
6. Latour 2004.

real. Nevertheless, the specific contours of any somatic behavior change with each set of technologies used to render it visible.[7] As such, over the course of classical antiquity, what was "manifest" [φαινόμενα] changed according to the shifting set of technologies that produced varying types of visibility. Different concepts of corporeality relied on different tools. Different tools displayed different bodies.

Along with these two vectors, a third remains: tracking the use of specific tools as heuristic analogies for individual body parts. What counts as "observable" is often very hazy, especially in ancient science.[8] More often than not, scholars react to what seems to be an astonishing blindness to certain corporeal features, asserting that ancient authors surely knew about such obvious physical behaviors as the pulse or respiration, mentioned above, and surely must have known about the vital role of the lungs or the heart. After all, people could have held up an ear to their chest and must have seen the rise and fall of the chest cavity during breathing. Moreover, animals were cut apart all the time in sacrifice, and butchers would have an intimate knowledge of the inner animal parts. Yet this book demonstrates that even ideas about basic somatic behaviors were assembled from a tangled web of theoretical commitments, medical observations, and in many cases, technological *comparanda*. In this regard, the phenomena described in medical treatises were always assembled from various places. It is because of this distinction that technologies can have a substantial impact on ideas about corporeal behaviors and processes, not least because of how technological environments can shape assumptions about what is possible and what is natural. To illustrate this dynamic, this book tracks the impact of water-delivery technologies and pneumatics on theories of respiration and the vascular system. Authors from Empedocles to Galen used devices such as the *clepsydra* wine server, the bellows, the force pump, pipes, and aqueducts to conceptualize the corporeal interior. As these technologies developed, theories changed accordingly, as did basic observations about the body's behaviors. Taking this approach reveals that even a formulation as simple as "the chest rises when we breathe in" surfaced through negotiation with technological comparisons and physical theories. Accordingly, this third vector will

7. See Kuriyama (1999), who illustrated the various ways that "the pulse" was construed as a phenomenon within Greek and Chinese medical approaches. For Greek authors such as Herophilus, the pulse was a singular action located in the arteries, while for Chinese physicians, the pulse manifested in disparate ways in different parts of the body depending on the state of health or illness.

8. Lehoux (2012) examines this point well.

emphasize how consequential ancient technological environments were for constructing basic corporeal assumptions and how shifts in these environments from the fifth century BCE to the second century CE contributed to changes in medical explanations.

The entanglement of these three vectors—structural, investigative, and analogical—reveals the deep interconnections between tools and bodies within ancient medical discourses and situates the question of technology at the heart of many Greek and Roman medical theories, beginning with the Hippocratics in the fifth century BCE and extending to Galen in the second century CE and beyond. This multilayered approach is necessary because countenancing the hybridity of investigative and observational techniques reveals how composite an object the body actually is, as each medical theorist created new corporeal configurations as objects of inquiry.[9] To be sure, ancient physicians paid close attention to real bodies that responded in physically determined ways, but "the body" discussed in each medical explanation cannot be considered identical or coterminous with these physical things. Medical theories explain conglomerate entities pulled together from exemplars both human and nonhuman, living and nonliving, that were viewed in different ways and articulated with different tools. This book explores how these multiple objects and observations transform into a single phenomenon. It attends to the messy business of making the objects of medical explanation and the tools used to complete this task.[10]

Part of this claim rests on the insistence that "the body" upon which medical theorists trained their intellectual focus is not a singular, stable physical object waiting to be witnessed by more acute viewers and comprehended by more accurate explanations.[11] On a basic epistemological level, objects of scientific comprehension are always assembled through multiple exemplars or are compiled from multiple cases of what is

9. See also Holmes 2010a, which tracks medical constructions of the body as the recipient of technical expertise.

10. In many ways, this threefold investigation therefore has much in common with the work of Mol (2002), whose anthropology of atherosclerosis examined how a patient experienced disease by privileging practices rather than theories. She notes that as a consequence of this approach, the phenomena under question start to fragment into multiple objects.

11. In this approach I follow Holmes (2010b), who calls the body a "conceptual object" (see esp. 18–19) following Rheinberger's discussion of "epistemic things" (1997: 11–23, 28–31); cf. Holmes 2010a: 84, which distinguishes between bodies as physical objects and "the body that emerges in the classical period as the object of expert knowledge and care."

deemed to be a single phenomenon. Determining what counts as a relevant observation in these contexts thus requires discrimination and selection. Theorists must choose what types of bodies count as informative (animals/humans, male/female/intersex, old/young, idealized/disabled, racialized, etc.), how many individual bodies to examine, what behaviors to include, in which ways they are to be observed or manipulated, whose reports can be trusted, and so on. The phenomena being explained therefore emerge from the apparatus that reflects these decisions, and a singular epistemic object such as "the body" is compiled from heterogenous elements. Different somatic behaviors and corporeal features are made visible by different technical means. In other words, the epistemic object of inquiry emerges only through its interactions with certain tools, skillsets, and larger intellectual apparatuses,[12] such that the body and the behaviors that it displays are the emergent properties of the explanatory apparatus that articulates it. Understanding how technologies impacted ancient medical theories therefore requires examining the dynamics of how these interfaces manifested corporeal phenomena and how these interfaces changed over time.

0.2 TELEOLOGY, MECHANISM, VITALISM, AND TECHNOLOGY

In some regards, the history of the organism can be seen as the introduction of teleology into accounts of the human whole. Indeed, when scholars have attended to how functional accounts of the body developed, they have generally done so through the lens of teleology, using this concept to ground their analysis of how authors attributed intrinsic goals and purposes to natural objects. The problem with merging the history of the *organism* with the history of teleology, however, is that the latter can denote many different relationships.[13] Teleology can be of the cosmic type,

12. Latour 2004.

13. The bibliography on teleology in ancient philosophy is extensive, although most of it focuses on Plato, Aristotle, and Galen because of the importance the concept plays for each of these thinkers. For Plato, most scholarship focuses on the external, cosmic teleology imposed by a divine demiurge on the world, which Lennox (1985) called "unnatural teleology," and the relationship it bears to eternal forms and material necessity (see, e.g., Balme 1987; Johansen 2004: 69–116; Scolnicov 2017). Since teleology and "final causes" play such an important part in Aristotle's metaphysics, physics, and biology, the bibliography extends even further, although two key discourses surround how teleology structures his account of living creatures (see, e.g., Gotthelf 1976; 1985; 2012; Lennox 1982; Gelber 2021)

wherein some divine figure arranges the world to certain purposes. In turn, this description itself can imply several meanings, including that God preordains every single event within some totalizing plan, that Nature assigns each species some role in a broader natural order, or that an Intelligent Designer has arranged some, but not all, features of the world toward particular ends. Yet teleology can also operate internally and immanently as well, implying that individuals should each seek to fulfill their own unique purpose in life, that each species possesses some innate mode of flourishing, or that animals all behave in certain ways because of respective inborn characters.[14] In addition, teleology can describe the operation of innate forces in matter, whether to guide nonliving substances to their proper places in the cosmos, initiate and direct growth in an embryo, or orient and compel the living body toward health. All these teleologies have nuances in turn and do not exhaust the way that "purpose" or "goals" can describe actions, configure the world, or structure its objects. Accordingly, rather than orient this investigation toward teleology, which might then bend our eyes to mythological cosmologies, Anaxagoras's Mind, Empedocles's Love, or theological reflections on divine design, it is more productive to establish the history of the organism on the emergence of a particular type of internal orientation—namely, the tool-like structural teleology that construed the body as composed of organa that each fulfilled some function. This tool-like teleology created an entity that "worked," insofar as its component pieces each completed

---

and whether he believed the cosmos as a whole was teleologically arranged such that individuals and regular natural phenomena served both internal and external goals (see Cooper 1982: 217; 1987; Furley 1985; Sedley 1991; 2007, esp. 194–203; Charles 1991; Wardy 1993; Sharples 2017). For general accounts of Aristotelian teleology, see also Johnson (2005) and Leunissen (2010). More recently, Rocca (2017) has edited a volume wherein teleology is discussed with a slightly wider frame, a discussion that shows how the term can be deployed in both highly specific technical contexts and diffuse broader modes. For example, contributors speak about teleology as benevolent creationism (Sedley 2017; cf. Sedley 2007), atemporal mathematical structure (Scolnicov 2017), Nous-arrangement of the world and its names (Tarrant 2017), arrangement to what is either necessary or merely for the better (Leunissen 2017; cf. Leunissen 2010), or a simple normative biological nature (Craik 2017).

14. It is common to distinguish between two types of teleology, one variously referred to as cosmic/external/intentional/unnatural and another described as individual/immanent/nonintentional/natural. This division is consequential largely insofar as scholars associate the former with Plato and the latter with Aristotle. See, for example, Lennox 2001; Ariew 2002: 8; Sedley 2010; Rocca 2017.

an individuated task. Understanding the physiology of such an object required knowing what each part did and what purpose [τέλος] it served.

To be sure, the emergence of structural, tool-like teleology cannot be divorced from the development of external, cosmic teleology, since the two historically arose together. Nevertheless, the two concepts cannot be collapsed. This book thus follows how ancient teleological biologies were built upon the model of tools. Such an analytic frame helps uncover the complicated role that technologies played in conceptualizing bodies and their components.

As these reflections on teleology suggest, even as this book proposes a new framework for assessing the interactions of technologies and the body in antiquity, it joins a longer legacy of scholars assessing the relationship between the artificial and the natural in ancient scientific discourse. One important thread of such scholarship evaluates whether ancient accounts of living things are "mechanistic" or "vitalist." Charles Wolfe has outlined how discourses within seventeenth- and eighteenth-century iatromechanism established two opposing modalities with which to explain biotic phenomena. On the one hand, the iatromechanists adopted "mechanistic" accounts that explained biotic behaviors solely with recourse to generic physical forces, such that living things abide by the same laws of matter as nonliving entities. On the other hand, "vitalist" arguments, culminating with the Montpellier school, attributed supposedly unique capacities to living tissues that abiotic substances do not possess, including an orientation toward the living whole irreducible to basic physical properties.[15] In this regard, living tissues possessed their own powers that could not be reproduced through technical ingenuity, even hypothetically. A mechanist might therefore explain the seemingly self-guided growth of an embryo by claiming that matter adheres to its inner structures by becoming enmeshed by some physical entanglement as it flows through its tissues. A vitalist might explain this same phenomenon by claiming that living tissues possess a special capacity to attract what they need at any given time, relative to the needs of the whole. For a vitalist, living things grow, whereas nonliving things merely accumulate in piles, while for a mechanist, nothing ontologically distinguishes living from nonliving entities.

Although Wolfe illustrates that these tidy divisions seldom remained unblurred in practice, the categories have proven attractive to scholars looking to interrogate what forces are at work inside living things. For example, scholars such as Iain Lonie have employed these basic categories

15. Wolfe 2021.

to assess ancient medical theories, asking whether certain Hippocratic authors propose explanations of one type or another.[16] Sylvia Berryman maintains the same categories but avoids an ahistorical rendering of them, instead asking whether and how ancient mechanics itself influenced theories of natural objects.[17] Jean De Groot applies a similar logic, arguing that ancient mechanical principles shaped Aristotle's account of animal motion in particular.[18] These investigations are often successful on their own terms, but they spend considerable time adjudicating boundary disputes between the living and the artificial which are sometimes not present in ancient sources.[19] Moreover, as a corollary effect, adopting the mechanist/vitalist frame can occlude certain ways that technologies influenced theories of the body, since this division arranges artificial technologies such as machines and tools on one side while placing biotic, natural things such as plants and animal tissues on the other. Such a dichotomy leverages the fact that gears and machines still operate as paradigmatic technological devices within modern imaginations (even as digital products are displacing this dominance), but it ignores the fact that living things and technologies are not completely separate sets. Rather, the vast majority of tools applied to human bodies, including drugs, foodstuffs, and other therapeutic tools, are made from biotic substances. Addressing how technologies and bodies interacted within medical theories thus requires adopting a definition of technology that does not restrict itself to mechanics and mechanistic devices alone.

The term "technology" has undergone a dramatic shift in common English parlance in the last decade, becoming ever more associated with digital tools and computer-based industries. The general usage of the term now privileges novelty and complexity, so that yesterday's technologies transform into today's infrastructure. Such semantic flux makes establishing an operational definition all the more necessary, and yet the attempts

16. Lonie 1981a; 1981b.

17. Berryman 2009; cf. Berryman 2002b and 2003. See also von Staden (1995; 1996; 1998; 2007), who examines the interactions of pneumatics, mechanics, and medicine.

18. De Groot 2014; cf. de Groot 2008.

19. By contrast, Mayor (2018) traces many Greek and Roman myths in which ancient craftsmen, divine or otherwise, constructed living creatures through technical means, although she does so without attending to the distinction between accounts involving a divinely imbued capacity for self-motion and those that present a living thing as a machine operating according to materially comprehensible forces.

to define technology have been astonishingly varied.[20] While continental European languages still treat "technology" as the science of applied arts, modern English references to "technology" tend to refer to the material products produced by these activities.[21] This book uses both definitions but highlights the latter, on the grounds that understanding how technologies structured ideas about the body in antiquity requires examining actual material tools. It therefore places considerable emphasis on substances, instruments, and artifacts themselves. That is, it treats technologies both as specialized knowledge sets and as the objects produced by the application of such expertise.[22] Technologies therefore include hammers, shovels, scalpels, cement, paper, glass, cranes, boats, and siege engines, but they also include human-transformed substances employed for some purpose, such as plant-based drugs, wine, beer, foods, and timber. Although this expanded definition of technology does not come directly from ancient actors' categories, employing it allows us to include material developments in mechanics alongside shifts in drug production and scientific implements. All of these phenomena were crucial in articulating different theories of the body and different types of corporeality. Understanding how technical products and bodies interacted in ancient medical discourses requires analyzing these substances, too.

An emphasis on the materiality of technologies also helps overcome an obstacle faced by using ancient categories as the sole analytic framework for this investigation. Many scholars examining the interaction of technology and living things in Greek and Roman antiquity ground their analysis philologically, scrutinizing the shifting definitions of the terms *techne* and *physis*. Heinrich von Staden, for instance, has outlined the agonistic

20. See Schatzberg (2006 and 2018) for an overview of the term and its uses in philosophical, industrial, and sociological contexts.

21. Schatzberg (2018) traces these semantic uses to the appearance of the term *technologia* in Latin in the sixteenth century, where it denoted a *logos*, or systematic account, of a *techne*, or technical discipline. He argues that the extension of the semantic range to include technically produced objects emerged at the beginning of the twentieth century under the influence of German discourses about *Technik*, which described the material methods and tools involved in the industrial arts. Yet *Technik*, along with the French *technique* or Italian *tecnica*, can also refer to systematized skills and methods (which thus correspond more closely to the English *technique*), even though all these words are more commonly rendered into English as "technology" by modern translators. Such difficulties make a clear linguistic or etymological analysis difficult to conduct and restricted in value.

22. See Franssen, Lokhorst, and van de Poel (2018) for philosophical accounts of the metaphysics surrounding such artifacts.

relationship between these two concepts in certain Hippocratic treatises, as authors employ their techne to force nature to reveal itself, while he also shows how other medical authors characterized a more complementary relationship between medicine and nature.[23] More recently, Maria Gerolemou has conducted an even wider analysis of these terms, examining how nature acts as an inspiration for technical reproduction of animal motion, exploring *techne* and *physis* within Homer, Hesiod, Greek tragedy, and beyond.[24] Multiple scholars have explored the polarity of these two fields in other contexts.[25] Although this philological approach can provide insight into how individual authors balance these perpetually paired terms, acknowledging the materiality of technologies helps connect bodies with the tools that articulated them both conceptually and physically. Moving beyond a semantic approach to techne also helps avoid another potential tautology. Since medicine is itself a techne that developed and applied various therapeutic techniques, the history of medicine can be written as the application of different technologies to the human form. Such an approach occludes a full grasp of how tools interacted with bodies at both a physical and an abstract level. That full grasp requires seeing technologies as more than systems of knowledge, but also as actual objects and substances, created through human intervention and artifice.

0.3 ANALOGIES, METAPHORS, AND MODELS

If artifacts provide one pillar of this study, comparison supplies the other. As Geoffrey Lloyd has illustrated, analogical arguments played a particularly strong role in ancient science, and comparison was one of the main techniques through which Greek and Latin theorists constructed explanations.[26] While this study does not set as its primary aim the elucidation of how different scientific traditions or different authors employed metaphors and analogies (or how these argumentative techniques related to other literary disciplines in antiquity), analogies do offer a powerful way to track the conceptual effect that developing technologies had on

23. Von Staden 2007.

24. Gerolemou, 2023.

25. See especially the contributions in Bensaude-Vincent and Newman (2007). Two forthcoming volumes will also continue to investigate the mirroring and intermingling of living and technical objects; see Gerolemou and Kazantzidis, 2023, and Gerolemou, Ruffell, and Burr, forthcoming.

26. Lloyd 1966; cf. Lloyd 2015.

ancient medical assumptions. Comparison is one of the main ways that tools for doing can become tools for thinking, and one of the primary ways in which authors integrated tools into their articulations of natural phenomena.

Over the last sixty years there has been a tremendous amount written about metaphors, analogies, and models and their use in both philosophy and science (as well as an increasing literature in cognitive science).[27] The greatest number of contributions have been attempts to rehabilitate these comparative modalities in the face of an analytic and empirical tradition that sought to rid science of the ambiguity and uncertainty inherent in comparisons.[28] Mary Hesse, in her seminal 1963 book *Models and Analogies in Science*, argued for the essential predictive function that comparison performs in theory formation. As part of her analysis, she argues that each comparison includes positive analogies (i.e., comparable relations that are present in both systems), negative analogies (i.e., comparable relations that are present in one but not the other system) and neutral analogies (i.e., comparable relations that may or may not be present in both systems). For instance, we can think of modeling the atom on the solar system. A positive analogy might be that both systems include a central mass surrounded by smaller, orbiting entities. A negative analogy might be that although the planets have color, electrons do not. A neutral analogy might be that planets have elliptical orbits and are kept in them by a force that the central mass exerts. Although these resemblances may or may not turn out to be true, it will originally be unclear whether these particular similarities hold. Hesse emphasizes that neutral analogies therefore provide a valuable and essential heuristic role in science, since they present hypotheses that can then be independently tested to determine which falls into the positive and which into the negative category.[29] Analogies can translate theories into predictions. Most philosophers of science arguing for the importance of models in science suggest something similar, stressing the heuristic role they play, whether only in the context of

27. Kövecses (2010) provides an overview of metaphor that incorporates much of the relevant scholarship.

28. Although Francis Bacon denigrated imagery in the sixteenth century and Boyle and Hooke rejected misleading images in the seventeenth, Pierre Duhem may have configured the modern distrust of comparisons in science by rejecting even mechanical models of phenomena and instead looking to model science on axiomatic-deductive logic (1906, esp. ch. 5); cf. Draaisma 2000: 53–55.

29. Similar sentiments can be found in Black (1962: 223).

discovery or in the application of generalized laws into individual instantiations.[30]

The problem with accepting Hesse's analysis for understanding ancient science is that authors in antiquity generally did not test the hypothetical claims predicted by analogies with independent empirical experiment. Instead, the analogies and models were *themselves* the argumentative support for physical claims. Consequently, any neat distinctions between positive, negative, and neutral resemblances evaporate such that making these evaluations relies on the arguments and assumptions of the author and audience. Yet analogies were used for more than a single purpose within scientific treatises and could be employed for both embellishment[31] and assurance,[32] as well as having didactic[33] and heuristic functions.[34] It is necessary to keep these varying roles in mind,

30. For contemporary interest in how analogies can function as heuristic models in science, see Frigg and Hartmann 2020, especially section 1, which supplies a relevant bibliography.

31. I define *embellishment analogies* as the comparisons often (dismissively) associated with literary, poetic, and rhetorical concerns. These comparisons seek to make the target system more vivid while highlighting certain relevant features, but they do not imply a specific causal mechanics. For instance, an embellishment analogy might claim, "lightning flashes like a sword glinting in the moonlight," even as the author attributes lightning to a stream of fire surging downward, not a reflection on a metallic surface.

32. *Assurance analogies* are comparisons that support how an entity can enact a particular physical behavior by pointing to other instances where such behavior occurs (in this regard, they overlap with comparative examples). In these comparisons the analogue and its target system generally display so little resemblance that it is hard to imagine the author intended the analogue to represent any isomorphic physical mechanics, although it cannot always be ruled out. Regardless, these comparisons primarily function to guarantee the validity of certain physical claims. For example, an assurance analogy might claim, "the cupping vessel draws flesh into its hollow, just as a magnet pulls fillings toward itself."

33. *Didactic analogies* are comparisons that reveal the physical mechanics of a target system, but only up to some determinate point. In this regard they can be useful teaching aids, while any unwanted implications are kept separate from the theory which they seek to make clear. Didactic analogies are therefore often accompanied by boundary work to specify where the comparison fails. Sometimes, however, it is up to the reader to determine when the comparison stops becoming useful or starts to misrepresent the author's position. For instance, a didactic analogy might claim, "the body is like a house, with each chamber designated for a certain purpose."

34. Like didactic analogies, *heuristic analogies* provide sufficient resemblance as to provide a potential physical correlate for the target system, but it is unclear

since doing so prevents us conflating the purpose of all comparisons in all ancient scientific contexts.[35] Nevertheless, the trouble with outlining these types is that they are no more than useful fictions, especially since, unless they are explicitly framed by a declaration of purpose, we can only adjudicate which functions these comparisons play based on clues in the text and our basic intuition. Analogic types exist on a spectrum of resemblance, with only potential authorial instruction and contested degrees of isomorphism separating the categories. Part of the argumentative strategy of this study is therefore to acknowledge the slipperiness of these categories and track how even comparisons that authors endeavor to limit can still produce unintended conceptual consequences. In other words, embellishment, assurance, didacticism, and discovery coexist, sometimes in the same comparison itself. Recognizing the porousness of these categories can help us see how analogies can perform multiple functions in ancient scientific arguments, sometimes for different readers, sometimes simultaneously, and sometimes despite any authorial declarations and delineations.

Analogic models do not simply present potentially testable correspondence; they direct our attention in certain directions, thereby excluding potentially troublesome information and occluding certain details as irrelevant. As Max Black states, a comparison "selects, emphasizes, suppresses, and organizes features of the principal subject."[36] We can call this implicit act of selection *cognitive focus*. The aspects of the phenomenon around the boundaries are neither completely invisible nor cut off entirely; they are simply blurry and demand little regard relative to the points on which our attention has been trained. For instance, we may think about

---

where heuristic comparisons fail, or whether a high level of similarity exists. In this, they supply greater explanatory opportunity, since heuristic analogies supply otherwise inaccessible possibilities and can carry predictive force. An example of a heuristic analogy might be "thunder occurs when the hollow clouds burst, just like an inflated bladder popping." These heuristics are more or less identical to analogic models, and the two terms can be used more or less interchangeably in the context of ancient science.

35. Since I am interested in generalized types, I omit perhaps the most systematic use of analogy in ancient science: the formalized biological analogies that appear in Aristotle's biological treatises, notably *Historia Animalia* and *Parts of Animals*. In these texts, Aristotle attempts to systematize animate creatures by means of analogic relationships between functional parts, e.g., lungs : mammal :: gills : fish. Since they are bound to Aristotle's particular interest in teleological causality, these analogies are not universally employed in ancient scientific systems.

36. Black 1962: 44–45. He is here speaking about metaphors, but his analysis later extends to scientific models.

the brain's "hardware" and "software" without asking whether it requires periodic "updates." And yet a marginalized feature such as updating can be made active simply by readjusting our focus—this is precisely why heuristic analogies can generate novel and successful scientific ideas. Using a comparison to create a heuristic model is thus both a creative and destructive act, producing new and potentially significant links, while excluding or at least obscuring other possibilities. It directs attention to a small number of details, ignoring other aspects of the phenomenon, both physical and functional.

As Jacques Derrida argues in "White Mythology," philosophical attempts to avoid metaphorical usage only uncover another layer of metaphors, and even the very concept of the "literal" or the *etymon* bases itself on metaphors of effacement and exclusion.[37] Philosophy itself, he argues, is in fact enabled by and generated in metaphor. Related claims have also been made for many years by psychologists and cognitive scientists, who have increasingly acknowledged the crucial role that analogical thinking plays in how we conceptualize and comprehend phenomena. Rather than seeing analogies and metaphors as mere teaching aids or artistic flourishes, researchers have argued that comparisons help make conceptual bridges between seemingly disparate fields, while also providing the very mechanism that allows us to categorize objects in the first place.[38] Lakoff and Johnson push this idea even further, articulating a number of simple "conceptual" metaphors that are rooted in basic corporeal experience (high/low or warm/cold as good/bad, etc.). These, they argue, underpin whole systems of language.[39] More recently, Hofstadter and Sander have gone so far as to claim that analogies are "the fuel and fire of thinking," or, in stronger terms, "thought's core."[40]

37. Derrida 1974.

38. Foucault (1970) discusses the historical construction and classifications of natural kinds through resemblance and comparison (see esp. ch. 1–5); cf. Bowker and Star (1999), who also examine how classification functions through arbitrated similarities.

39. Lakoff and Johnson 1980. They also tie these conceptual metaphors to embodied experience (i.e., privileging human orientation, so that the human embodied experience of up/down underpins the metaphorical correlation with happy/sad).

40. Hofstadter and Sander 2013: 3, 18. Many other works have focused on the importance of analogies in cognition, notably the seminal discussions of Black (1962) and Ricoeur (1978). For various discussions concerning the use of analogies in science, see the early contributions of Oppenheimer (1956), Hesse (1966), Kuhn (1979), Gentner (1982), Gentner and Clement (1988), and Gentner and Jeziorski

We need not accept a poststructuralist account of language or accept that analogy is the foundation of cognition to recognize that analogic models do not have sharp boundaries where the comparison stops and the literal begins (indeed, that is why analytic philosophers have gone to so much work to locate or create such borders). This fuzziness is especially true in the context of ancient science. In fact, assumptions derived from comparisons often become so naturalized that they lose their status as metaphors and models. For example, in ancient treatises, digestion, fermentation, and ripening are all described by the same word: *pepsis* [πέψις], or "cooking." This basic categorization often attends ancient assumptions that heat must be involved in all four processes.[41] Yet rarely do ancient authors acknowledge that these processes are metaphorically linked, understood by an implicit comparison. By contrast, we presume that one of these words, "cooking," is the primary use of this term (i.e., the tenor), while others, such as "digestion" or "ripening," are the comparative targets (i.e., the vehicles), rather than allowing for the possibility that some Greeks might simply have considered all these activities to be instances of the same phenomena and thus represent no metaphor at all.

So many of our assumptions about the natural world come from basic technological behaviors that it is often difficult to discern where our assumptions come from in the first place. We cannot isolate or quarantine analogy so neatly, since comparisons provide structural frameworks through which we understand and explain the world. When looking for the impact that technologies had on scientific theories in antiquity, we must therefore start with explicit analogic models but can then examine whether these assumptions appear without any analogic justification, naturalized and transformed into frictionless assertions about the world. Indeed, sometimes technological models are tacit, active, and influential without being explicitly invoked, since they have worked their way down into our basic beliefs about the world and its operations.

(1989). For a discussion about metaphors in ancient science, see Pender 2003. I am particularly indebted to Black's "interactive" view of metaphor, which discusses how metaphors and analogies can organize our conceptions (see esp. Black 1962, ch. 4), and Ricoeur's emphasis on the "predicative" force of analogies, which creates meaning in the interaction of two terms, rather than via the substitution of one for another.

41. See Lloyd 1996: 83–103 for a fuller explication of this concept within Aristotelian biology; cf. Jouanna 1999: 314, 320. The verb πέσσειν can also refer to fruit ripening (cf. Totelin 2009: 148–49) or embryological maturation (cf. Arist. *Hist. an.* 6.2, 560b17–18; 6.4, 562b19; 6.7, 564a2; see Lehoux 2017 for a discussion of embryology and spontaneous generation, both of which involve "concoction").

0.4 CHAPTER OVERVIEW

Ancient theories were built in worlds composed of objects, implements, and certain material realities, and these environments shifted dramatically from the fifth century BCE to the second century CE (and beyond). The following chapters thus trace how these changing technological landscapes facilitated new theories of the body and its operations, while also showing how new theories of the body produced and were sustained by new sets of tools. Sometimes these new tools were technologies that directly intervened in corporeal processes, while at other times they were used as cognitive models for the corporeal interior. Since these issues are interdependent—tools do not supply easy models for corporeal functions if theorists do not think the parts function in the first place—the chapters will proceed chronologically rather than thematically, revealing how these interwoven ideas changed together over time.

Chapter 1 will discuss early Greek medicine of the fifth and early fourth centuries BCE, examining how Hippocratic authors used gruels, purgative drugs, and cupping glasses to understand the nature of the corporeal interior. It will show how these technologies formed recursive interfaces with the very bodies that they revealed, acting as physical and conceptual extensions of the body's essential substances and ideal composition. For instance, the author of *Ancient Medicine* considers the flavors in foods administered by the physician to be the literal substances balanced in the body, and the consistency of his ideal medicalized comestible—barley gruel—was thought to be the ideal composition of a healthy body. By contrast, *Nature of the Human* put forward a slightly different theory, this time using purgative drugs to make the four inner humors manifest to the theoretical eye. As with *Ancient Medicine*, the therapeutics overlap with the corporeal theories, insofar as an emblematic form of these emetic drugs resembles the pathological substance that they supposedly pulled from the body. Other treatises, such as *Diseases* 4, privileged a wider array of therapeutic technologies that included material instruments along with food and drugs, and this orientation mirrors the greater emphasis this treatise places on structures and their mechanisms within the body. This same tendency once again occurs in Hippocratic gynecological treatises, which argue that women's bodies are essentially different from men's insofar as they are wetter and need to purge this excess moisture through menstruation. In this light, the author of *Diseases of Women* 1 compares female flesh to wool, before going on to use wool as the primary vehicle for vaginal and uterine pessaries where irritating substances are introduced as emmenagogues to induce menstruation. In short, different

Hippocratic authors privilege different therapeutic technologies as their primary investigative tools, and these discrepancies align with divergent theories of the body.

Chapter 2 will start by returning to the fifth century BCE to look at a parallel development in theories of corporeality. It starts with a prehistory of organs in the pre-Socratic *physikoi*, who largely ignore the body as a conceptual unity until Empedocles begins to explain certain corporeal features such as breathing. He does so with a tool called the *clepsydra*, which had already become a totemic technology for understanding the power of air and the void. His account is remarkable insofar as it maps quite poorly onto human anatomy as it is now understood, implying that we breathe through both our nose and our skin and that air is drawn inward through the advance and retreat of blood within our body. This chapter argues that this is no mistaken interpretation but reveals how corporeal processes had yet to be housed in individual parts. In this regard, the clepsydra analogy came first and subsequently helped facilitate the idea in later authors that individual "organs" were responsible for corporeal functions. Indeed, the second part of the chapter turns to Plato, who adopts features of Empedocles's account in his own theory of respiration, which appears within his cosmogonic account in the *Timaeus*. It emphasizes that Plato was the first to call the interior parts *organa* and that doing so was crucial to his broader teleological view of the world as crafted by a divine demiurge. Yet Plato's version of the organism remains different from our own, insofar as it is oriented toward rationality, not life, as its highest purpose. The chapter ends by examining how this theory of respiration fits with an accompanying account of blood distribution in the body, an account that seems to reflect contemporary irrigation technologies of the fourth century BCE. It weaves together the structural, tool-like teleology that came with a cosmic creator and the ways in which the surrounding technological environment shaped the features and behaviors attributed to this new epistemic object.

Chapter 3 examines how Aristotle adopted tool-like teleology as the foundation of his biological research program, which attempted to outline a teleology of difference that explained why the parts of each animal were suited to their own particular lifestyle and how changes in one part would be balanced by alterations in others. This approach foregrounded parts and organs as crucial divisions within the body and made anatomy a privileged investigative practice. Moreover, he arranged the parts and their operations around a new goal: the maintenance of life. His interest in explaining how this organized body worked promoted what we might now term "physiology," insofar as he described both the function and mechanism of corporeal tasks. He thus merged form and function within individuated organs, and in so doing articulated a new theory of the body. The remainder of the chapter examines some of these mechanisms, including pulsation, respiration, and blood delivery, to illustrate the complex role that tool analogies played in his theories. They were not determinative, but neither were they inconsequential, and they could smooth the edges between the different demands of various corporeal tasks. The final section of the chapter extends this analysis to the way that automata feature in his account of animal motion to argue that his assertion that sinews, not muscles, provided the generative force relies on assumptions derived from the use of these animal tissues in motion-generative machines.

Chapter 4 will follow the rise of anatomical practices in the Hellenistic era and look at how they supported a mode of corporeality that placed emphasis on tool-like parts and their activities. It shows how vessels, formerly passive channels, were assigned active functional roles, as the organization of the body spread to multiple parts and structures. Although this chapter highlights this broader trend, it spends more time with Herophilus and Erasistratus as the two most important anatomists of the era, both working in Ptolemaic Alexandria. This newly founded city saw a swell of intellectual energy, which led to significant developments in mathematics, lexicography, aesthetics, astronomy, mechanics, and pneumatics. Scholars have long identified some conceptual similarity between Erasistratus's model of the heart—which he takes to be a mechanism of propulsion for the first time—and Ctesbius's contemporary pneumatic invention, the force pump. This chapter assesses these claims but argues that the impact of these new devices moved beyond a single instance of correspondence. Instead, it emphasizes that Erasistratus attributed disease to the "infiltration" of one substance (air or blood) into passageways that contained the other and argues that this etiology reflects the operations of pneumatic devices performing marvelous tricks through the carefully calibrated separation of air and water into various chambers.

Chapter 4 examines these challenges to the organized body from the second century BCE to first century CE. The success of the organism as a conceptual object did not eliminate its detractors, who launched epistemological arguments against anatomy as a mode of knowing inner corporeal activities. The Empiricists, Asclepiades of Bithynia, and the Methodists all critiqued the physiological approach to medicine that the organism had facilitated and presented their own rival forms of medical practice. Even though they did not foreground, or even include, functional parts, their medical theories and their reception were still impacted by their technological environment. Roman aqueducts and public baths were now a crucial part of the sprawling infrastructure of empire. Not only do certain features of Asclepiades's theory about blockage and flow in the body seem indebted to specifically Roman water technologies, but his popularity can also be attributed in part to the comprehensibility of his claims to a patient base that was now tending to health within bathing complexes fed by aqueducts under constant threat of blockage. This chapter stresses that technologies can create cultural heuristics and that we should examine the reception of medical ideas with these environments in mind. The last section of the chapter returns to female bodies to show how Soranus of Ephesus decoupled female corporeality from the biotechnical interfaces of Hippocratic gynecology and used Methodism to articulate women's bodies instead. As a consequence, he accepted a far wider range of menstrual volume and did not even see the absence of menstruation as itself pathological. In this regard, this chapter examines what the absence of certain technologies meant for theories of female corporeality.

The sixth and final chapter tracks the resurgence of anatomical investigations in the late first and early second centuries CE, when they appeared in both public performances and the medical texts of Rufus of Ephesus. With the return of anatomy came a greater interest in tool heuristics to understand the body. This interest can be seen most clearly in the works of Galen of Pergamon, who held that the body was a perfectly engineered object, with each part playing a functional role. As part of his arguments, Galen insists that each corporeal organ possess its own unique natural capacity to perform its primary task for the body. These forces have often been described as "vitalist" powers, which commentators then contrast with the "mechanistic" forces at work in artificial instruments. This chapter argues that instead of representing a step away from tool heuristics, Galen's vitalism follows logically from his assertion that the corporeal parts are tools responsible for certain somatic activities. Moreover, it emphasizes that both the existence of natural faculties and their specific properties are demonstrated by a series of technical interventions into animal bodies. It therefore argues that these vivisection experiments enact Galen's vitalist powers, which cannot truly be separated from the technologies used to disclose them. Finally, it examines the specialized tools required for these vivisections and notes that Galen implicitly establishes himself as a demiurgic figure, one who reverse engineers the body that Nature has built. The specialized tools that he designs to articulate the body mirror the specialized parts that they reveal. In sum, all chapters highlight the profound impact that tools and technologies had on Greek and Roman theories of corporeality.

CHAPTER ONE
Hippocrates and Technological Interfaces

1.0 INTRODUCTION

In the fifth century BCE, multiple Greek physicians began to write about medicine and defend its practices as a principled and effective discipline, designating it as an art, or *techne*. This development was certainly not the birth of healing in the Mediterranean world and its environs. To the south, Egyptian medical papyri attest to surgical care, gynecological treatments, and general therapeutics from the early second millennium BCE onward,[1] while to the east, Assyro-Babylonian cuneiform tablets from the eleventh century BCE list the sign sets used to determine which disease assailed a sick person and which treatments to apply as a consequence.[2]

1. For overviews of Egyptian medical and the papyrological sources, see Strouhal, Vachala, and Vymazalová 2014 and Nunn 1996. In the eighth century BCE, Homer casts Egypt as the homeland of the god of healing Asclepius, as well as the source of powerful drugs (*Od.* 4.231–232). In the fifth century BCE, Herodotus (*Hist.* 2.84.1) still marvels at Egypt as a place of medicine and medical knowledge; cf. Theophr. *Hist. pl.* 9.15.1.

2. See Scurlock (2014). These medical documents align with broader semiological practices that organize social and cosmic orders by means of a set of predictive signs (cf. Rochberg 2016). For an overview of Babylonian medicine, see Geller 2010 and Zucconi 2019: 15–56. Medicinal substances from both these geographical locations appear in early Greek sources, which attests to some level of interaction, and more connections between Egyptian, Assyro-Babylonian, and Greek medicine continue to be uncovered. It has been difficult to locate specific doctrinal influences that these traditions had on Greek medical notions, even if broader intellectual tendencies can be found (see van der Eijk 2004a and 2004b). Nevertheless, medicinal substances appear in Greece from both Egypt and the territories of Assyro-Babylonian culture (see Totelin 2009: ch. 4), which suggests ongoing interchange in the classical period, and more recent work has shown the influence and interchange between medical cultures of the ancient world (see Rumor 2016).

Within Greece itself, Homer's *Iliad* and *Odyssey* attest to the presence of wound treatment and plague-fighting paeans around the eighth century BCE,[3] and there is evidence of an ongoing pharmacological tradition in the centuries that followed,[4] with funerary statues and inscriptions even providing the names of a few Greek doctors from the sixth and early fifth centuries.[5] Nevertheless, the late fifth century BCE represents a major watershed in the development of medicine as a self-conscious literary discipline within Hellenic discourses. It was at this moment that Greek medical authors started to define their principles and theories in writing and defend their practices and methodologies in opposition to other traditions, notably magic and natural philosophy, even as they sometimes adopted ideas and techniques from both these fields.

The major evidence for this early Greek medicine comes from the so-called Hippocratic Corpus (or Collection), a group of approximately sixty texts (depending on inclusions and text divisions), composed by various authors but collected under the single moniker "Hippocrates" during the Hellenistic period.[6] The texts themselves display considerable diversity in terms of style, content, purpose, and doctrine, with some treatises presenting polished defenses of medicine, others detailing regimental and gynecological practices, while yet others preserve what appear to be notes. Many attempts, stemming back to antiquity, have been made to determine the "authentic" treatises of the historical Hippocrates,[7] or, if not

3. Hom. *Il.* 1.473–474, 5.396–403, 5.899–904; *Od.* 19.455–459.

4. See Totelin 2009.

5. A marble statue from the sixth century BCE was found in Megara Hyblaea, Sicily, with the inscription "Som[b]roditas, the doctor son of Mandrocles." A marble disc, likely the lid of a funerary urn, found in Athens and dated to around 500 BCE, displays the inscription "wise skill of Aineios, best of doctors," and a copper tablet from 499–498 BCE or 477–475 BCE, found at Idalion in Cyprus, honors Onesilas and his brothers, all doctors, with a bilingual Cypriote/Greek text detailing their service to Citium and their rewards. See Lane Fox 2020: 43–46 for these references; cf. fig. 3 on p. 61.

6. Craik (2015) lists fifty-one treatises. Considerable controversy concerns the date at which the Hippocratic Corpus was first collected. Although the current grouping of texts did not appear together in the 1526 Aldine edition, there are intimations that these treatises appeared in various clusters from as early as the fourth century BCE. The strongest case can be made that the compilation of these texts occurred within the bibliographic practices of Hellenistic Alexandria. Jouanna (2018) presents a full textual history of the corpus, although Totelin (2009: 4–5) and King (2020: 19–42) provide useful and concise overviews.

7. Hippocrates himself was reportedly born around 460 BCE on the island of Cos in the eastern Aegean, where he is said to have founded a medical school. He

that, to distinguish and group different "schools" of thought in the texts (the so-called Coan and Cnidian).[8] Most such approaches have been challenged and abandoned, so what we are left with are some relatively uncontroversial text groupings and some broadly accepted dates, with the bulk of the treatises originating in the late fifth and early fourth centuries BCE.

Even with long-standing debates surrounding textual authenticity and affiliations, there is wide consensus on a key doctrinal issue. That is, it is all but common knowledge that the Hippocratic texts located the nature of the body in its "humors" [χυμοί], "moistures" [ὑγροί], or other such constitutive substances, sometimes simply called "the things that" [ἅ] comprise human nature. Blood, phlegm, and bile (both black and yellow) are perhaps the best-known liquid candidates, but authors also proposed water, breath, air, and fire. In recent years, scholars have given more weight to how divergent these various Hippocratic theories can be, perhaps exemplified by Philip van der Eijk's suggestion that we abandon the collective title "Hippocratic" altogether and instead simply speak about the texts as representing "early Greek medicine."[9] Despite the increased acknowledgment of theoretical diversity in the Corpus, less attention has been given to differentiating Hippocratic authors' individual therapeutic practices. Because no treatise provides a full therapeutic manual, reconstructing Hippocratic treatments requires drawing evidence from across multiple texts. The therapeutic tool kit thus generated will include interventions into diet, exercise, sleep, and sexual activities, paired with the application of drugs, enemas, fumigations, venesections, cauterizations, joint relocations, bone settings, and a few instances (although not many) of soft-tissue surgeries, with another major division between general care and women's reproductive care. To be sure, there are many shared strategies and methods, even if nothing approaching a standard of care existed. Nevertheless, this chapter emphasizes that Hippocratic authors privilege different therapies in their writings, most notably when they describe which tools are the most epistemologically valuable for revealing the nature of the body. Tracking these varying therapeutic priorities can help

achieved notoriety as a physician within his own lifetime, providing the example of the doctor par excellence for both Plato and Aristotle by at least the early fourth century BCE (cf. Pl. *Prt.* 311b–c; *Phdr.* 270c–d; and Arist. *Pol.* 1326a15–16). Beyond these accounts, biographical information remains apocryphal and highly debated, but for full evidence, see Jouanna 1999; Craik 2015: xx–xxiv; Flashar 2016; and Lane Fox 2020. For an overview of the "Hippocratic question" and why it has been abandoned, see Lloyd 1975b and Craik 2018.

8. See Lonie 1965a, 1978a, and 1978b; Jouanna 1974; and Grensemann 1975.

9. Van der Eijk 2016.

reveal the intimate connection between developing medical technologies and emerging Hippocratic notions of corporeality. As was outlined in the introduction, different tools built different bodies.

It is difficult to stabilize the objects of scientific knowledge. Indeed, in establishing medicine as a techne, early Greek physicians had to make the body [σῶμα] and its "nature" [φύσις] their primary objects of expertise.[10] As Brooke Holmes has revealed, doing so was no minor task. Prior Greek healing practices had placed the patient in a network of relations between themselves and divine actors, such that the etiology of corporeal afflictions could lead either inward or outward, and often did both. By contrast, the Hippocratic authors located the source of disease in the body itself and, accordingly, needed to conceptualize and define corporeality as it pertained to medicine.[11] Yet the body's opaque interior presented immediate epistemological hurdles. We can neither see our insides nor directly witness the dynamic corporeal processes that seem to take place under the skin. Even if we could, visibility does not equate with comprehensibility, and the messy, oozy innards of an animal body do not disclose their systematic relationships in any straightforward way. Hippocratic authors therefore relied on various strategies to make our corporeal contours manifest to the theoretical eye. Some strategies included producing philosophically inflected arguments about first principles. Others involved cataloging what we would call symptoms (although they are perhaps more productively thought of as sign sets).[12] Beyond passively watching the distressed body, however, or theorizing about its insides, early Greek physicians also used active interventions to reveal what they thought was the

10. For Hippocratic use of the term "nature" and its variance, see Bourgey 1980 and Gallego Peréz 1996; cf. Craik 2017: 207. Von Staden (2007) notes that someone or something's *physis* is identified through a set of recurrent, stable characteristics that can only be known via its *dynamis* or "power," which itself can only be known through its effects upon and reactions to other entities and substances; cf. H. Miller 1952. Since *physis* is such a fundamental concept in Hippocratic medicine, but also a term that has incredible multivalence—as it can be used in the context of the "nature" of the human body in general, an individual's body, a body part, an internal substance, an external substance, etc.—reflections upon or discussions involving it are widespread; von Staden (2007: 43 n5) presents a useful bibliography on the subject.

11. See Holmes 2010b, which deals with these developments at length and examines how these developments implicated corporeality within a broader ethics of taking care of one's own health. See also Holmes 2014 and 2018, which present additional reflections both on this transformation and on the emergence of the Hippocratic body as an epistemic object.

12. For the semiotic approach to inner states, see *Art.* 11.1–12.1 = 6.18–24L.

essential nature of the human whole. There is evidence for some small degree of animal dissection in the fifth century BCE (although very little for human dissection), and this corresponded with an interest in mapping interior corporeal topography. Nevertheless, the most powerful investigative tool for Hippocratic authors was their own therapeutic techniques. Here, technologies directly intervened in the body, which was then structured around its reactions to these interventions. In short, Hippocratic theories were not simply abstract ideas or quasi-philosophical commitments. They were ideas and explanations that were demonstrated with, crafted alongside, and sustained by certain tools and practices. As a result, when therapeutic technologies changed, so too did the physical substances and conceptual objects they articulated.

As Heinrich von Staden has noted, many scholars have noted the mimetic relationship that Hippocratic authors establish between the techne of medicine and the physis of humans.[13] Authors can claim that "natures are the physicians of disease" [νούσων φύσιες ἰητροί], as appears in *Epidemics* 6,[14] or argue that all *technai*, medicine included, imitate aspects of the body, as in *Regimen*.[15] It is crucial to understand the material implications of such assertions. That is, mimesis often stretched beyond the conceptual, insofar as many early therapeutic technologies were physically integrated into Hippocratic theories of corporeality, sometimes being treated as the actual components that composed our physis. At other times, these therapeutic tools were affixed to the body, which then absorbed them as models of the very behaviors the tools aimed to treat. In other words, Hippocratic authors articulated and built their epistemic object in part through establishing therapeutic interfaces that often merged the physical and conceptual. In this regard, Hippocratic corporeality developed in the interactions between these medical tool kits and the flesh they aimed to treat. Yet to understand how these interfaces operated, and how this blending of prosthesis and mimesis worked, it is first necessary to understand some basic notions about how Hippocratic ideas about the body differed from our own and how they went about articulating and assembling the human whole.

13. See von Staden (2007), who analyzes both the ways that the medical techne is seen as an extension of physis and the moments that the techne forces physis to reveal or "confess" its secrets.

14. Hipp. *Epid.* 6.5.1 = 5.314L. For this author, physis can act to fend off illness by stimulating tears, moisture in the nose, yawning, coughing, urination, etc.

15. Hipp. *Vict.* 1.11–12 = 6.486–488L might be emblematic in this regard. For a discussion of these passages, see page 36–37.

1.1 CORPOREAL COMPOSITION WITHOUT ORGANS

One striking feature of Hippocratic notions of corporeality is that their body seems to lack a certain structural teleology that undermines our basic assumptions about what consolidates a living thing and makes it a unified living entity.[16] That is, modern explanations of the body tend to start with our parts, which are then understood as enacting some function for the operation of the living animal, and we understand bodies by asking how they "work." Yet for Hippocratic authors functional parts barely feature, if at all, in human physis. As mentioned above, early Greek physicians tended to describe corporeal liquids as the essential factors that maintained health and produced disease. Even when authors disagreed on what these liquids were, most still treated health as a homeostatic balance of these substances that could be disturbed by either environmental or behavioral factors. Authors placed far less emphasis on interior structures, most often treating them as receptacles and reservoirs for the moistures that flowed through and around them.[17] As Beate Gundert has demonstrated, there are many parts (ligaments, muscles, vessels, glands, hollow spaces) that individual Hippocratic authors treat as serving some function (closing the mouth, facilitating perception, creating the conditions for sounds, voice, sight, etc.).[18] Yet these references arise as offhanded comments or are only inferred from pathological states, and rarely are the viscera treated as functional components. Certainly, some authors treat structures as more important than others do, but even these authors do not foreground such parts when discussing the essential nature of the body. For example, "the cavity" [ἡ κοιλίη] that provides the site of digestion—so crucial within many Hippocratic systems of humoral and nutritive balance—cannot be pinned to a single organ but instead operates as the entire digestive tract, from stomach to rectum,[19] and several Hippo-

16. Craik (2017) has recently written about the "incipient" teleology found in Hippocratic texts. For where my stance differs, see note 31 on p. 32.

17. See Holmes (2014), who investigates these fluid dynamics relative to the emergence of "sympathy" as a physical force within the body (and cosmos). See also Joly (1966: 136), who describes the parts as passive and governed by "une physique du récipient."

18. Gundert (1992) catalogues the instances where Hippocratic treatises do examine structures. Nevertheless, she agrees that fluids are still more directly identified with the "nature" of humans.

19. For an overview of the semantic meaning of the term "cavity" [κοιλία], see Bubb 2019: 131 n11.

cratic authors even distribute digestion throughout the flesh as well.[20] In short, functional parts were not absent, but they neither structured nor consolidated corporeality.

One of the consequences of this view is that medical theorists of the fifth century BCE did not take "organs" as an essential corporeal subdivision. That is, even though authors had names for the heart, liver, kidneys, and so on, they did not generally treat these interior structures as responsible for individuated tasks. The Hippocratic treatise *Places in the Human* can illustrate one version of this view. The treatise is primarily pathological in outlook, insofar as it discusses the body in illness, not in health, and its dominant pathological category is "flux," wherein disease overwhelms individual sites—the so-called places [τόποι] that the title seems to reference[21]—and then cascades into other areas.[22] To put this in other terms, this treatise characterizes illness not as the malfunction of bodily components but as the inappropriate accumulation of moisture in various locations, such that the disease maintains its identity even as it travels around the body.[23] Yet even while promoting such a pathological theory, *Places in the Human* still expends considerable energy articulating the interior geography that hosts such fluids. The author documents the sensory orifices, the network of vessels transporting the corporeal moistures, and the skeletal system, including joints and tendons, all of which are characterized as pathways for illnesses to travel, sites in which disease can take hold, or the points of therapeutic intervention.[24] None of the

20. See discussion about Hipp. *Diseases* 4 on p. 54–67; cf. Arist. *Juv.* 469b6–13.

21. The text itself does not use this term to describe these sites, and Craik (2017: 206) argues that it was a late title.

22. Cf. Hipp. *Loc. Hom.* 1.2 = 6.276L. The moisture moves in two main ways, both insofar as the wet moves by nature to what is dry and by simple gravity flow. He also offers a general principle of fluid dynamics, that "everything flows toward what yields" (Hipp. *Loc. Hom.* 9 = 6.292L). These fluxes can arise from excessive or deficient phlegm, overcooling (which tightens vessels in the head and squeezes out moisture), overheating (which liquifies moistures and opens passages [e.g., Hipp. *Loc. Hom.* 9 = 6.292–294L]), and excessive dryness (which causes dry parts to attract moisture from elsewhere), as well as from other conditions that cause excessive humors to collect in inappropriate areas.

23. For instance, even when function might be implied, such as when blockage in the bronchial tubes arrests the flow of air (Hipp. *Loc. Hom.* 14 = 6.306L), the author provides no indication of what breathing does for the body or why it might be essential.

24. Hipp. *Loc. Hom.* 3 = 6.280–282L describes the vessels proceeding downward from the crown of the head to the sensory parts (most often providing moisture), as well as those that continue on to the lower places. Even the tendons [νεῦρα] in

structures that we are more apt to call "organs," including the heart, liver, lungs, kidneys, or brain, are given functions. In fact, this text describes a body that maintains an intrinsic unity down to the level of its smallest substances, where the parts therefore do not fulfill specific vital tasks but operate as homogenous representatives of the larger whole.[25]

Another text, *On Fleshes*, presents a related, albeit slightly different, view of corporeality. It begins by proclaiming a set of lofty goals—namely, to explain what the soul is, what being healthy and being ill are, what is bad and what is good in a human, and, lastly, from what one dies. In the text that follows, however, the author spends his time establishing the physical parameters that create the various tissues of the body, describing how heat, an immortal, all-perceiving, and motion-producing substance, acts on three additional components: the cold [τὸ ψυχρόν], the fatty [τὸ λιπαρόν], and the gluey [τὸ κολλῶδες]. He describes the formation of bones, sinews, and vasculature, the hollow digestive tract—which includes the pharynx, esophagus, stomach, and intestines—and then the brain, heart, liver, spleen, kidneys, flesh, joints, nails, teeth, and hair. Throughout his accounts, the author of *On Fleshes* remains most concerned with explaining how membranes form and how hollows emerge, while paying little attention the parts' shapes beyond this. Although he occasionally explains the unique physical properties of individual tissues, *On Fleshes* almost fully neglects function.[26] Even moments that at first may seem to assign function do so by describing the parts as sources or thruways, not

---

the latter case are described as recipients of fluid diseases (Hipp. *Loc. Hom.* 4 = 6.282–284L), although the text does supply the opaque comment that they "press" the bones, which offers a glimmer of function (Hipp. *Loc. Hom.* 5 = 6.284L).

25. Hipp. *Loc. Hom.* 3 = 6.278L; cf. Hipp. *Alim.* 23 = 9.106L; Hipp. *Vict.* 1.2 = 6.468L. This is not to say that there are not parts that can be described in terms of function. Indeed, *Places in the Human* begins by describing how the sensory orifices pass along sensory flux and how various vessels supply moisture. In fact, the author describes the eyes, ears, and nose first, describing their structures and implying that their particular material composition facilitates the transmission of sensory flux to the brain (Hipp. *Loc. Hom.* 2 = 6.278–280L). There are also some other brief indications of function—e.g., the little vessels supplying the purest moisture to the eyes (Hipp. *Loc. Hom.* 2–3 = 6.280L) and mucus lubricating the joints (Hipp. *Loc. Hom.* 7 = 6.290L). These functions, however, seem primarily to be transporting fluids.

26. For example, the author only describes the formation of the lungs, not their role in respiration or the intake of heat or cold. In fact, at Hipp. *Carn.* 18 = 8.608L, he even mentions that humans draw air into their whole body, even if it collects largely in the hollow parts. In other words, just as digestion was distributed throughout the body, so too was respiration.

active organs. The treatise consolidates the body as a dynamic system that balances the hot and the cold, the moist and the dense, where the activities of these essential substances are primary, and the parts of the body simply host their powers and effects.[27]

A third example from *On Glands* provides further clarifying evidence. This text describes the glands of the body as spongy [σπογγώδης], loose-textured [ἀραιαί], fatty [πίονες], crumbly [ψαφαρά], and wool-like [οἷον εἴρια] to the touch.[28] It states that these glands tend to occupy the "hollow" parts of the body (armpits, groin, and under the ear), around the joints, and wherever there is ample moisture and blood. Because of their texture and the hollow veins running through them, the glands attract the excess moisture arising from exertions and/or collect the moisture running down into them from above. In this way, they act as a buffer and help ensure that no surplus moisture builds up in the body.[29] On the one hand, the text explicitly describes this type of activity as a "task" [χρέος/χρείη] that the glands engage in or perform.[30] On the other, the glands do

27. For example, the author calls the brain the "mother city" [ἡ μητρόπολις] of the cold and the gluey, while he names heat the "mother city" of the fatty (Hipp. *Carn.* 4 = 8.588L). Elsewhere he insists that the heart and the "hollow vessel" [κοίλη φλέψ] together possess the most heat and "dispense *pneuma*" [ταμιεύει τὸ πνεῦμα] to the rest of the body (Hipp. *Carn.* 5 = 8.590L). However vivid these metaphors may be, calling the brain the mother city of the cold and the gluey marks it as the location where these two substances primarily collect, whereas by stating that the heart distributes pneuma, the author seemingly implies little more than that it acts as a passive distribution hub, rather than implying that it plays some active role in determining when, where, or how much pneuma leaves this site—especially since he immediately proceeds to describe the pathways of the vessels extending from this juncture and not how the heart might control or restrict pneuma.

28. Hipp. *Gland.* 1 = 8.556L.

29. Hipp. *Gland.* 2 = 8.556L.

30. See Hipp. *Gland.* 7 = 8.560L, where he describes how the "paristhmia" glands of the throat serve as an overflow reservoir, which prevents damage being done, so long as the flow into them is commensurate and small enough for the glands to overrule it, although if a large, sharp amount flows into it, illness follows. He explicitly states that serving as a reservoir in this way is the glands' "function" [χρείη]. At Hipp. *Gland.* 4 = 8.558L, he describes how the glands and the hair both "take up the same task" [ἄμφω χρέος τωὐτὸ λαμβάνουσιν], the former with regard to the moisture flowing into them, the latter in growing up from the surplus. Potter (1995) translates χρείη as "function" but χρέος as "office" in the above sentences. These nouns are used very sparingly elsewhere in the Hippocratic Corpus, only appearing at Hipp. *Flat.* 4 = 6.96L, which speaks about the great "necessity" [χρείη] of wind for life, while Hipp. *Decent.* 1 = 9.226L; 4 = 9.230L;

not supply this benefit for the body because they operate as teleologically oriented functional parts, but because they are seemingly looking out for their own interests.[31] The author states that they use the moisture that enters them as nutriment and thus "derive profit" [κέρδος ποιεύμεναι] from their surroundings.[32] The glands in the omentum "graze" in the intestines by squeezing out moisture for themselves [νέμονται αἱ ἀδένες ἐν τοῖσιν ἐντέροισιν ἐκπιεζόμεναι τὸν πλάδον],[33] while the glands in the kidneys "draw to themselves" [ἕλκουσι πρὸς σφέας] whatever moisture they can so as to profit from it [ἀποκερδάνωσιν].[34] In each case, the glands and their allies are not parts of an integrated functional machine but individual agents attending to their own interests. We can thus think of the glands as analogous to salt marshes, which lessen the damage from storm surges, but not because they are designed to do so or consciously attend to this purpose. Thus, the glands do not truly function and malfunction. They sometimes provide "benefits" [ἀγαθά] by absorbing moisture, but other

---

and 7 = 9.236L describe the "usefulness" or "need" of certain wisdom [χρέος], and elsewhere (Hipp. *Decent.* 17 = 9.242L) "what is needed" [τὸ χρέος]; cf. Hipp. *Iusi.* 5 = 4.628L, which uses χρέος as a synonym for "money." It also appears in the title of the treatise *On the Use of Liquids.*

31. Craik does not acknowledge this difference, instead suggesting that the text "has an Aristotelian tenor" (2017: 215). She argues more broadly that there are multiple moments of proto- or "incipient" teleology in the Hippocratic Corpus and—incorrectly, in my view—blends Hippocratic notions of physis as normative or just with Platonic and Aristotelian notions of functional organization.

32. Hipp. *Gland.* 4 = 8.558L.

33. Hipp. *Gland.* 5 = 8.560L.

34. Hipp. *Gland.* 6 = 8.560L. Similarly, those in the intestines greedily want more than their share [μᾶλλον πλεονεκτεῖν ἐθέλουσα], but none is able to "gain advantage" [ἐπαυρισκόμεναι] at the expense of the others (Hipp. *Gland.* 9 = 8.564L). Hair, too, which grows from the moisture, likewise "seizes the opportunity" presented by the glands [τὴν ἀπὸ τῶν ἀδένων ἐπικαιρίην] (Hipp. *Gland.* 4 = 8.558L), while the brain, which is like a gland, "steals [ἀποστερέει] moisture, thereby providing vengeance [τιμωρέων] against the moisture (Hipp. *Gland.* 10 = 8.564L). Similarly, the breasts are glands that "steal the surplus from the rest of the body" [ἀποστερίζουσι τὴν πλεονεξίην τοῦ ἄλλου σώματος] (Hipp. *Gland.* 17 = 8.574L). The nostrils, too, "are able to avenge themselves" [ἱκαναὶ τιμωρέειν σφίσιν] when moisture enters them from the brain, because they are wide (Hipp. *Gland.* 13 = 8.568L). At the beginning of the treatise, the brain is treated as comparable to the glands in both texture and behavior, but by the end of the discussion, the author refers to the brain and "the other glands" [αἱ ἄλλαι ἀδένες] as though there is no real distinction between them (Hipp. *Gland.* 15 = 8.570L).

times produce "harms" [κακά] by acting as repositories for disease.[35] In other words, in *On Glands* the body is an ecology of self-interested parts.[36]

The point to take from these examples is not that function *never* appears or gets implied in the Hippocratic Corpus but that explicating the nature of a human does not privilege integrating these functional parts within an organized structure, even in texts about the body's inner landscape. Function can appear at various registers, but it is not a fundamental structuring principle of corporeality. Indeed, even the essential liquids themselves are not generally discussed relative to their role in keeping humans alive, and it is not clear why phlegm or bile exist other than to produce disease.[37] Humors maintained health through their balance, but their collective presence did not generally explain life.[38] Physis, not function, consolidated the body.[39]

This basic structural assumption is widespread across the Hippocratic Corpus and remains true for the other attested contemporary medical

35. Hipp. *Glan.* 2 = 8.556L. He states that the glands produce both benefits and harms [τὰ δ'αὐτά οἱ δοκέει παρέχειν ἀγαθὰ καὶ κακά] (Hipp. *Glan.* 8 = 8.562L), and the brain, too, provides the same benefits to the head as the glands [ταὐτὰ ἀγαθὰ τῇσιν ἀδέσι ποιεῖ τὴν κεφαλήν] (Hipp. *Glan.* 10 = 8.564L).

36. To illustrate further that an ecology of interests, rather than a functional teleology of parts, structures this author's account of corporeality, I can note that even though the brain absorbs excess or surplus moisture that flows into the head, thereby providing an added buffer against disease, it is unclear why the head attracts moisture in the first place. Similarly, the brain can send off the moisture it "steals" into one of seven outflows: the eyes, ears, nose, pharynx/lungs, esophagus/stomach, the vessels of the spinal marrow, and the vessels leading to the hips (Hipp. *Glan.* 11 = 8.564L). If it sends them through the nose to the exterior, no disease generally results, but flowing through any of the other orifices generally produces pathologies. Nowhere does the text ask why six of the seven outflows attached to the brain are disease-producing.

37. For example, Hipp. *Genit.* 3 = 7.474L lists the four humors, which the author claims are the "inborn kinds" [ἰδέας συμφυέας] of moisture, but instead of discussing what purpose these might serve for the body, only states "and from them diseases arise." The same author states that semen is derived from these four humors (and therefore forms the essential matter of a human) but supplies no indication that these liquids perform discrete tasks for the body.

38. Some authors make fire responsible for motion and water for nourishment (see Hipp. *Carn.* and Hipp. *Vict.*), while Hipp. *Flat.* 4 = 6.96L calls breath the "cause" [αἴτιος] of life. Still, Hippocratic authors do not create systems in which moistures cooperate to animate humans and instead focus on the roles of these liquids in disease creation.

39. Cf. Hipp. *Loc. Hom.* 2 = 6.278L.

authors. Alcmaeon of Croton (fl. early 5th c. BCE), a physikoi with tentative links to the Pythagoreans, reportedly defined health as the equal rule of "wet, dry, cold, hot, bitter, sweet, and the rest," whereas disease was the dominance or "monarchy" of one such opposite arising in the blood, marrow, or brain.[40] Euryphon of Cnidos (fl. 5th c. BCE) reportedly held that diseases are caused by the residues left in the cavity rising to the head.[41] Alcamenes of Abydos maintains a comparable theory but adds that the residues reaching the head are then supplemented and sent out all over the body.[42] Herodicus of Cnidos describes how residues themselves break down into acid and bitter moistures, and each produces different diseases, as do the different "places" [τόποι] that these moistures occupy, whether the head, liver or spleen.[43] In all instances, illness arises in inappropriate matter occupying certain sites, but there are few indications that these places perform any explicit individuated function for the body before such illness takes hold, other than to hold the potentially pathogenic fluids themselves. Indeed, other evidence illustrates that medical theorists could be interested in the interior sites of the body without implying an interest in what task each part completed for the body.[44] For example, Aristotle describes how three physicians from the fifth century BCE outlined the vascular system: Diogenes of Apollonia, Syennesis of Cyprus, and Polybus of Cos (the son-in-law of Hippocrates and author of the Hippocratic treatise *Nature of the Human*).[45] Their descriptions track the main vessels around the body, generally describing one major vessel on the right and one on the left (although Polybus describes four main vessels), extending from the head to the feet, rectum, or penis, and passing through various viscera in sequence. Each physician presents different

40. DK24 B4 (D30 LM) = Aët. 5.30.1

41. Anon. Lond. 4.31–40 Manetti.

42. Anon. Lond. 7.40–8.10 Manetti.

43. Anon. Lond. 4.40–5.34 Manetti. Nutton (2004: 74) notes a parallel between Herodicus of Cnidos and the Hippocratic treatise *Places in the Human* insofar as both works attribute various diseases to the places in which fluxes settle. See also the entry for Timotheus of Metapontum (Anon. Lond. 8.10–34 Manetti), who argues that when the passages in the head are blocked, residues rise into its "places" [τόποι] but have no passage and so stagnate into salt or acrid moisture, and this moisture then floods into various other "parts" [μέρη], which can each lead to different diseases.

44. We might compare Osborne (2011), who argues that fifth- and early fourth-century BCE artists were interested in delineating parts of the body, not exhibiting functional musculature.

45. Arist. *Hist. an.* 3.2, 511b11–513a7.

pathways for these vessels, but they still cast the viscera only as sites or stations along these channels, not operative parts. Syennesis and Polybus fail even to mention the heart, while Diogenes and Polybus both mention the vessels as sites where lancing and venesection should occur. In short, even as authors were interested in describing inner structures and their connections, they seemingly did so to map the interior geography, potentially as it related to points of therapeutic interest, not to create an operational model of the body.

In many ways, the basic assertions above are largely uncontroversial, but their consequences can get overlooked. For instance, if the body is not "organized" into tool-like functional parts, it risks mischaracterization to assume that the body "works" as a primary conceptual framework to understand physiology.[46] Moreover, when Hippocratic authors draw comparisons with tools to conceptualize the corporeal interior, some of the most obvious ways we use tools as heuristic analogies fit less obviously as explanatory modes. Ideas that we take to be obvious or perhaps even natural (e.g., that the liver acts as a filter, the lungs as bellows, the heart as pump) only truly emerge once the parts are already thought to be tool-like functional organs. As such, it is worth examining how mimetic affiliations were developed within early Greek medicine, especially as therapeutic tools were being used to both treat and conceptualize the corporeal interior. These recursive relationships happened with multiple medical technologies, including some that we might not think of immediately as tools, such as food or plant-based drugs.

1.2 REGIMEN AND THE BODY

Along with a new sense of (admittedly fraught) disciplinary boundaries surrounding medicine, the middle of the fifth century BCE also witnessed the emergence of a novel type of therapeutic practice as part of this shift: the use of regimen-based medicine and lengthy dietetic prescriptions. This approach regulated food intake, exercise, exertion, baths, and sex, as physicians sought to preserve the balance of health and treat illness along what was now considered the trajectory of a disease over a numerically significant number of days. It is a matter of debate where such practices came

46. There is one place this vocabulary appears, but only in a cryptic assertion in *Nutriment*, which states, "All things according to the whole, and the parts in each part in regard to the task [πρὸς τὸ ἔργον], according to the part" (Hipp. *Alim.* 23 = 9.106L). Other unmarked examples occur, such as at Oss. 5 = 9.170L, where the kidneys filter off urine from blood.

from, whether Pythagorean religious practices or the training of athletes as part of the rise of gymnasia in the late sixth century BCE.[47] Regardless, the mid-fifth century saw the development and expansion of regimen-based medicine, complete with robust theoretical frameworks for the first time.[48] Herodicus of Selymbria, a trainer [παιδοτρίβης], is reported to have originated such practices, and his extreme regimens designed to prolong life made him the target of Plato's scorn in the *Republic*.[49] Despite its paramedical origins, however, regimen-based care abounds within the Hippocratic Corpus, and multiple authors characterize it as their primary therapeutic intervention. For these same authors, dietetic techniques greatly structure notions about the corporeal interior and its processes. Indeed, dietetic practices and accompanying theories of the body seem to have emerged in tandem, as the techniques used in medicalized regimens formed recursive interfaces with the bodies that they treated.[50]

The Hippocratic treatise *Regimen* provides the lengthiest account of dietetics and provides a good example of these dynamics. This text holds the human body to be comprised of two constituent substances, fire and

47. Zhmud (2012: 347–79) attributes dietetics to Pythagorean practices. Dean-Jones (2013) shows that athletic trainers and physicians had remarkably different aims; cf. Bartoš 2015: 29–36, which argues for a distinction between the practices. Bartoš (2015, esp. ch. 1), argues that dietetics developed in the mid- to late fifth century; cf. Lonie (1977), who holds the same view. For the rise of gymnasia and the history of Greek sports, see Papakonstantinou 2021.

48. For overviews of dietetics and regimen-based medicine in the fifth century, see Bartoš 2015. Grant (2018) provides an even broader survey of regimen in antiquity but does not develop a coherent picture of regimen as a specifically medicalized mode of consumption.

49. Pl. *Rep*. 3.406a–c; cf. Pl. *Prt*. 316d. Plato later displayed a more positive view toward regimen; see *Ti*. 44b. In *Plt*. 294d4–e1, Plato mentions gymnastic trainers who must provide general rules of regimen for everybody, starting and stopping classes at specific times, rather than treating individuals, since they are all different; he also elides them with physicians. For other ancient references to Herodicus, see Anon. Lond. 9.20–36 Manetti; Hipp. *Epid*. 6.3.18 = 5.302L; Soranus *Vita Hippocr*. 2.175, 7–9 (which casts Hippocrates as a pupil of Herodicus); Porph. *Iliad* 11.514 (which casts Herodicus as the inventor of dietetics and Hippocrates, Praxagoras, and Chrysippus as subsequently perfecting it). For full testimony and confusion between Herodicus of Selymbria and Herodicus of Cnidos, see Grensemann 1975: 15–20; cf. Manetti (2005).

50. The Hippocratic treatises *Regimen*, *Regimen in Health*, and *Regimen in Acute Diseases* are all devoted to the topic; regimental practices and theories are discussed in *Airs Waters Places*, *Ancient Medicine*, *Nature of the Human*, and *Affections* (esp. 47–60); and therapeutic dietetics appear in the prescriptions found in *Diseases* 1–3 and *Diseases of Women* 1 and 2.

water, which function as the kinetic principle of motion and the principle of nutriment, respectively. Each requires the other for its continued existence, since fire consumes water, and water slows to a stop without fire.[51] As rooted as these ideas may be in pre-Socratic physics,[52] these mutually dependent and mutually antagonistic substances also reflect the text's two main categories of therapeutic intervention: exercise (fire/movement) and food (water/nutriment). The author's two therapeutic vectors thus mirror the composition of the body on a broad structural level.[53] *Regimen* also insists upon smaller-scale mirroring as well, insofar as the treatise claims that a wide array of artisans imitate [μιμεῖσθαι] the processes of their bodies even without knowing it. Ironsmiths, fullers, cobblers, carpenters, house builders, musicians, cooks, curriers, basket weavers, and gold workers all imitate the production or maintenance of human life in their technical activities, and in this way they all imitate dietetic trainers.[54] He claims that these practitioners are unaware of these connections and "do not know how to see the invisible from the manifest" [Οἱ δὲ ἄνθρωποι ἐκ τῶν φανερῶν τὰ ἀφανέα σκέπτεσθαι οὐκ ἐπίστανται].[55] In this formulation, the "manifest" refers to information embodied in technologies and technical processes, while the "invisible" is the hidden nature of the body. These comparisons highlight the ways in which the body was transforming into a technically produced object for Hippocratic physicians, as well as underlining the ways in which authors' own tools could imitate and inform the nature that they make visible to the theoretical eye.[56] Still, the conceptual impact of regimen-based medicine moved beyond structural patterns and included the corporeal integration of specific foodstuffs and food-processing techniques. This integration is most evident in another regimen-based treatise, *Ancient Medicine*.

51. Hipp. *Vict.* 1.3 = 6.472–474L.

52. See Bartoš 2015 and Preus 2020 for a discussion of these associations.

53. Hipp. *Vict.* 1.3 = 6.472L even begins to speak about the function [χρῆσις] of fire and air in the body, showing a close connection to his therapeutic interventions; cf. Bartoš 2015: 76–79 and H. Miller 1959: 148–49.

54. Hipp. *Vict.* 1.11–24 = 6.486–496L; cf. Gal. *San. Tu.* 2.11 = 6.155K, who picks up and expands these passages. Hipp. *Vict.* 1.22 = 6.494L mentions building bodies with the same "tools" [ὀργάνα] as potters. Lefèvre (1972: 210–11) and Demont (2014) both argue that this passage refers to bodily "organs," but the text clearly refers to the dry and moist, which suggests clay and fire, or other substances from pottery making.

55. Hipp. *Vict.* 1.11 = 6.486L.

56. Bartoš (2015) argues that "imitation" [μίμησις] does not here (if ever) indicate copying but rather a dual-direction mirroring.

Ancient Medicine begins by establishing some antagonists, whom the author names as those engaged in the "inquiry into nature" who establish a "postulate" or "hypothesis" [ὑπόθεσις] as the foundation of their account of human nature and human disease. This postulate can be "hot, dryness, moistness, cold, or whatever else they want."[57] Multiple attempts have been made to determine whether this author has a specific target in mind and who it might be, but it seems more likely that he takes issue with the general class of pre-Socratic physikoi and their explanatory methods.[58] The author criticizes these thinkers on the grounds that their failure to acknowledge medicine's true origin and discovered method [ἀρχὴ καὶ ὁδὸς εὑρημένη][59] has led them into vain speculation about "empty" [κενή] substances. By contrast, he repeatedly casts physicians as artisans [δημιουργοί] and "handworkers" [χειροτέχναι] engaged in a practical techne, not theorists engaging in guesswork.[60] That is, the author explicitly distances himself from abstract speculation, and he instead attempts to root his own arguments in applied therapeutic tools and techniques. Foods, not philosophy, will be his epistemological bedrock.

As support for this methodological claim, *Ancient Medicine* outlines a (historically dubious) account of the origins of medicine.[61] The author argues that medicine arose from the basic observation that sick people could not eat the same foods as healthy people, nor could humans eat the same foods as animals without pain and consequent illness. By necessity, then, humans gradually discovered techniques to avoid discomfort by applying various methods of food preparation. Ill patients could re-

57. Hipp. *VM* 1 = 1.570L. See Holmes 2010b: 162–71 for scholarly approaches to this text.

58. Lloyd (1963) argues that Philolaus is the text's primary target; Vegetti (1998) suggests instead that it is Empedocles, whom Hipp. *VM* 20 = 1.620L cites by name as he rejects the approaches of previous philosophers and physicians. Other scholars pay more attention to the related question of the text's philosophical debts and influences. Diller (1952) sees Hipp. *VM* as drawing from Plato; Edelstein (1967: 195–203), as influenced by Protagoras; while Longrigg (1963 and 1983) suggests that *VM* itself influenced the philosopher Protagoras; cf. Schiefsky (2005: 1–4), who likewise considers the *physikoi* the general targets. For philosophical influences on *VM*, see Bartoš 2015: 75.

59. Hipp. *VM* 2.1 = 1.572L. Schiefsky (2005: 75) translates this sentence as "a principle and a discovered method," but it is clear from the author's arguments that the origin of medicine *is* the method on which true medicine rests.

60. See Hipp. *VM* 1–2 = 1.570–574L for examples.

61. See Bartoš 2015: ch. 1 for the origins of dietetics in the second half of the fifth century; see also Lonie 1977: 242.

duce food intake, but they could also weaken foods by mixing and cooking them to produce gruels [τὰ ῥυφήματα].[62] Early humans also learned how to soak, crush, grind, sift, mix, and bake grains. From wheat they made wheat bread [ἄρτος], from barley they made barley cakes [μᾶζα].[63] They boiled some things to make them moister, while baking others to make them drier.[64] Lastly, they "mixed and blended the strong and unmixed with weaker foods, molding everything to the nature and power of the human."[65] In sum, humans started cooking, breaking down and mixing their comestibles in order to ameliorate the deleterious effects of raw, unmixed, and overly powerful foods.[66] While other Hippocratic treatises describe how each foodstuff, whether meat, fish, or grain product, can have its own nature and "power" [δύναμις] that must be matched with the constitution and nature [φύσις] of the patient so as not to overwhelm his system.[67] *Ancient Medicine* emphasizes how to produce both healthful and therapeutic regimens not simply by selecting different types of ingredients but by employing techniques to weaken food types.[68] In fact, technical processes, not curated ingredient lists, form the core of both its therapeutic and investigative practices. The author goes on to argue that these ancient practices and techniques of transforming food deserved the title of "medicine" insofar as they were discovered for the sake of health, protection, and nourishment. Physicians eventually recognized that humans have a nature that differs from animals and that each individual also possesses a nature [φύσις] with its own particular power [δύναμις] to

62. Hipp. *VM* 5.22–24 = 1.582L.

63. Hipp. *VM* 3.5 = 1.576L; cf. Hipp. *Nat. Hom.* 9.22–24 = 6.54L, where wheat bread and barley cakes are also treated as the paradigmatic foods of regimen and represent opposite poles on a spectrum.

64. Cf. Hipp. *Salubr.* 1 = 6.72–74L, which puts this schema into action, while Hipp. *Vict.* 2.56 = 6.564–566L lays out this logic.

65. Hipp. *VM* 3.12–15 = 1.578L; cf. Hipp. *VM* 13 = 1.598–600L, where exposing food to fire and water changes its power and nature by mixture and combination.

66. Hipp. *VM* 3 = 1.578L.

67. See *Regimen* and *Affections* for applications of this notion.

68. See esp. Hipp. *VM* 14.1–2 = 1.600L, which describes how the nature and potency of bread will change according to how much water is added to the dough, whether the wheat is sifted, how much the dough is kneaded, etc.; cf. Hipp. *Vict.* 1.2.13–14 = 6.468L, which claims that foods have a power both "according to nature" [κατὰ φύσιν] and "through compulsion and human technique" [δι᾽ ἀνάγκην καὶ τέχνην ἀνθρωπίνην]; cf. Hipp. *Vict.* 2.39.1 = 6.534L, which describes the powers of individual foodstuffs as well as the capacity of the medical art to strengthen or weaken these substances.

master [ἐπικρατεῖν] and assimilate food.[69] Foods and individuals can thus be plotted within a single commensurate spectrum.

What type of body hosts a "nature" so defined or is constructed as the recipient of these dietetic practices? What body do these food-processing and therapeutic technologies articulate? As it turns out, one that absorbs and mirrors the medical tools that treat it. For example, the actual substances flowing in the body are not simply commensurate with powers in foodstuffs, they are in fact the same "humors" or "flavors" [χυμοί] that undergo processing.[70] These constitutive flavors in the body include the "salty, bitter, sweet, acidic, sour, bland, and countless other things having all kinds of powers in both quantity and strength" [ἁλμυρὸν καὶ πικρὸν καὶ γλυκὺ καὶ ὀξὺ καὶ στρυφνὸν καὶ πλαδαρὸν καὶ ἄλλα μυρία παντοίας δυνάμιας ἔχοντα πλῆθός τε καὶ ἰσχύν].[71] That is, *Ancient Medicine* identifies the flavors found in foods as the primary potencies that enter the body, provide nutriment, and potentially cause harm. The flavors present in food are also "present in a human" and thus they are the powers directly altered, reduced, and transformed through the practices of cooking, as well as subsequently introduced into the body through consumption.[72] For instance, when the author describes why eating lunch can produce illness for those who generally skip the meal, he suggests that the "cavity" did not have sufficient time to take benefit from the food, master it, be emptied and then rest; instead, new substances were added to a gut that was "boiling and fermenting" and needed more time to cook/digest [πέσσουσι] its food and recover.[73] When the author seeks to explicate what it means in corporeal terms for a human physis to be mastered by foods, he uses the same words to describe the digestive processes as he uses for the type of preparation techniques that regimen-based medicine employs to mas-

69. Cf. Hipp. *Loc. Hom.* 43 = 6.336L, which provides a similar notion of the body "ruling" or "being ruled" by its food. This language of "ruling" [κράτειν] is etymologically related to "mixing" [κρᾶσις].

70. The word "humor" [χυμός], which he uses to describe these substances, refers to "juices" or "flavors." It only later becomes the standard vocabulary to refer to the inner moistures or "humors."

71. Hipp. *VM* 14.33–35 = 1.602L.

72. That said, not all potencies are strictly identified with flavors. In one instance, Hipp. *VM* 15.17–18 = 1.606L mentions a potency that is "disturbed/disturbing" [ἄραδος], which thus denotes it by the effects it produces in the body, not its flavor; cf. Schiefsky 2005: 264.

73. Hipp. *VM* 11.9 = 1.594L.

ter food outside of the body.[74] Dietetic tools and medicalized bodies are literally made of the same things and involve the same basic technical processes.

In general, food-processing techniques, which include heating/cooling, moistening/drying, disintegrating, and mixing, are all actions conducted within the body.[75] Nevertheless, the comparison to food processing and digestion in fact moves in two directions: the Hippocratic author understands the internal actions of the body on the model of cooking and medicalized techniques, but only insofar as these medical techniques have themselves been envisioned as prosthetic extensions of the body's inner processes.[76] In fact, he explicitly advocates looking at the behavior of foodstuffs outside of the body in order to understand its internal processes, functionally establishing external proxies to imitate the actions of inner substances.[77] Later in the treatise, the author again picks up this thought to suggest that we can understand how the humors behave inside the body by witnessing their transformations outside of it:

74. Schiefsky (2005: 29) notes that "just as the task of the cook is to bring about *pepsis* in foods outside the body, so the task of the doctor is to bring about *pepsis* of the humors inside the body."

75. Cf. Hipp. *Vict.* 1.20 = 6.570L, which describes processing grain as a type of technical mimesis of digestion.

76. As he states, ancient cooks "molded everything in relation to the nature and power of the human" (Hipp. *VM* 3.33–35 = 1.578L), and yet, "we must learn what a human is in relation to the things eaten and drunk" (Hipp. *VM* 20.18–22 = 1.622L).

77. He states: "Completely opposite things result from each of these [sc., hot and cold substances that are either sour or insipid, etc.], not only in a human, but also in leather and in wood and in many other objects that are more insensate than a human [ἀλλὰ καὶ ἐν σκύτει καὶ ἐν ξύλῳ καὶ ἐν ἄλλοισι πολλοῖσιν ἅ ἐστιν ἀνθρώπου ἀναισθητότερα]. For the hot is not that which possesses great power, but the sour and the bland and all the others I mentioned, *both in the human and outside of the human*, things that are eaten and drunk, and things that are applied externally as ointments and plasters." (Hipp. *VM* 15.22–30 = 1.606L, emphasis added). It is not clear precisely whether the author means that sour and bland substances have effects *on* wood or leather, or that they demonstrate certain powers (or undergo certain processes) when contained *in* leather or wood vessels (such as wine stored in leather skins). The latter seems more plausible given the most common reading of the Greek preposition ἐν, especially when compared to a passage where Theophrastus establishes a similar thought but uses the preposition πρὸς to indicate the effect that medicinal plants have *on* lifeless things: "Herbs and shrubs, as has been said, have many virtues which are shown in their effects not only on living bodies but on lifeless ones" [Αἱ δὲ ῥίζαι

> Concerning the powers, it is necessary to examine what each of the humors themselves can do to a human—just as I indeed said earlier—and how they relate to one another in kind. I mean the following: if a sweet flavor should change into another form, not from mixture but altering all by itself, what form would it first become? Bitter, salty, sour or acid? Acid, I think. Therefore, if sweet flavor is the most suitable of all to administer, an acid flavor would the most suitable of the rest. *Thus, if someone should be able to happen upon the best treatment while investigating outside of the body*, he would also be able to always select the best treatment of all.[78]

This particular example seems to be referencing the transformation of grape juice into acidic wine, or sweet wine into vinegar. Indeed, these were some of the liquids that physicians administered to patients as part of therapeutic interventions. These therapeutic products therefore serve as heuristic comparisons for the very corporeal substances that they physically interact with—that is, both analogical models and physical guarantors of the body's fluids—and in so doing they form a recursive interface with the body that they articulate.

This interface establishes other ways in which dietetic foods and their recipient bodies mirror each other. Regimental prescriptions can be recondite, varied, and complex, used both to maintain health and to treat manifold illnesses across the corpus. Despite the multiple gradations of food potencies, Iain Lonie details a broad structural pattern present in dietetic prescriptions, a broad typology that categorizes foodstuffs by their consistencies rather than their ingredients. That is, Hippocratic dietetics divides comestibles into three categories: foods [σῖτα], gruels [ῥυφήματα], and drinks [πότα]. Food generally refers to any solid comestible but can especially imply bread or other cereal-based solids.[79] Drinks refer to liquid, such as water, wine, vinegar, and *oxymel* (vinegar with honey). Gruels thus occupy a middle position, insofar as they are quasi-liquid preparations of grains such as barley, most notably a concoction called *ptisanē* [πτισάνη]

καὶ τὰ ὑλήματα, καθάπερ εἴρηται, πολλὰς ἔχουσι δυνάμεις οὐ πρὸς τὰ ἔμψυχα σώματα μόνον ἀλλὰ καὶ πρὸς τὰ ἄψυχα] (Theophr. *Hist. pl.* 9.18.1). This reading also makes more sense in light of the author's comments at Hipp. *VM* 24.1–3 = 1.634L that seem to reference his earlier passage.

78. Hipp. *VM* 24 = 1.634–636L, emphasis added.

79. Lonie 1977. See Hipp. *VM* 5= 1.580–582L and Hipp. *Acut.* for this distinction, while Lonie also notes that this same structure exists for the Hippocratic treatises *Diseases* 2, *Internal Affections*, *Affections*, *Diseases of Women* 1, *Diseases of Women* 2, and *On Fractures*.

or "barley gruel," prescribed frequently throughout the corpus.[80] Despite divergent and diverse regimental recommendations, Hippocratic texts all agree that drinks and gruels alone should be given to a patient still suffering from acute fever, with solid foods administered only after the patient has reached a "crisis" or "decision point" in the trajectory of their illness. The ubiquity of this foods-gruels-drinks division is widespread enough that Lonie claims it to be "deeply established in Greek medicine."[81]

As common as it is, there are some clues that such a schema was relatively new in the fifth century BCE. The term "gruel" [ῥυφήματα] and its related verbal forms [ῥυφάνειν/ ῥοφέειν] do not appear outside of early medical texts, and the items to which they refer within the corpus bear little resemblance to regular Greek cuisine.[82] All this, as Lonie argues, suggests that the descriptions develop as technical terms to denote a now medicinally marked therapeutic food.[83] In short, gruels, although certainly not invented at this time, first transformed into a medicalized category of food in the mid-fifth century BCE.

The use of gruels in health practices transformed a humble substance into a potent and emblematic medical tool that left its mark on the Hippocratic theories of the body it helped support. For instance, *Ancient Medicine* characterizes the safest, most healthful foods as those that are well balanced with a composition that is physically well mixed. These descriptions align with his characterization of bread, barley cakes, and gruel, which, by virtue of their uniform consistency, are "simple, single, and

80. *Regimen in Health* begins with a discussion of this substance, and this discussion could be why the treatise circulates in the manuscript tradition with the alternate title *On ptisanē*. Hipp. *Ep.* 21 mentions a Hippocratic treatise called *On ptisanē*; cf. Cael. Aur. *Acut. Pass.* 1.11.99–1.12.103. By the next century, ptisanē can be seen as an emblematic therapeutic substance. Arist. *Pr.* 1.37, 863b states: "The gruel [τὸ ῥόφημα] which is given to a fevered case should be such as to provide nutriment in small amount and to have a cooling effect. Barley gruel [ἡ πτισάνη] has this property" (cf. Lonie 1977: 243 n39). Gal. *Alim. Fac.* 1.28.22 = 6.541K (= Diocles fr. 190a van der Eijk) mentions both Diocles and Philotimus using ptisanē, which evinces its growing popularity. Galen will later devote a full treatise to discussing this substance and its application; see Gal. *Ptis.*

81. Lonie 1977.

82. See Davidson 1997, Garnsey 1999, and Darby 2003. It should also be noted, however, that both Aristophanes and Nicophon included ptisanē in lists of what appear to be rustic foodstuffs in the late fifth or early fourth century BCE, around the same time as their appearance in the Hippocratic texts; see Ar. Fr. 159, 412; Nicophon, Fr. 15.

83. Lonie 1977: 243.

whole" [ὅλον ἕν τε γέγονε καὶ ἁπλοῦν].[84] Dangerous foods, by contrast, have imbalanced flavors/powers in them, which get separated off by themselves. Yet the author also extends this logic to the internal composition of the healthy body, claiming that just as healthful foods are well concocted, well mixed, and unmastered by a single flavor or potency, a healthy body should likewise be well concocted and well mixed, without any flavor separated off by itself:

> These things, when they have been mixed and mastered by one another, are neither apparent nor harm the human. But whenever one of them is separated off, and becomes isolated by itself [ὅταν δέ τι τούτων ἀποκριθῇ καὶ αὐτὸ ἐφ' ἑωυτοῦ γένηται], this then becomes clear and harms the human. And with respect to this, of all the foods that are unsuitable for us and harm the human when they enter, each one is either bitter and unmixed or salty, acidic or anything else unmixed and strong, and because of this we are disturbed by them, *just as we are disturbed by things separating off in the body* [ὥσπερ καὶ ὑπὸ τῶν ἐν τῷ σώματι ἀποκρινομένων].[85]

The author goes on to state: "A human is in the best of all conditions when the whole is concocted and in rest, showing no individuated power" [μηδεμίαν δύναμιν ἰδίην ἀποδεικνύμενον].[86] As with gruel, bread, barley cakes, and other "safe" foods, the interior of a healthy body should quite literally taste the same all over, such that the body created within this medical interface not only incorporates the techniques of cooking and medicine but also absorbs the composition of the least harmful foods.[87]

There are, to be sure, other cultural practices that support this denigration and distrust of unmixed foods. For example, unlike a few surrounding cultures, the ancient Greeks drank their wine mixed with water rather than "unmixed," which the author mentions as a dangerous practice.[88]

84. Hipp. *VM* 14.57 = 1.604L; cf. *Hipp. VM* 19.53–57 = 1.620L. Although foods are not healthy in themselves but in how they interact with a patient's nature (cf. Hipp. *VM* 9 = 1.588–590L), the author does characterize weak and mixed foods as the safest and best.

85. Hipp. *VM* 14.37–44 = 1.602–604L, emphasis added.

86. Hipp. *VM* 19.53–57 = 1.616–620L.

87. Interestingly, insofar as raw meat presents the paradigmatic dangerously strong food, the healthy bodily composition imagined by this author is *least like* the animal flesh that it consumes and *most like* the highly processed foods and drinks.

88. Hipp. *VM* 20.30–34 = 1.622L.

Ethical prescriptions likewise lurk behind these regimental prescriptions, insofar as luxuries such as sweets and salty snacks become unsuitable and hazardous. Moreover, the philosophical interest in harmony and mixture extends back to Parmenides, Empedocles, and Alcmaeon, and scholars most often emphasize this tradition as the intellectual lineage of the basic notions underlying *Ancient Medicine*'s conception of the body in health. Nevertheless, it is medicalized cooking techniques that the author himself establishes as his epistemological foundation. More importantly, the Hippocratic treatises that discuss the concept of corporeal mixture [κρᾶσις, σύγκρασις] are all treatises that deal with dietetics (*Airs Waters Places*, *Nature of the Human*, and *Regimen*).[89] In other words, only the texts that employ food-processing techniques as a core practice of medicine formulate corporeality on the model of these techniques, even though this terminology will later become far more widespread in humoral medicine. The body becomes the opposite of raw, unmixed, and overly powerful by incorporating the food-processing techniques of cooking, mixing, and watering down, the same tools that the author sets up as the proper investigative method native to medicine.[90] The body and the gruels that treat it themselves integrate, mix, and blend.[91]

1.3 HIPPOCRATES'S *NATURE OF THE HUMAN*

Although no other Hippocratic text fully replicates the specific theories advanced in *Ancient Medicine*, another treatise, *Nature of the Human* (often

89. Smith (1992: 270–71) notes twenty-seven uses of the terms κρᾶσις/σύγκρασις in the corpus: nineteen in *Regimen*, five in *Ancient Medicine*, two in *Nature of the Human*, and one in *Airs Waters Places*; cf. Bartoš 2015: 39.

90. Parmenides DK28 B16 (D51 LM) = Arist. *Metaph.* 3.5, 1009b22–25/Theophr. *Sens.* 3 claims that the mixture of the limbs affects the quality of thought. Empedocles DK31 B105 (D240 LM) = Porph. in Stob. *Anth.* 1.49.53; DK31 108 (D244 LM) = Arist. *Metaph.* 3.5, 1009b20–21 describes cognition as reliant on a harmonious mixture around the heart. Tracy (1969: 32–66) discusses these passages relative to the philosophical concept of the mean. Similarly, Lonie (1977: 241) notices what he calls the "kinship" between the powers in our bodies and those outside our bodies that affect us, but he attributes this kinship to cosmological theories, not technological integration.

91. Something quite similar happens in the integration of the Hippocratic body and its environment, whereby the external weather physically alters the inner environment while also providing a model and epistemological guarantor of the medical theory into which it is integrated; see Le Blay 2005 for a reflection on this methodology; cf. Bartoš 2015, ch. 2.

attributed to Polybus, Hippocrates's son-in-law),[92] closely mirrors several of its views regarding the healthful composition of the body, insofar as it, too, identifies health as a thorough and uniform mixture of the internal substances. This Hippocratic treatise adopts many of the same ideas about regimen-based treatment as *Ancient Medicine*. Despite considerable similarities, however, *Nature of the Human* privileges a different therapeutic technology as its primary investigative and conceptual tool: purgative drugs.[93] This distinction accompanies a different set of ideas about the body and its constituent substances.

Drugs and pharmacological substances have a long history in the Mediterranean and surrounding cultures. Homer describes a restorative drink called a *kykeon*, often a mixture of wine, honey, barley meal, and cheese, but he also mentions healing *pharmaka* that can reduce the pain of wounds, as well as drugs that make people forget.[94] As mentioned above, Greek authors long viewed Egypt as a land of drugs and drug lore, and medicinal substances from Egypt, Cyrene, Arabia, Persia, and elsewhere flow through Greek recipes, which evinces a certain level of cultural exchange and commerce.[95] Doctors were not alone in producing and administering drugs. Authors of the fifth century BCE likewise mention "root-cutters" [ῥιζοτόμοι] and "drug makers" [φαρμακοπῶλαι], while magical practitioners also employed pharmaka and amulets.[96] Despite this wide range of pharmacological traditions and treatments (on which the corpus draws), Hippocratic authors of the fifth century BCE

92. Polybus is generally identified as the author of *Nature of the Human*, since the description of the vascular pathways that Aristotle ascribes to him at *Hist. an.* 3.3.512b11–513a2 (cf. *Hist. an.* 3.2.511b24–30) appears at Hipp. *Nat. Hom.* 11 = 6.58–60L. It is also later included at Hipp. *Oss.* 9 = 9.174–176L.

93. Hipp. *VM* 19.31–32 = 1.618L also mentions purgative drugs as a more extreme intervention to remove unmixed, separated humors from the body but does not discuss purgatives as revealing the nature of the body. Bartoš (2015: 37–46) discusses the comparable use of dietetics in both *Ancient Medicine* and *Nature of the Human*.

94. For *kykeones*, see Hom. *Od.* 10.234; *Il.* 11.624–641; cf. Hom. *Hom. Hymn Dem.* 210. For topical, pain-reducing drugs, see Hom. *Il.* 4.191; 4.218; 5.401; 5.900; 11.515; 15.394. At Hom. *Il.* 22.94, a snake eats "terrible drugs" (which suggests that the term extends to the plants themselves).

95. See esp. Totelin 2018.

96. Nutton (1985) provides a vivid account of how many different groups were involved in the drug trade, although the bulk of his evidence concerns imperial Rome. Lloyd (1983: 119–35) outlines how Theophrastus relied upon the rootcutters for certain sets of plant knowledge, while Scarborough (2010) also provides an overview of the practices of ancient Greek and Greco-Roman pharmacology.

generally—although not universally—restrict the term *pharmakon* to refer to purgative medicaments in particular.[97] To be sure, the gynecological treatises—which contain the vast majority of pharmacological recipes in the corpus—describe many remedies that could be classified as drugs in a modern sense, including ointments, plasters, enemas, and fomentations.[98] Moreover, insofar as dietetic practices medicalized foodstuffs, no strict line can be drawn between food and medicalized substances.[99] Nevertheless, when Hippocratics used the term *pharmakon*, or recommended "applying drugs" [φαρμακεύειν], they generally meant vomit- and feces-inducing substances meant to "cleanse" or "purge" [καθαίρειν] the patient, either "upward" or "downward."[100] It is these purgative pharmaka that the author of the Hippocratic treatise *Nature of the Human* establishes as his main therapeutic and investigative technologies, and this shift in therapeutic emphasis coincides with subtle changes in his medical theories about the essential corporeal substances and inner processes. Different tools here once again construct a different body.

Nature of the Human sets out, as the title suggests, to explicate the essential "nature" [φύσις] of the human. To do so, the author begins—

97. Totelin (2018: 208) argues that the Hippocratics employ this term to avoid confusion with foods; cf. Totelin 2015; Artelt 1968; Touwaide 1996; Goltz 1974: 297–302; Lonie 1977: 245; von Staden 1999: 257–58; Thivel 1999: 35–37; and Holmes 2010b: 79 n161.

98. Totelin (2009) presents a thorough overview of Hippocratic pharmacology, arguing that the extant material evinces a living oral tradition on which these texts and practitioners drew; see also Stannard 1961; Goltz 1974; Scarborough 1983; Hanson 1991, 1998, 1999; King 1995a, 1995b, 1998; Laskaris 1999; Totelin 2009. Although most Hippocratic pharmacological recipes are found in the gynecological texts, Hippocratic authors refer multiple times to a *Pharmakitis*, which may either be a lost treatise on drug lore or simply each physician's own notes (Craik 2006: 17; cf. Totelin 2018: 204 n10 for references and bibliography).

99. Totelin 2015; Bartoš 2015: 101. Multiple passages distinguish emetic *pharmaka* as distinct from regimen; see Hipp. *Mul.* 1.11 = 8.42L; 1.17 = 8.56L; 1.66 = 8.136L; 2.115 = 8.248–250L; 2.118 = 8.254–258L; Hipp. *Aff.* 22 = 6.232–234L; Hipp. *Loc. Hom.* 23 = 6.314L; Hipp. *Nat. Hom.* 9 = 6.54L; Hipp. *Prog.* 15 = 2.146–148L; *Hipp. Acut.* 56–57 = 2.508–512L; *Hipp. Morb.* 1.14 = 6.164L; cf. Pl. *Rep.* 459c.

100. Purgative drugs were generally used alongside dietetics as a more immediate and powerful way to evacuate harmful matter when regimen alone did not suffice; see Hipp. *Nat. Hom.* 9 = 6.54L; Hipp. *Morb.* 2.72 = 7.109–110L; Hipp. *Mul.* 2.115 = 8.248–250L; and Hipp. *Vict.* 3.67 = 6.592–594L. Even *Nature of the Human* is followed in the manuscript tradition by *Regimen in Health*, which many scholars suggest is either an extension of the same treatise or at least by the same author, and this latter treatise has a section on the use of emetics and enemas. Food and emetic drugs were viewed as two tools within the same therapeutic programs.

just as in *Ancient Medicine*—by denigrating the methodologies of the pre-Socratic physikoi. Those theorists, according to the author, hypothesize about the fundamental substances of the body using philosophical argumentation, while he, once again, seeks to ground his theories more solidly in observation derived from practice, stating, "I do not assert that a human is entirely air, nor fire, nor water, nor earth, or anything at all that is not manifestly present [μὴ φανερόν ἐνεὸν] in a human."[101] Unlike *Ancient Medicine*, however, *Nature of the Human* does not consider the body's fundamental constituents to be flavors and potencies, and instead it presents what became, thanks to Galen, the most standard form of humoral theory from antiquity, characterizing the four basic substances of the body to be phlegm, yellow bile, black bile, and blood.[102] These four substances collectively form our physis, so that the unity of the human whole relies on their combination and cohesion.[103] Health occurs when they are "measuredly disposed" [μετρίως ἔχῃ] to one another in respect to balance, power, and amount [κρήσιος καὶ δυνάμιος καὶ τοῦ πλήθεος], and their physical intermixture is especially important. In other words, *Nature of the Human* attributes health to the body being physically well mixed and identifies diseases as arising when a single substance separates off.

In what way do these four constituent humors qualify as "manifestly present" in humans, as he states? The author is clear that he does not mean that these substances appear naturally, either in a healthy body or through spontaneous eruptions in illness, but that they can be *made* manifest through particular therapeutic interventions. He states that phlegm-attracting drugs cause patients to vomit phlegm, and that the same happens with yellow and black bile–attracting drugs. In addition, blood is drawn from the body through the act of venesection:

> You would recognize in these following things, that these things are not all one, but each of them has its own power and nature. For if you give any person any drug that draws out phlegm, he will vomit phlegm for you, and if you give any drug that draws out bile, he will vomit bile for you. According to the same things, black bile is also purged, if you give

101. Hipp. *Nat. Hom.* 1.4–7 = 6.2L. The author explicitly follows these comments with reference to Eleadic arguments regarding the unity of being and then mentions Melissus.

102. Hipp. *Nat. Hom.* 4.1–3 = 6.38–40L.

103. He never refers to these substances as "moistures" [ὑγρότητες] and only once refers to them as "humors/tastes" [χυμοί] (Hipp. *Nat. Hom* 15.29 = 6.68L). Instead, he discusses them more abstractly as the "things" of the body, or the substances that the body possesses, as in the above passage.

> any drug that draws out black bile; and if you cut anything on his body so as to create a wound, his blood will flow.[104]

In short, his confidence in the self-evidence of the four humors cannot be separated out from his application of particular therapeutic practices and tools, since the nature of each specific humor only becomes manifest when drawn off by different purgative drugs or extracted through therapeutic wounding.[105] Elsewhere he claims that exercise and diet affect the balance of these humors on a quotidian basis, while the seasons affect the balance over the course of a year.[106] Some of these fluctuations are apparent from disease-states, but he claims that the clearest evidence of these "manifest" humors is that if you give the same purgative drug to a person four times in a year, "he will vomit the most phlegm stuff in the winter, the most moist in the spring, the most bilious in the summer, and the blackest in autumn."[107] This author consistently suggests that the effects of emetic drugs reveal the nature of humans and their primary liquids.[108] Drugs and venesection articulate and reveal the unseen corporeal interior by drawing out its inner substances and making them manifest to the theoretical eye.

How does this author think that these purgative pharmaka work? The

104. Hipp. *Nat. Hom.* 5.19–28 = 6.42L, emphasis added.

105. He also adds that they have different names, and so are separate substances "by custom" [κατὰ νόμον], as well as being manifestly different in feel, so must be separate in nature [κατὰ φύσιν] (Hipp. *Nat. Hom.* 5.1–9 = 6.40–42L).

106. As with Hipp. *Loc. Hom.*, this text likewise describes how the pathological consequences of these seasonal fluctuations land on "places" and "parts," with a vascular system seemingly transporting the humors (and diseases) between these locations (see esp. Hipp. *Nat. Hom.* 10–11 = 6.56–60L). Hipp. *Nat. Hom.* 9.1–9 = 6.52L provides a general programmatic statement about treating all therapy/healing as allopathy, whereby opposites cure opposites, and establishes two etiological categories: diseases caused by regimens affecting individuals and those caused by air affecting communities.

107. Hipp. *Nat. Hom.* 7.67–71 = 6.50L. This passage seemingly contradicts his earlier arguments that we can be certain of the four humors, since phlegm-attracting drugs attract phlegm at any age at any time of year (and same with other moistures); the author here suggests that the same emetic will elicit various humors in different seasons.

108. Cf. Hipp. *Nat. Hom.* 6.1–13 = 6.44L, where he imagines that even his mono-humoral opponents rely on their interactions with drugs when formulating their theories. Compare Hipp. *Art.* 12.4–5 = 6.24–26L, which describes the ways that the physician can force certain substances to emerge from the body (phlegm, liquids, etc.), which then "inform against" the hidden inner affections.

basic logic appears to be that like substances attract like substances, insofar as purgative drugs draw out whatever is especially in accordance with their nature [κατὰ φύσιν].[109] First they evacuate the most unmixed, unintegrated individual humor, then the mixed substances. He likens the power that they exert on the body to the power displayed by (drug-producing) plants in the ground:

> For just as with plants and things that are sown, whenever they enter the earth, each draws that which accords with its own nature in the earth, and acid and bitter and sweet and sour and many sorts of substances are present. And so first it draws the majority of whatever especially accords with its own nature to itself, but then draws the rest. Drugs do this sort of thing in the body. The ones that draw bile clean out the most unmixed bile first, then mixed; and the phlegm drugs draw out the most unmixed phlegm first, then mixed. And the hottest and reddest blood flows first from those [animals] who have had their throats cut, then the more phlegmy and bilious blood flows.[110]

On the surface, this looks like a straightforward heuristic analogy, as discussed in the introduction, whereby the author uses the known behavior of the plants (which supply purgative pharmaka) to understand how purgative drugs operate inside the body. Yet how does the author know that plants first attract the flavors/humors to themselves most in accordance with their nature? He assumes this to be evident because it is the way that purgative drugs work in humans. In other words, he uses the model of plants to understand the body, and he uses the model of the body to understand plants. In this recursive loop, pharmaka-producing plants and the bodies that they treat mutually determine each other.

Beyond using plants to conceptualize the inner workings of the body (and vice versa), a closer look at what emetic drugs would have looked like in the fifth century BCE shows another potential instance of a mimetic relationship between a therapeutic technology and the body that it treats. It is difficult to reconstruct emetic pharmaka of the era exactly,

109. *Nature of the Human* seems to make a distinction between the natural substances in plants (bitter, sweet, salty, acid, etc.) and the congenital elemental humors of humans (phlegm, blood, yellow bile, black bile). Unlike in *Ancient Medicine*, these plant substances can be *like* human humors, but do not seem to be identical to them.

110. Hipp. *Nat. Hom.* 6.28–41 = 6.44–46L.

since Hippocratic gynecological texts, which contain the vast majority of pharmaceuticals in the corpus, describe how to prepare other types of drugs with detailed lists of substances, dosages, and preparation instructions, but the prescriptions for purgative medications are generally far less explicit, often consisting solely of the recommendation for an emetic drug to "clean upward" or a laxative to "clean downward," as though any such pharmakon would do.[111] Theophrastus's *Enquiry into Plants* thus provides a better picture of what these drugs might have looked like, notably perhaps the most powerful emetic: the so-called driver [ἐλατήριον], a purgative prescribed or used within the Hippocratic corpus sixty-seven times. This emetic consisted of the "juice" [ὀπός] of the squirting cucumber plant, which purged both upward and downward.[112] Other purgatives include the juice of spurge, which mainly purged downward;[113] the juice of cyclamen, which cleaned the head;[114] and the juice (or root) of thapsia, which purges both upward and downward.[115] In all these instances, the medicalized juice of each plant seems to refer to moisture collected from medicinal plants via small incisions in either their roots or stalks (although a small group of purgative plants emit such medically potent liquid naturally). Theophrastus describes this technique:

> Juice collection [ὁ ὀπισμὸς] occurs either from the stalks, as with spurge, wild lettuce, and many plants, from the roots, or from the head, as with poppies. For this alone is extracted in this way, and it is unique in this. For others, the juice [ὁ ὀπὸς] arises spontaneously, as a certain tear-like substance... but the majority comes from incision.[116]

The techniques here discussed are those of the so-called root-cutters [ῥιζοτόμοι], the collectors and producers of medicinal substances who would have competed for patient attention alongside the Hippocratic

111. Cf. Pl. *Rep.* 406d–e, which presents the same description of the effects of "drinking a *pharmakon*" [φάρμακον πιὼν], which likewise suggests a liquid substance.

112. See Theophr. *Hist. pl.* 9.9.4; 9.9.5; 9.14.1–2; 9.15.6. The driver is also used as an emmenagogue at *Mul.* 1.74 = 8.154L and elsewhere.

113. Theophr. *Hist. pl.* 9.11.8.

114. Theophr. *Hist. pl.* 9.9.3.

115. Theophr. *Hist. pl.* 9.20.3.

116. Theophr. *Hist. pl.* 9.8.2. Theophrastus further explains how "juice" can also be produced by soaking bruised vegetation and then straining it, but states that doing so results in an inferior product (*Hist. pl.* 9.8.3).

physicians. In addition to juice extraction, root-cutters collected "herbs" [πόα], leaves, fruit, and whole roots (not to mention non-plant-based substances).[117] That said, the very moniker "root-cutter" seems to reference the juice-extraction process and suggests the particular importance of this method in their practices and identities.[118] Moreover, these extracted liquids appear as emblematic emetic drugs in popular culture more broadly. For instance, Aristophanes includes them as part of a reducing regimen that will make Aeschylus's verses thinner through the application of small poems, walks, white beets, and "nonsense-juice filtered off from books" [χυλὸν διδοὺς στωμυλμάτων ἀπὸ βιβλίων ἀπηθῶν].[119] Thus, when the author of *Nature of the Human* references pharmaka that draw out phlegm or bile, extracted purgative juices would have functioned as common, perhaps even iconic, examples of emetic drugs.[120]

If these are the tools that the author of *Nature of the Human* uses as his foundational demonstrative technologies, it is worth noting that there is a superficial resemblance between the extracted juice of certain plants and the humors they are explicitly said to make manifest. In some ways this resemblance must have been normative. When patients vomited "phlegm and bile" or passed them through their stools, the substances evacuated

117. Hipp. *Epid.* 2.2 = 5.106L mentions several drug preparation techniques, including drying, crushing, and boiling, but then states, "I pass over most."

118. See Theophr. *Hist. pl.* 9.1.5–6, which describes incision techniques for extracting the "gum" or "tears" of trees. The pseudepigraphic Hippocrates *Ep.* 16, addressed to the "best root-cutter," Crateuas (perhaps an allusion to the famous pharmacist and physician of Mithradates VI of Pontus), mentions gathering plants from high mountains (which are stronger, moister, and more bitter), and flowers from beside rivers, ponds, and springs (which are weaker, slacker, and sweeter). He continues, "All that have flowing juice or liquid [χυλοί τε καὶ ὀποὶ] should be brought in glass jars" (Hipp. *Ep.* 16.25 = 9.344L = W. Smith p. 71–73). See also Dioscor. *MM* 9 Beck.

119. Ar. *Ran.* 939–943. Sophocles also wrote a play called *The Rootcutters* (see Soph. Fr. 534), which illustrates how well known these practitioners and their methods might have been.

120. These extracted juices could be employed to multiple medical ends, but they certainly served as purgatives. Indeed, there are other purgative substances employed in Hippocratic treatises, but these do not appear to be called *pharmaka*. For example, Hipp. *Mul.* 2.115 = 8.248–250L prescribes hellebore as an emetic and then recommends a "pharmakon to purge downward" as though these are two distinct pharmaceuticals; cf. Hipp. *Morb.* 2.72 = 7.109–110L. By contrast, Theophrastus makes no such distinction (e.g., *Hist. pl.* 9.15.4). Hellebore later becomes the paradigmatic purgative; see, for example, Hipp. *Ep.* 18–21 = 380–392L = W. Smith p. 92–101.

from the body would have been, to say the least, messy. There are no doubt instances where we can vomit out the contents of our stomachs with a mucus-like foam, while at other times we can run out of food to eject and end up heaving out yellow gastric acids. Yet singling these instances out as privileged observational moments occludes a huge body of observations that would be less convincing instances of a single humor exiting our system. Indeed, black bile has no obvious physical correlate, yet the author indicates that it too is equally as "manifest," coming directly before blood, as we vomit ourselves to death.[121] This interdependence of therapeutic practices, technologies, and theories prohibits attributing these arguments to straightforward empiricism, since the observational boundaries are constructed by these codependent forces. The notion that purgative drugs make phlegm and yellow and black bile patently observable is in no way clear cut. The physical appearance and extraction process of these emblematic purgative pharmaka thus no doubt implicitly helped support a humoral framework for human physis. If a healthful body looks like gruel (a healthful food), a diseased body looks like a purgative drug, separated off in a storage vessel.

It is important not to overstate the importance of these resemblances. Purgative drugs did not create the idea that essential humors were responsible for disease, which predates this treatise, and the author himself invokes "custom" [νόμος] as justification for the existence of these humors, in addition to sensory evidence, which reveals the potential longevity of the concept.[122] Moreover, these drugs do not demand that there be *four* humors in particular (a number that seems instead to be influenced by the philosophical precedent of Empedocles's four-substance ontology). Nevertheless, the author of *Nature of the Human* repeatedly invokes purgative drugs as his foundational demonstrative technology, and his essential internal substances resonate with the physical appearance and production of emblematic examples of these pharmaka. In some sense, the extracted juice in a vial sits as a visible proxy for illness in the body it treats in a way that resembles the logic underpinning sympathetic magic, whereby a substance can produce effects that resemble it in some way (e.g., a red stone can increase blood flow, or reattaching two parts of a broken stick can mend a broken bone). Drugs thus play a similar role in providing an image of the same excessive liquids they are trying to extract.

121. See Stewart 2018 for Galen's theory of black bile and the debts that his conceptions owe to Hippocratic precedents.

122. Hipp. *Nat. Hom.* 5.1–9 = 6.40–42L; see note 103 on p. 48.

1.4 MEDICAL IMPLEMENTS AND HIPPOCRATES'S *MORB. 4/GENIT./NAT. PUE.*

Up until this point, we have been examining the corporeal fluids and the role that they play in pathogenesis. Despite the emphasis placed on these substances in theoretical explications of disease, the Hippocratic authors do speak about problems caused by internal arrangements as well. In fact, near the end of *Ancient Medicine*, discussed above, the author sets up a dichotomy between two types of diseases: those that arise from "powers" and the separation of the essential humors of the body, and those that arise because of internal "structures" [σχήματα]. He clarifies that by "structures" he means the shapes (broad, long, round, tapering, etc.) and textures (dense, fleshy, porous, etc.) of each inner part, which affect how fluids are attracted, collected, or repelled.[123] Within this broader framework, he makes an assertion that is relatively commonplace as a physical assumption in later treatises—namely, that broad, hollow, and tapering structures draw fluid into themselves. He claims that cupping vessels [αἱ σικύαι] illustrate this phenomenon, and that the bladder, head, and womb are all the same shape [σχῆμα] such that they all "manifestly attract and are always full of moisture brought into them."[124] In the final sections of his treatise, then, *Ancient Medicine* opens up a second type of interface, now between bodies and medical tools that we are perhaps more accustomed to calling "technologies"—that is, physical implements and instruments. Yet, if *Ancient Medicine* ends by indicating that these implements might operate as heuristic tools, it is the Hippocratic treatise *Diseases* 4 that adopts this project more fully. In so doing, it comes closer than other Hippocratic treatises to characterizing the body as a functional, organized system.[125]

Diseases 4 seems closely related to both *Generation* and *Nature of the Child*, insofar as all three treatises propose the same humoral theory otherwise unattested among the Hippocratic Collection, and all three make similar cross-references to *Diseases of Women* 1, which the author identi-

123. Hipp. *VM* 22 = 1.626–634L.

124. Hipp. *VM* 22.29–31 = 1.628L; cf. Hipp. *Morb.* 4.35 = 7.548L. Hipp. *Morb.* 1.15 = 6.168L states that the cavity attracts liquids to itself when heated, an assertion that implicitly uses the cupping glass as a model (cf. Hipp. *Morb.* 2.8 = 7.16L; 2.11 = 7.18L). Similarly, Hipp. *Nat. Pue.* 8 = 7.506L mentions that newly created hollows attract substances into developing bones. See Dean-Jones 1994b: 65–66.

125. Excluding those Hippocratic treatises dated to the Hellenistic period. An expanded version of this section can be seen in Webster 2023.

fies as a text that he also wrote.[126] A survey of this last text and other gynecological treatises reveals just how many implements and medicaments were involved in women's care. Far beyond the simple foodstuffs or emetics present in the regimen-based therapeutic treatises, the gynecological tool kit included cupping vessels, sitz baths, fomentations, suppositories, birthing implements, and menstruation-inducing medicaments. Perhaps it is no surprise, then, that the corporeal interior imagined by this author likewise teems with an expanded technological set. Indeed, *Diseases* 4 has received much scholarly attention for its proclivity for using vivid analogies to imagine the body's interior fluid movements, many of which involve tools.[127] It is crucial to note that how many of these comparisons are medical technologies, many described within the gynecological treatises. In other words, as this author's preferred medical tools changed, so too does the body that he articulates.

The author of *Diseases* 4 presents some ideas that resemble the claims of the Hippocratic treatises discussed so far. He advances a somewhat modified four-humor theory that attributes illness to the imbalance of phlegm [φλέγμα], blood [αἵμα], bile [χολή], and water [ὕδωρ, ὕδρωψ].[128] Yet whereas both *Ancient Medicine* and *Nature of the Human* characterized health as the uniform mixture of their respective essential substances, this author assigns specific locations for each humor in the body. That is, *Diseases* 4 identifies health with the correct and appropriate humoral separation rather than with unified mixture. Four sites, called "springs" [πηγαί], provide these primary humoral stations. The head provides the spring for phlegm, the heart for blood, the gallbladder for bile, and the spleen for water. Despite the connotations of the term "spring," which

126. Hipp. *Genit.* 4 = 7.476L; Hipp. *Nat. Pue.* 4 = 7.496L; and Hipp. *Morb.* 4.26 = 7.612L. In each case, the passages alluded to can be found in Hipp. *Mul.* 1 (see Hipp. *Mul.* 1.1 = 8.10L. Moreover, passages at Hipp. *Mul.* 1.44 = 8.102L and Hipp. *Mul.* 1.73 = 8.152–154 both claim, "How milk is produced, I have explained in *Nature of the Child in Childbirth*," which seems to reference the *Genit./Nat. Pue./Morb.* 4 cluster. For the intimate connections between these treatises and the other gynecological texts, see Lonie 1981b: 51–54 and Dean-Jones 1994b: 10–13.

127. See esp. Lonie 1981a. Lonie asks whether this reliance on structural comparisons evinces a commitment to a "mechanistic" account of inner fluids, which move according to universal physical forces, rather than moving themselves by a type of "vitalistic" power and agency only found in living things. Yet the same Hippocratic writer also employs comparisons between embryos and plants in *Generation/Nature of the Child*, and these seem to suggest that living entities possess unique capacities for self-directed growth.

128. Hipp. *Pharm.* 31–34 Schöne, a later Hippocratic treatise that does not appear in Littré's edition, adopts this same four-humor theory.

might seem to suggest that these parts supply a source or production point for these humors, the springs instead provide internal reservoirs for the four innate fluids, which get replenished (and perhaps replaced) through our food and drink. When we eat, phlegm-like, bloodlike, bile-like, and watery substances first land in the "cavity" [κοιλίη], which we can loosely identify with the hollow digestive tract stretching from stomach to rectum.[129] From this channel, each spring draws off its own respective humor through interconnected vessels or "veins" [φλέβες] according to the principle "like with like." The springs then pass on their individual humors to the "body" [σῶμα], which we can loosely identify with the flesh surrounding the cavity and springs. When we consume food, the springs draw off new humors from our cavity. This process pushes the older humors, which have been broken down, concocted, and made thinner by the body's heat, back into the cavity, where they are expelled through the orifices, generally in the form of urine and stool (although they can also exit through the mouth and nose).[130] In a healthy body, this humoral procession moves with numerical regularity, enacting a three-day cycle for fluids and a two-day cycle for solids.[131] Disease arises when one of the humors dominates the body as a whole, when an empty spring draws moisture back from the body in the wrong direction,[132] when an individual spring attracts too much of its humor, or when the humors stagnate too long in one location.[133] The springs thus provide reservoirs that moderate excessive or insufficient humors, which must proceed in their basic downward flow through the body (see fig. 1).

More than any of the Hippocratic texts already mentioned, *Diseases* 4 thus concerns itself with articulating the specific forces at work inside the body. When describing how each spring attracts its own proper humor from the cavity, however, the author invokes the same basic comparison that *Nature of the Human* has already used to explicate the action of emetic drugs in the body, except that *Diseases* 4 uses the attractive capacity of plants to explicate the innate capacity of inner fluids to attract

129. Cf. Hipp. *Nat. Pue.* 6 = 7.498L. See also Holmes (2010b), who uses "cavity" to refer to the unseen corporeal interior, although this author uses a far more restricted meaning.

130. Hipp. *Morb.* 4.11 = 7.564L.

131. See esp. Hipp. *Morb.* 4.11 = 7.562–564L.

132. Hipp. *Morb.* 4.2 = 7.544L; cf. Hipp. *Morb.* 4.8 = 7.556L.

133. For the latter, see Hipp. *Morb.* 4.14 = 7.570L.

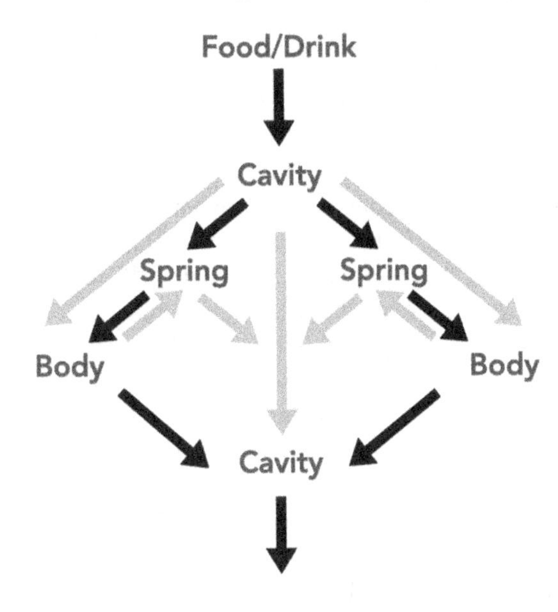

FIGURE 1. Flowchart of four humors in *Morb.* 4, with healthy flow lines in black, and pathological or dangerous flow lines in gray.

what is similar to themselves (not just purgative drugs introduced into the system):

> For in this way the earth contains countless manifold potencies. And for everything that grows in it, the earth provides a moisture that is similar to what each plant already contains in itself, similar according to its kind. And each plant draws nourishment from the earth that is like itself. For the rose [τό ῥόδον] draws the moisture [ἰκμὰς] from the earth that is of one sort, and garlic [τὸ σκόροδον] draws moisture from the earth of a different sort, such as it itself already is in its potency. And all other plants likewise draw each thing from the earth according to its kind.[134]

His argument is that just as these plants each attract their own proper moisture from the earth in accordance with their kind [κατὰ συγγενὲς], each of the four springs must likewise attract its own particular moisture from the cavity, according to the principle "like moisture attracts like"

134. Hipp. *Morb.* 4.3 = 7.544–546L. He also describes the plants as deriving *potencies* [δυνάμιας] from the earth, which highlights their medicinal aspects; cf. Hipp. *Vict.* 1.2 = 6.468. Dioscorides and Galen both employ the language of "potency" [δύναμις] to discuss the medical effects of drugs.

[ἕλκει ἡ ὁμοίη ἰκμὰς τὴν ὁμοίην].[135] Yet it is important to remember that plants do not simply represent some intrinsic biotic life force; they were also used as medical technologies in the form of food and drugs (as we already saw in *Nature of the Human*).[136] In fact, the particular plants that this author invokes have particular uses in women's care. Roses have a particular association with uterine treatments, and their petals and leaves are used in vaginal suppositories,[137] uterine fumigations,[138] potions to clean out the lochia after birth,[139] fomentations,[140] sitz baths,[141] and topical remedies.[142] There are also forty-seven uses for "rose unguent"[μύρον ῥόδινον] and twenty-five for "rose oil" [ἔλαιον ῥόδινον] in the gynecological texts.[143] Garlic appears in Hippocratic corpus seventy-two times, of which forty-two occur in the gynecological treatises, where it is used as the sole ingredient in a uterine fomentation,[144] to treat uterine dropsy,[145] and to correct a twisted uterus,[146] among other such therapies.[147] The author also invokes a third pharmaceutical, silphium [τό σίλφιον], to illustrate how plants attract their own unique moistures, since silphium famously would only grow in certain moisture-specific soil found only in Libya. Silphium is another plant used for multiple medical purposes (although it can be used as a simple condiment too), and forty-six of its seventy-five references appear within gynecological recipes, where it helps expel a stillbirth and induce abortion.[148] Thus, invoking these particular plants as heuristic aids aligns with the established tendency of early Greek medical authors to merge therapeutic technologies and the bodies

135. Hipp. *Morb.* 4.2 = 7.544L.

136. For the prevalence of plant analogies in the *Morb. 4/Genit./Nat. Pue.* Cluster, see Lonie 1969; Lloyd 1978: 45–49; Holmes 2017.

137. Hipp. *Nat. Mul.* 84 = 7.406L; Hipp. *Mul.* 2.87(2) = 8.380L.

138. Hipp. *Nat. Mul* 34(11) = 7.372L; Hipp. *Mul.* 1.51 = 8.110L; Hipp. *Mul.* 2.86(7) = 8.378L; Hipp. *Mul.* 2.97.2 = 8.398L.

139. Hipp. *Nat. Mul.* 109(23) = 7.428L.

140. Hipp. *Mul.* 2.57 = 8.344L.

141. Hipp. *Mul.* 2.58 = 8.346L; Hipp. *Mul.* 2.101 = 8.406L.

142. Hipp. *Mul.* 1.49 = 8.108L.

143. According to a Thesaurus Linguae Graecae search of these terms.

144. Hipp. *Mul.* 2.133 = 8.284–286L.

145. Hipp. *Nat. Mul.* 2 = 7.314L.

146. Hipp. *Nat. Mul.* 8 = 7.322L.

147. For the use of garlic as both a foodstuff and a medicinal remedy, see Totelin 2015. It also appears at Ar. *Plut.* 716–721 and *Eccl.* 403–407 in treatments within the Asclepieia.

148. For a discussion of silphium, its medicinal use, and scholarship on the issues surrounding it, see Totelin 2009: 158–60.

that they treat, even as it evinces a slightly different set of therapeutic commitments linked to the expanded pharmacology of Hippocratic gynecology.[149]

In addition to the plant-like capacity housed in the moistures of each spring, the text further explicates how structural arrangements affect these dynamics. The author first discusses the head, attributing its capacity to draw phlegm upward to its basic hollow form, and, like the writer of *Ancient Medicine*, he compares this form to a cupping vessel:

> I say that when whatever is phlegm-like in food or drink enters into the cavity, some the body attracts to itself, and some the head attracts, insofar as the head is hollow and set on the body, just like a cupping vessel [ὥσπερ σικύη]. And the phlegm, insofar as it is gluey, follows one part after another into the head. And as new phlegm, arising from the food, occupies the head, an equal amount of old phlegm is pushed out by it, and because of this, whenever someone drinks or consumes anything phlegm-like, he will cough out phlegm.[150]

Cupping vessels worked by being heated (or having a flaming piece of lint placed under them) and then pressed against the skin. As they cooled, they formed a seal and drew flesh (or pus, blood, etc.) up into their interior. They could also be placed over a wound, or beside small incisions to encourage greater flux.[151] They are the sole medical tools surviving from the fifth century BCE and were among the most iconic elements of medicine, appearing as an icon on the *Peytel Aryballos* of the so-called Clinician Painter, located at the Louvre (fig. 2); on a grave stele of a doctor from 480 BCE (fig. 3); as well as on the founding stele of the temple of Asclepius in Athens.[152] Different cup sizes and shapes were thought to have vary-

149. If different tools articulate different bodies, different bodies likewise articulate different tools. Whereas *Nature of the Human* invoked purgative plants to understand the inner processes involved in purgative drugs, the substances that those plants contained were not identical to the humors found in the body. For *Diseases* 4, the plants physically provide the "phlegm-like" and "bloodlike" moistures that enter the cavity and flow through the body (Hipp. *Morb.* 4.2 = 7.548L; cf. 4.5 = 7.550L). In this way, the comparison is an analogy, but one where the conceptual and physical continuums actually blend together.

150. Hipp. *Morb.* 4.4 = 7.548L.

151. Hipp. *Morb.* 2.55 = 7.86L; Hipp. *Int.* 51 = 7.296L.

152. "Telemachus Relief," founding stele of the Sanctuary of Asklepios, Athens, ca. 420/419 BCE, Acropolis Museum, EAM 2490. In this case, the cupping vessel appears alongside forceps and another, indeterminate implement.

FIGURE 2. *Peytel Aryballos,* 480–470 BCE, clay, 8.6 × 8.8 × 8.8 cm, Louvre, Paris, Louvre CA 2183, photo via G. Dagli Orti.

ing effects and thus could be used to multiple therapeutic ends.[153] They were applied topically to move humors around the body, including application to the breasts to control menstruation, or to the head to treat headaches.[154] As such, when this author compares the phlegm-attracting head to a cupping vessel, he is describing an implement that could be

153. Cf. Hipp. *VM* 22 = 1.628–630L and Hipp. *Medic.* 7 = 9.212L.

154. For application to the head to treat headaches, see Hipp. *Aff.* 4 = 6.212L. For application to the breast, see Hipp. *Mul.* 2.110 = 8.236L; cf. Hipp. *Epid.* 2.6.16 = 5.136L; Hipp. *Aph.* 5.50 = 4.550L. *Diseases of Women* includes thirty references to

FIGURE 3. Greek relief, ca. 480 BCE, stone, Antikenmuseum Basel und Sammlung Ludwig, Basel, inv. BS 236.

physically affixed to the head in therapeutic use. This example presents another instance where a prosthetic appliance is absorbed into the conceptual fabric of the body.

The author then moves on to describe how each of the other three

cupping glasses. Of the seventy-five mentions of cupping glasses in the corpus, forty-five appear in gynecological contexts. For general Hippocratic use of cupping vessels, see Bliquez 2015.

springs attracts its own proper moisture. The (hollow) spleen attracts water down into it through gravity flow, while water can be drawn upward from the legs when vessels become dry and attract water back into them.[155] The gallbladder attracts bilious moisture because of the tendency for like to attract like and because the thin vessels that lead to it prevent thicker moistures from arriving.[156] Lastly, the heart attracts bloodlike moisture to itself, presumably because it abounds in blood already, which thus activates the principle "like humors flow to like."[157] These explanations thus incorporate the homogenous attractive capacity of the fluids themselves along with other types of forces, including suction and gravity flow, presumably because the author sees no real difference between these forces as types of physical explanations.

After these individuated accounts, the author then generalizes, stating that each of the springs will supply moisture to the body when it is full, but will draw from it when empty (as will the cavity). To illustrate this point, he draws a comparison to another set of implements, this time three interconnected bronze dishes:

> For it is just as if someone should pour water into three or more bronze vessels [χαλκεῖα], having positioned and fit them together on level ground, and should arrange them as nicely as possible, and then, having fit pipes into openings in them, should gently pour water into one of the bronze vessels until they were all full of water. For water will flow from one into the other bronze vessels until the rest are also filled. And whenever the bronze vessels are full, if someone draws off water from a single one, the water would flow backward and give back to that one vessel. And the bronze vessels will be emptied again in the same way as they were also filled. This is also the case in the body. For whenever foods and drinks enter the cavity, the body draws off [moistures] from the cavity and is filled along with the springs. But when the cavity is emptied, moisture is given back again, just as one bronze vessel received [water] from the others.[158]

Scholars often treat the three bronze vessels as though they illustrate how the four humors balance each other in each of the reservoirs, but the three vessels likely represent the cavity, a spring, and the body in each individ-

155. Hipp. *Morb.* 4.6 = 7.554L.
156. Hipp. *Morb.* 4.9 = 7.560L.
157. Hipp. *Morb.* 4.7 = 7.554–556L.
158. Hipp. *Morb.* 4.8 = 7.556–558L.

FIGURE 4. *Footbath with Stand,* late 5th or early 4th c. BCE, bronze, 21.5 cm high, Metropolitan Museum of Art, New York, accession no. 38.11.5a, b.

ual reservoir system.[159] Because of the vivid description, many commentators have taken this comparison as a quasi-experiment completed by the author (and potentially to be completed by the reader). Yet, we might ask, why does the author insist on *bronze* vessels in particular? These would have been especially difficult to drill through and expensive to ruin, as an example from the late or early fourth century BCE illustrates (fig. 4). Moreover, *Diseases* 2 describes boring holes in pot lids, presumably terra-cotta, thereby illustrating other, less costly, options.[160]

The *Peytel Aryballos* (fig. 2), mentioned above, suggests a possible answer, insofar as it depicts a very similar bronze vessel as a prominent medical implement alongside the cupping vessel. In the vase painting, the

159. See Lonie 1981a: 130; Lloyd 1978: 45–49; and Schiefsky 2005: 292–93 for the former interpretation. Yet the author explicitly introduces the bronze vessels as a way to conceptualize the fluid dynamics of these three components, and he refers to them in explicating the comparison. Moreover, the four springs do not directly communicate or contain the same moistures. For instance, Hipp. *Morb.* 4.9 = 7.560L describes how the gallbladder contains only bile, while the head, spleen, and heart contain all four humours, although they each contain a majority of their own natural moisture.

160. Hipp. *Morb.* 2.26 = 7.42L

vessel sits directly in front of the patient, perhaps to collect the blood produced through the venesection performed directly above the dish.[161] That is, the bronze vessel acts as a tool for imagining the inner reservoirs insofar as it already serves as an external reservoir in the medical intervention.

The author of *Diseases* 4 invokes bronze vessels in other comparisons as well, namely when he explains how bile can nourish fever. He accounts for this phenomenon by comparing the process to boiling water in a bronze vessel with unguent oil mixed in with it: the water evaporates, while the majority of oil remains.[162] This analogy, as specific as it is, references a production practice seen exclusively in the gynecological treatises, whereby a physician can make salves, plasters, eye medications, and other such medicaments by leaving the ingredients in a bronze vessel over low, sustained heat or simply by leaving it in the sun.[163] Unguent appears again in *Diseases* 4, when the author explains why swelling occurs after something strikes the flesh: the collected fluids block their own exit passages, just as unguent can block its own way out of a leather oil vessel [λήκυθος σκυτίνη].[164] In other words, the behaviors of the implements and technologies used in his therapeutic practices become a large part of his conceptual framework for understanding the motion of humors within the body.

Indeed, the remainder of the text contains numerous similar analogies to medical devices and medicalized substances. Sometimes these are medicalized foodstuffs, such as when he explains why hot environments disturb the body and produce disease by comparing this phenomenon

161. Krug (2012) presents an alternative interpretation.

162. Hipp. *Morb.* 4.18 = 7.580L.

163. Hipp. *Ulc.* 12 = 6.412L describes making a topical salve of grape juice, wine, honey, myrrh, and flower of copper in a bronze vessel; Hipp. *Haem.* 3 = 6.438–440L uses a bronze vessel to make a salve; both involve leaving the contents of the vessel out in the sun. *Mul.* 1.98 = 8.224L uses one to make a plaster for gout; Hipp. *Mul.* 1.102 = 8.224L and 1.104 = 8.226L both describe an eye treatment that involves drying the contents of the vessel in the sun. Hipp. *Mul.* 1.105(3) = 8.228L slowly boils an eye medication. This process also appears in a related comparison in the sibling treatise *Generation*, which explains how sex determination relies on the amount of the female "weak" and male "strong" seed from each parent mixed together (Hipp. *Genit.* 6 = 7.478L): "It is just as if someone were to mix wax and fat together, making the amount of fat greater, and melt them beside a fire: as long as the mixture is liquid, it is not obvious which one predominates. But when they solidify, then it becomes clear that the fat exceeds the wax in amount. And this is how it is with the male and female seeds too."

164. Hipp. *Morb.* 4.20 = 7.588L; cf. a similar comparison at Hipp. *Mul.* 1.33 = 8.78L.

to churning horse milk into cheese and butter [ἱππάκη],[165] or how cold causes illness in the same way that cold fig juice coagulates milk into cheese.[166] Other instances involve medicalized implements, such as when he claims the bladder stones coalesce just as precipitate collects together in a swirling wine cup [κύλιξ] or a bronze vessel,[167] even as wine is one of the treatments described elsewhere in the corpus when children get bladder stones.[168] Even comparisons that may not at first appear medical often still come from the medical realm. For example, he compares how bladder stones can dissolve to slag melting out of iron ore during the process of refinement.[169] This comparison does not directly reference any therapeutic tool, but physicians were likely directly involved in the design and manufacture of their metal instruments,[170] and bladder stones, for which this comparison is invoked, was one of the few conditions treated by invasive surgery in early Greek medicine (which would have

165. Hipp. *Morb.* 4.20 = 7.584L. This mention of Scythian cheese might reflect borrowing from Hecataeus (see Lonie 1981b: 70), but since it also appears at Hipp. *Aer.* 18, it may be considered part of Hippocratic medical tradition, if only as a topos of exotic foodstuff.

166. Hipp. *Morb.* 4.21 = 7.590L. Hom. *Il.* 5.902–905 compares the speed of fig juice coagulating milk to the speed at which Paiēon healed Ares, which suggests that this particular comparison had longstanding medical overtones. By comparison, Emped. DK31 B33, ln. 4 (D72 LM) = Plut. *De amic.* 5.95A compares the effects of fig juice on milk to love's capacity to bind and fix, while Arist. *Gen. an.* 1, 729a12–14, compares the effects of semen on menstrual blood in the formation of the embryo to this interaction of fig juice and milk; cf. Arist. *Gen. an.* 2, 737a14–16; 3, 771b19–27. Perhaps this comparison supplies an even closer connection to embryology, a connection that fits neatly with the Hippocratic author's use of it in describing diseases forming in utero. For the relationship of Aristotle's *Generation of Animals* to the *Genit./Nat. Pue./Morb. 4* cluster, see Byl 1980 and Oser-Grote 2004. Figs were also used in fertility medicaments, probably because "fig" was a slang term for "vagina."

167. Hipp. *Morb.* 4.24 = 7.600L.

168. Hipp. *Aer.* 9.51–54 = 2.40L.

169. Hipp. *Morb.* 4.24 = 7.602L.

170. Bliquez (2015: 14–16) emphasizes the large role physicians played in the design and production of their own implements in antiquity. This account accords with descriptions in the Hippocratic treatises (see esp. *Haem.* 2 = 6.436L; *Fract.* 3 = 3.516–518L), as well as with later comments by Galen (*Ind.* 4–5 Pietrobelli), who laments how a fire destroyed not only tools that he had designed but wax models he used to replicate these instruments; cf. Celsus *Med.* 8.20.4; Gal. *Hipp. Art.* 18A.338–339K. Chapter 2.4 will examine Galen's expanded, purpose-built dissection tools and how those tools helped construct a body that reflected their intricate and thoughtful design.

FIGURE 5. Theagenes Spring/fountainhead at Megara, Greece, ca. 600 BCE.

required metal tools).[171] Even "springs" themselves should likely be seen as the human-made stone reservoirs that served as catch basins from which waters could be drawn (fig. 5)—waters that formed a key component of medical attention in texts such as *Airs Waters Places* and *On the Use of Liquids*.

Indeed, the treatise mentions that these springs do, in fact, serve an important corporeal function in this latter regard, although not through performing some physical task. Instead, the author claims that "even before sensation, they always each interpret their own potency [κατὰ τὴν ἑωυτοῦ δύναμιν ἑρμηνεύει] for the rest of the body from the things being eaten and drunk, that is, whatever is bile-like, phlegm-like, bloodlike, and water-like."[172] In this way, the springs operate like surrogate physicians inside the body, intimating which foods will be healthful and dispensing

171. Bladder stones are described in multiple places in the Hippocratic Corpus, and while *The Oath* 4.17L explicitly prohibits lithotomy, this proscription, to my mind, suggests that it was practiced by at least some physicians (otherwise, there would be little need to forswear it). This surgical intervention later became a more common practice (see Celsus *Med.* 7.26.2), and physicians such as Meges of Sidon (first century BCE) produced specialized medical tools for this surgery; see Bliquez 2015: 15, 98–102, 181–83; cf. Jackson 2010.

172. Hipp. *Morb.* 4.8 = 7.558L.

moistures as needed. This assertion harmonizes with the claim in *Epidemics* 6, noted above, that during illness our physis is our primary physician, insofar as it is an innate and self-regulating vital force dynamically guiding the body and its processes, whether to maintain balance or to restore health when it is injured.[173] On the one hand, such a conception implies that a body's own physis can maintain it in a healthy state without medical intervention—at least up to a point—such that the medical skill only intervenes when natural processes cannot overcome disease. On the other hand, this formulation explicitly describes the intrinsic capacity of corporeal physis as though it were an internalized medical power. Nature is not identical to the medical art, but its actions are modeled as therapeutic interventions. Here, too, the reservoirs mirror a type of broad technical expertise, such that the springs act as internal administrators of the external therapeutic tools. To be sure, the physician can still correct the patient's course, but the body has absorbed his practices to the degree that the medical techne functions as an extension of the body's nature, though only insofar as the body and its parts have been simultaneously modeled on the technologies of the physician.

Out of all the Hippocratic texts from the later fourth and fifth centuries BCE, *Diseases* 4 comes closest to articulating the body parts in terms of their function. Nevertheless, even the humors themselves, which provide the elemental substances responsible for health, serve no explicit function for the body other than to remain in balance.[174] As such, the body parts do not truly "work" or "function" to keep the human alive, even if they are part of a system that maintains health-producing homeostatic balance. Instead, what the text offers is a body punctuated with medical technologies, where the emphasis is not on individual tools but on the fluids flowing through and around them. Even these fluids themselves transform into mimetic reflections of the physician and his art.

1.5 FEMALE CORPOREALITY AND GYNECOLOGICAL TECHNOLOGIES

We have been speaking about Hippocratic bodies without addressing one of the key divisions within corporeality as they construct it in their

173. See Holmes 2018: 85; cf. Bartoš 2015: 268.

174. Hipp. *Genit.* 3 = 7.474L mentions that blood, bile, water, and phlegm are *innate* humors but says "the human has this many innate kinds [of moisture] in him, and *diseases arise from these*" [τοσαύτας γὰρ ἰδέας ἔχει συμφυέας ὁ ἄνθρωπος ἐν ἑωυτῷ, καὶ ἀπὸ τούτων αἱ νοῦσοι γίνονται].

texts: the distinction between male and female. As multiple scholars have emphasized, the primary distinction between sexes that the Hippocratics highlight is not different genital presentations but distinct types of flesh.[175] Men and women, according to the Hippocratic authors, are simply composed differently, such that male bodies are denser and drier, while female bodies are more porous and wet.[176] As a result, women supposedly draw excessive blood into their flesh, which then needs to be discharged, lest it cause illness. Menstruation acts to purge this blood and is required to maintain health. Pregnancy fits within this same structure, insofar as blood still flows into the uterus after conception, but instead of being expelled through menstruation, it gets consumed as nourishment by the growing fetus. In fact, Hippocratic gynecological authors insist that pregnancy facilitates future menstruation by stretching the uterus and the body's vessels, thereby easing future flow and making the body accustomed to accommodating more blood.[177] For this reason, women who have given birth get sick less often and less severely than those who have not. Menstruation and pregnancy thus supply the key behaviors around which Hippocratic authors structure female corporeality, and dysregulated menstruation forms the essential gynecological pathology, such that almost all female ailments are attributed to the inability to expel excess blood and moisture from the body.[178]

As part of this general view of female corporeality, *Diseases of Women* lists more than two dozen emmenagogues,[179] therapies for excessive or deficient menstruation, and remedies to align the cervix correctly to en-

175. Sex distinctions start in utero, where Hipp. *Genit.* 6–7 = 7.478–480L asserts they are caused by the relative proportions of strong (male) and weak (female) semen and result in differing speeds of fetal development; see Dean-Jones 1994b: 166–70 for other accounts.

176. See Lloyd 1983: 58–111; Gourevitch 1984; Hanson 1991 and 1992; Dean-Jones 1994a and 1994b; King 1998; Flemming 2000; Holmes 2012; and Connell 2016 for general treatments of ancient gynecological thought. While ancient medical authors agree on the relative densities of male and female flesh, they disagree about whether male or female bodies are warmer and colder, respectively, or vice versa.

177. E.g., Hipp. *Mul.* 1.1 = 8.10L.

178. See, for example, the comments at Hipp. *Genit.* 4 = 7.476L; Hipp. *Loc. Hom.* 47 = 6.344–348L; Hipp. *Mul.* 1.2.62 = 8.14L; and *Diseases of Girls*. Menstrual difficulties can occur through a number of means, including, quite frequently, uterine displacement that misaligns the cervix and vagina, thereby preventing menstrual flow.

179. Hipp. *Mul.* 1.74 = 8.154–160L lists twenty-eight. These are overwhelmingly suppositories applied to the cervix, although two are unguents used to anoint the cervix and genitals.

able menstrual discharge from the uterus. Moreover, since the pathologies created by menstruation create manifold symptoms, many of which are initially subtle, and the consequences of not acting early can be fatal, it seems likely that physicians intervened after only a single skipped, deficient, or excessive period.[180] If we add these practices to fertility treatments, the remedies applied to ensure lochial discharge, and the near-constant cycles of pregnancy expected of ancient Greek women of the fifth century BCE, we can see how closely therapeutic interventions were integrated into notions of female bodies, where natural and medically induced "purgings" or "cleanings" [καθάρσεις] lack a great deal of distinction.[181] Indeed, *Diseases of Women* 1 claims that menstruation purges blood, but that the physician and woman must carefully balance the inflow of food (which creates blood) with exertion (which consumes blood) and the dilation of interior vessels throughout the entire body.

No internal organ (such as the uterus) is responsible for regulating menstruation as it would in a functionally arranged organism, and instead the excess blood can come out through the anus or be extracted through other orifices. Even still, the uterus is no piece of static infrastructure, but a dynamic unit that can twist or move to seek moisture upward in the cavity when it is too dry, thereby putting harmful, sometimes pathogenic pressure on other interior parts or restricting menstruation by creating misalignment between the cervix and the vagina.[182] At times

180. See Hipp. *Mul.* 1.2–4, 8.14–24L. Dean-Jones (1994b: 95) argues that Hippocratic doctors would not have intervened in the menstrual cycle of sexually active women without waiting considerable time for fear of inducing abortion. To my mind, the number of emmenagogues described in the corpus and the demonstrable anxieties about absent or insufficient menstruation suggests that they would be willing to intervene as soon as a disturbance was noted, but likely only after a fully missed cycle, rather than a delay of a few days. Indeed, comments such as those at Hipp. *Mul.* 1.11 = 8.44L, where the author claims that there is no need to provoke menstruation that is already occurring, suggest some hesitancy to intervene when it is not necessary, but even the recommendation itself potentially also implies that physicians might be tempted to do so.

181. Hippocratic authors do not carefully delineate the effects of normal menstruation and induced menstruation, and the same is the case for lochial cleaning/*catharsis* (see, e.g., *Hipp. Mul.* 1.36 = 8.86L). Dean-Jones (1994b: 143), however, suggests that there are some minor differences in efficacy between natural and induced menstruation.

182. Hipp. *Mul.* 1.2 = 8.14–22L; cf. Hipp. *Mul.* 1.7 = 8.32L. For this reason, Hippocratic gynecological authors recommend intercourse, such that a husband's semen moistens the uterus and prevents displacement; see, for example, Hipp. *Genit.* 4 = 7.476L.

its behaviors seem almost animalistic, able to be attracted upward or repulsed downward by the application of pleasant or unpleasant smells.[183] Yet the Hippocratic remedies exploit and treat this tendency, too.[184] In short, the tools and treatments of the techne seem requisite parts of maintaining health and were closely assimilated into basic notions of female corporeality. This assimilation opened up new opportunities for the mirroring and interface of therapeutic tools and the bodies that they treat and conceptually support.

When it comes to explicitly supporting these theories of female bodies, therapeutic techniques again make their appearance, although somewhat obliquely. For example, *Diseases of Women* 1 starts by insisting, as mentioned above, that female flesh is more porous; it then supports this claim with a vivid comparison:

> For example, if someone should place clean flocks of wool and a clean, tightly woven cloth, the latter exactly equal in weight to the flocks, over water or a moist place for two days and nights, he will find upon picking them up and weighing them that the flocks are much heavier than the cloth. This is because [moisture] always moves away from the water in a wide-necked vessel into that which is above, and the flocks, insofar as they are less dense and softer, will take up more of what is moving toward it, while the cloth, insofar as it is compact and woven, will become quite full without having taken up much of what moves toward it. In the same way, a woman, being more porous, will draw into her body more of what is being exhaled from her cavity, and more quickly, than a man does.[185]

183. Plato's *Timaeus* later develops this idea more imaginatively into the so-called "wandering womb" that could travel widely throughout the body, but within the Hippocratic gynecological texts, the uterus seems to be a part that hosts reproduction and moves according to basic physical behaviors, including the general principle that dry things seek out moisture. Scholars disagree as to how far the Hippocratic authors think the uterus can travel in the body. Hanson (1991), King (1993 and 1998: 35–37), and Dean-Jones (1994b: 69–77) suggest that they believed it could relocate entirely, traveling as far up as the head, perhaps via the veins, whereas Adair (1996: 159) disagrees, arguing that the uterus merely turns toward these parts. We might also consider that *Diseases of Women* 1 describes relatively small displacements, as the uterus shifts upward, toward the liver, hips, etc., while *Diseases of Women* 2 includes dislocations that affect locations as distant as the heart and head (see Hipp. *Mul.* 2.14–15 = 8.266–268L), so it could be that different authors had different opinions about how far the uterus could travel.

184. Byl (1989) provides a survey of examples.

185. Hipp. *Mul.* 1.1 = 8.12L.

Wool working was the paradigmatic female labor in Greek antiquity, and both spinning and weaving were essential activities in the idealized performance of feminine domesticity.

This comparison no doubt draws on these cultural associations, such that women's bodies adopt physical characteristics of the raw material most closely associated with womanhood. The "wide-necked vessel" also clearly reflects long-held connections between the womb and a jar, which Helen King has traced back as far as the Pandora myth.[186] It is important to note, however, that wool possesses not only cultural resonances but also medical ones, and it is used extensively as a vehicle for medicaments applied as suppositories.

Diseases of Women 1 describes several different treatment types, including a handful of dietary suggestions, but direct treatment of the uterus is dominant, whether to induce menstruation or to remedy uterine displacements and other ailments. Some treatments involve fomentations and fumigations, others cleansing and douching the uterus with liquid medicaments forced in through a pipe inserted through the cervix, here called the "mouth" of the uterus. That said, uterine and vaginal suppositories feature highly, being used as emmenagogues, to encourage postpartum lochial flow, and attend to other uterine disorders.

These suppositories are often applied directly to the cervix and involve substances that can occasionally be molded into a ball and applied. More often, however, the medically active ingredients are infused into a vehicle, either a wheaten plug or, most commonly, a flock of wool, which soaks up this substance. For example, to promote pregnancy, the author says, "having finely ground up an astringent and dissolved it in an unguent, soak it up in a piece of wool [εἰρίῳ] and apply it as a suppository, and let her keep it there for three days."[187] Elsewhere, to clean the lochia, he recommends, "Having taken two drafts of Cnidian berries and pepper, grind them finely and mix into white Egyptian oil and the finest possible honey, then soak this into wool, having wound it around a feather, and apply it for a day and a night."[188] To soothe an ulcerated uterus, he says to "mix myrrh, goose fat, white wax, and frankincense with the hairs from the belly of a rabbit, and, having made this smooth, apply it in wool."[189] To induce menstruation, he suggests having her "mix goose oil, bitter almond, and resin, and taking it up in wool, apply as a suppository."[190] Such prescriptions

186. King 1998: 23–27, 34–35.
187. Hipp. *Mul.* 1.19 = 8.58L; cf. Hipp. *Mul.* 1.23 = 8.62L.
188. Hipp. *Mul.* 1.37 = 8.90L.
189. Hipp. *Mul.* 1.49 = 8.108L.
190. Hipp. *Mul.* 1.74(3) = 8.156L.

abound in this and other gynecological texts. Indeed, the above comparison between wool and female flesh describes weighing the flocks, and this process, too, parallels the measuring and weighing of ingredients when infusing these woolen plugs. Moreover, *Diseases of Women* 1 describes how suppositories should be placed on the cervix, which thereby draws liquid downward. His comparison—wherein flocks of wool are placed on the mouth of the vessel to draw moisture upward—provides an inverted image of the wool suppository drawing fluids downward, but otherwise looks quite similar. The analogy he uses to explain the essential nature of women's flesh—the analogy that supports the entire apparatus of pathologies associated with female corporeality—is drawn from the treatment of this same flesh and mirrors its behaviors.

Beyond wool suppositories and female flesh, other therapeutic technologies articulate the features of the uterus itself, even providing normative behavioral and physical claims that seem to emanate from features of the tools rather than from independent observation of the body. For example, *Diseases of Women* 1 structures much of its pathological system around normal menstruation, which it claims should be around two Attic cotyles in volume, discharged over two or three days.[191] This equates to about one pint, which greatly exceeds how much menstrual fluid we now think emerges during menstruation. Lesley Dean-Jones suggests that this normative claim comes from the use of uterine douches, which are often two cotyles in volume—presumably giving rise to the notion that the uterus holds this much liquid.[192] In other words, the basic features of menstruation and uterine structure are as much artifacts of the technologies with which they are interfaced as they are of the body and its behaviors. Ideas about the tool and uterus are thus mutually dependent.

This tendency to integrate therapeutic tools into the essential conceptualization of the female interior and the processes that it hosts occurs in another treatise likely written by the same author as both *Diseases of Women* 1 and *Diseases* 4. *Generation/Nature of the Child* is a paired set of treatises (which were likely a single text originally) that explain how male and female seeds coalesce in the womb and how the embryo then develops through various stages of fetal growth, according to the sex of the child. Sex difference relies on the relative quantities of "strong" (male) and

191. Hipp. *Mul.* 1.6 = 8.30L.

192. Dean-Jones 1994b: 90–92; cf. King 1998: 30. For douches of two cotyles, see Hipp. *Mul.* 1.78(53, 56) = 8.188–190L. See also Hipp. *Mul.* 1.78(60) = 8.192L and Hipp. *Mul.* 1.80(2) = 8.200L, which both list four-cotyle prescriptions (and one five-cotyle), which would presumably involve two sequential douches.

"weak" (female) seed, which can both come from either the mother or the father. If both parents contribute strong seed, the embryo will be male. If both contribute weak seed, the embryo will be female. If one parent contributes strong seed and the other weak, the sex will be determined by which of the two dominates. He compares this last case to mixing wax and fat [κηρὸν καὶ στέαρ], which are hard to distinguish when liquefied.[193] In some ways, this comparison is strange, insofar as the relative dominance of one substance over another hardly needs explication, whereas the hiddenness of the sex determination is nowhere at issue outside of the analogy. It must be noted, then, that wax and fat are common ingredients within gynecological treatments, whether melted together and collected in wool flocks for a suppository or used as fertility-promoting unguents to anoint the cervix.[194]

Elsewhere in *Generation/Nature of the Child*, the douche apparatus comes into view again, no longer to determine uterine capacity but to help illustrate the physical principles according to which embryos grow. The text describes how breath entering the embryo through the umbilical cord articulates the body and forms its parts through its catalytic tendency to move like to like. He then says you can see these principles at work, if you "attach a reed pipe to a bladder and put in earth, sand, and fine scrapings of lead through the pipe into the bladder; then, having poured in water, blow through the pipe."[195] He claims that lead, sand, and earth will all eventually separate, and if you cut the bladder open, you will see that like has moved to like. This, the author claims, is how the breath flowing through the umbilical cord causes different components within the blood to collect and create various internal parts, whether bones, internal structures, limbs, or sinews. Lonie noted that the device described in this comparison looks very much like a repurposed douche. Indeed, *Nature of Woman* describes curing a particular uterine displacement by "attaching a pipe to a bladder" so as to "blow into the uterus" [αὐλίσκον προσδήσας πρὸς κύστιν, φυσῆσαι τὰς ὑστέρας] before cleaning it with a vapor bath, flushing it out and fumigating it with foul-smelling substances.[196] Even the ingredients that he inserts into the bladder—dirt, sand, and lead—can potentially be seen in the manure that he applies

193. Hipp. *Genit.* 6 = 7.478L.

194. See, for example, Hipp. *Mul.* 1.49 = 8.108L, which saturates wool with goose grease, white wax, and other ingredients. The two ingredients are also commonly paired in Hipp. *Ulc.* 21–22 = 6.424–428L.

195. Hipp. *Nat. Pue.* 6 = 7.498L.

196. Hipp. *Nat. Mul.* 14 = 7.332L.

during fertility treatments, the lead sounds that he uses to open the cervix so as to introduce the douche's nozzle, and the sand that he elsewhere seems to clean from the uterus.[197] In short, the tools and substances that comprise uterine therapies simultaneously operate as theoretical models of that same uterus.

One last analogy from this set of treatises reveals the integration of therapeutic technologies into female corporeality, one that occurs in *Generation*, where the author explains that when a mother gives birth only to weak children, the problem must arise from the insufficient size of her uterus. He then supplies a comparison with the techniques used to train the growth of a bottle gourd [σίκυος]:

> It is like this, as though if someone should put a bottle gourd that has already finished flowering, but is still young and attached to the stem, into a cup; the bottle gourd will grow equal to the inside of the cup and the same shape. But if someone places it into a large vessel that is sufficient to contain it, but not that much larger than the nature of the bottle gourd, the bottle gourd will become equal to the insides of the vessel and similar in shape. For it will strive in its growth to the inside of the vessel. This is almost to say that even plants are all disposed however someone constrains them.[198]

On the one hand, this comparison participates in a series of comparisons in this treatise between the growing fetus and plants, which have led both Lonie and Holmes to evaluate whether this author truly believes that the growth of the embryo obeys mechanical physical principles or requires specific living capacities. On the other, the plant that he chooses to demonstrate the potential of the uterus to restrict growth, the bottle gourd, is a uterine purgative, used to clean the uterus after birth.[199] More-

197. For manure in fertility medicaments, which come as part of so-called *Dreckapotheke*, found as part of female-focused pharmacology, see von Staden 1991 and 1992d. For examples of lead sounds, see Hipp. *Mul.* 1.11 = 8.46L and *Mul.* 1.37 = 8.90L. Hipp. *Mul.* 1.74(2) = 8.154L describes an emollient that expels water and sand [ψάμμος] from the uterus, which might provide a corollary for the sand in the bladder. That said, Hipp. *Mul.* 1.84(1–2) = 8.204–206L describes two similar cleaning agents that are said to expel water, pieces of skin [δέρματα] and bloody serum (as well as mucuses in the second instance). It is hard to tell what to make of this transposition.

198. Hipp. *Genit.* 9 = 7.482L.

199. E.g., Hipp. *Mul.* 1.37 = 8.90L; cf. *Mul.* 1.79 = 8.198L; *Mul.* 1.80(2) = 8.200L; *Mul.* 1.109(3) = 8.232L.

over, these recipes twice ask for a specimen the width of a hand,[200] which might lead us to ask whether the mold-grown bottle gourd relates to the medical standardization required for precisely these medicaments. Lastly, the Greek name, σίκυος, for the plant we call the "bottle gourd" closely relates to the word for "cupping vessel" [σικύη], so called because of the implement's resemblance to a gourd. As mentioned above, the author of *Ancient Medicine* mentions that the uterus bears this shape too.[201]

1.6 CONCLUSION

What has emerged, then, is a broad conceptual integration of the tools with which Hippocratics used to treat the body and the parts that receive this intervention. While it is interesting, it is perhaps not entirely surprising that the technological environment in which the Hippocratics immersed themselves provided many of the analogies that helped them imagine and express ideas about the corporeal interior. After all, what are explanatory comparisons if not the use of something familiar to explain something less so? Yet what appears in these treatises does more than draw information from one realm to explain another. Instead, the physical interfaces provided by certain therapeutic technologies help guarantee the basic notions of corporeality that support the use of these technologies. When women menstruate (or simply bleed) after the application of a wool-flock suppository, it supports the notion that women possess wool-like flesh in need of cleansing. When the uterus is cleaned through the action of a two-cotyle-sized bladder squeezing its contents through the cervix by means of an attached pipe, it helps confirm that the actions that take place in that uterus operate according to these same mechanisms. Conception inside the body is understood with the same medicaments used to foster fertility through their application. Although not totalizing in their force, these technological interfaces structure bodies around different sets of tools and create different notions of corporeality in the process.

As Heinrich von Staden states, "Whether ancient Greek medicine depicted the relation between "art" and "medicine" as mimetic or semiotic, as one of dependence or interdependence, as adversarial or as harmonious, it reveals the remarkable extent to which the concepts of *technē* and *physis* became reciprocally defining."[202] As far as this statement is

200. See Hipp. *Mul.* 1.69 = 8.194L; 1.80(2) = 8.200L.
201. Hipp. *VM* 22 = 1.626–628L.
202. Von Staden 2007: 42.

true, it also requires acknowledging its material implications and recognizing that the technical practices and associated implements were not identical for all Hippocratic physicians. To be sure, there is broad overlap between the therapies presented across the corpus and many shared Hippocratic notions of corporeality. Nevertheless, the technologies inherent in regimen-based approaches to medicine—namely, food preparation techniques—created a body whose interior cooked, mixed, and integrated flavors, and became gruel-like when healthy. The author who privileged emetic drugs constructed a body whose interior liquids resembled iconic versions of these drugs. The expanded tool kit of the gynecological treatises helps construct bodies, both male and female, that teem with more devices and multiform natures, whether cupping vessels, interconnected bronze vessels, or wool suppositories.

CHAPTER TWO

The Origins of the Organism

2.0 INTRODUCTION

In his 1952 lecture "Machine et organisme," published later in *La connaissance de la vie*, Georges Canguilhem put forward a strong case for seeing tools as extensions of the body, declaring that "tools and machines are kinds of organs, and organs are kinds of tools or machines."[1] In the next two chapters, we will spend more time examining how this statement amounts to a type of tautology or linguistic truism, insofar as understanding the parts of the body as "organs" already means conceptualizing them as biotic tools that complete functional tasks. As we have seen, ascriptions of function certainly occur within the Hippocratic Corpus, which describes the vessels conducting humors, the uterus attracting semen, the head collecting phlegm, and more. Nevertheless, the Hippocratic authors more often cast liquids, heat, air, and other such substances as the active agents, not the corporeal structures or parts, and even when these authors privileged the body's liquid substances, they rarely discussed the functions that such humors fulfilled. No extant text offers a full physiology of how the body functions, because, as I have argued, they conceptualized corporeality in slightly different terms. As was stated in the last chapter, humors maintained health through their balance, but their presence did not generally explain life. This chapter will show how functional parts became the essential divisions of corporeality and how function became the key concept around which the body was consolidated. To this end, it follows the emergence of tool-like corporeal components in the works of the Greek physikoi of the fifth century BCE, illustrating how this mode of thinking started as a way to describe sensory parts before expanding to become a heuristic for conceptualizing the body in its totality. In so doing, it presents the prehistory of organs within early natural philosophical

1. Canguilhem 1952: 143, my translation; cf. Hacking 1998.

discourses before tracing this concept's development in the work of Plato and then, in the next chapter, Aristotle.

Even as it attends to these broader intellectual movements, this chapter also pays attention to the minutiae of the body constructed under this functional structure to illustrate how particular material technologies shaped discourses and expectations about corporeality and changed the physical contours of the somatic interiors imagined by early authors. It was Empedocles who first compared the sensory parts to organa and expanded this analogic mode to address respiration, which he explained with a lengthy comparison to a pottery vessel called a *clepsydra*. This device utilizes vacuum pressure and the mutual exchange of water and air to trap liquid and release it when desired. Despite our intuition that, if the clepsydra provides a model for the mechanism of breathing, it must stand in for the lungs, Empedocles seemingly held no such opinion. Instead, he held that air entered the body through our vessels more generally, and it is very difficult to see how his extended simile conforms to human anatomy at all. This chapter therefore argues that we cannot take even major anatomical divisions to be crucial to early authors and that these divisions became important largely because of a reliance on technologies to delineate corporeal mechanisms. That is, by using the clepsydra to articulate the mechanics of respiration, Empedocles helped lead later authors to locate breathing specifically in the lungs as a type of tool. Tools helped create human anatomy as a medically relevant concept even before they served as heuristic aids to explain anatomical functions.

The second part of this chapter turns to Plato, who presented what might be the first full account of human physiology within his cosmogonical narrative in the *Timaeus*. That is, Plato is the first not only to present an explanation of the mechanics of the human interior, but also to illustrate how these processes are required for the maintenance of life. As opposed to the disease-focused discussions within Hippocratic treatises, or Empedocles's description of respiration's mechanics, the *Timaeus* is the first source to outline both corporeal operations and the functions that inner processes accomplish for the body. Moreover, insofar as he locates these functions in corporeal tools, Plato presents something that begins to resemble our organism. Yet his construction of corporeality differs substantially from our own, linked as it is to his conception of a tripartite body that derives its unity through imitating the cosmos as a whole. The functions he assigns to his corporeal tools thus sometimes diverge substantially from our expectations of what "organs" in biotic systems do, especially since he does not conscript all our parts into systems operating to maintain life. This chapter thus tracks the specific notions that attended the birth of the organism to illustrate that we cannot simply project our expectations about anatomy, physiology, or corporeality back onto ancient authors, even when some of their ideas start to feel more familiar. They propose bodies that contain different types of parts, and different parts imply different wholes.

As novel as it is in some regards, Plato's account certainly draws on the medical tradition before him, especially when discussing diseases.[2] Moreover, he also adapted and significantly altered Empedocles's explication of breathing. Although Plato does not explicitly invoke the clepsydra, his predecessor's account and the material tool it employed still inform his physiology of respiration to some degree, since he, too, explains it as a type of air exchange across a membrane perforated with holes, all taking place in a vessel that must remain rigid for the process to work. In fact, while the rise and fall of the chest seems fundamental to our experience and conception of respiration, Plato's account prohibits this and instead requires the chest to stay rigid, maintaining the same volume in its inner cavity at all times. This chapter thus illustrates how corporeal activities as basic as breathing were not clearly delineated physical phenomena that authors explained in different ways but behaviors whose contours emerged only through negotiations between lived experiences, material devices, and theoretical commitments. Even the processes we seem so intimately familiar with are compound constructions whose relevant details, constituent parts, and functional boundaries emerge both within and from an explanatory apparatus.

The nature of the corporeal complex becomes even clearer when placing Plato's account of respiration next to another of his physiological explanations, that of blood distribution, which he describes as a type of irrigation. As simple as this image may seem, treating the distribution of nourishment through the body as akin to gravity-flow pipes makes it hard to understand how blood travels upward in our body. To be sure, Plato makes suggestions to overcome this complication, but these alternate explanations start creating tensions. The demands he places on the body—and sometimes the same body parts—shift from one explanatory frame to

2. Plato discusses diseases according to another tripartite schema at the end of his discussion of the construction of the primordial human body (*Ti.* 81e7–86a10). For the relationship of this material to his predecessors, see Rivaud (1925: 114–15), who evaluates it relative to the Hippocratics; Taylor (1928: 587–610), who argues for the influence of Philistion, Alcmaeon, and Empedocles; Mugler (1958), who argues for Alcmaeon's influence; and H. Miller (1962), who is more skeptical of substantial Empedoclean influence.

the next, such that air and blood need to flow through the vascular system in different ways, perhaps at the same time. To elucidate some of these difficulties, the final section of the chapter suggests that features native to irrigation pipes from fourth-century BCE Greece help us understand why Plato and his contemporaries were so comfortable imagining that air and blood traveled through the same vessels in the body.

Tracking all these interactions reveals how technologies are not totalizing heuristics but can work modally with one another, in partnership or opposition, and illustrates how using tool analogies can transform incidental material features of these devices into physical attributes seemingly native to the body. They highlight the incongruous and multifaceted ways that the organism, its components, and their behaviors were built as objects of inquiry and explanation. Moreover, if the last chapter illustrated how different therapeutic technologies endogenous to medicine articulated the body by creating recursive interfaces with it, this chapter emphasizes the use of exogenous tools as heuristic models. It shows how these instruments, too, merged with the body in their own way and left traces of their material configurations. Technologies not only explained *how* the body behaved as it did but also altered the basic behaviors that authors thought demanded explanation in the first place.

Another vector of this investigation must simultaneously be acknowledged. Plato's *Timaeus* describes how a divine demiurge crafted the entire world, including the human body, which he manufactured so that it imitated the living cosmos and, ultimately, the rational good to the greatest extent possible. As was outlined in the introduction, the emergence of a teleologically oriented body comprised of tool-like parts cannot be separated from this notion of divine design, since the two were so closely wedded for so much of ancient biomedical discourse. Nevertheless, the question for this chapter is not when teleology got introduced into physical explanations or discussions of the body, but when the body became consolidated around function as its organizing principle, structuring its internal relationships and providing the dominating concept with which to understand corporeality. More than anything, this chapter emphasizes that the shift toward function does not simply reflect a different epistemology, as though tool-like teleological accounts of the body and its parts simply altered and expanded what counted as a sufficient scientific explanation.[3] Instead, it argues that this shift created a new type of epistemic object.

3. For an example of this type of claim, see Rocca (2017: 1–22), who frames teleology in purely epistemological terms.

This developing corporeality had different expectations placed upon it, demanded new types of attention, and displayed different behaviors.

It takes work to recover the process by which the organism was made the default paradigm of corporeality and to defamiliarize the behaviors associated with it, such as breathing, blood circulation, and the pulse. This is especially the case since we live with the naturalization of the organism such that Canguilhem's assertions can seem provocative rather than the expression of a historical merging of conceptual forms.

2.1 EMPEDOCLES'S CLEPSYDRA AND THE CORPOREAL INTERIOR

As the last chapter illustrated, by the beginning of the fourth century BCE, the Hippocratic physicians established a medical interest in the interior structures and processes of the body. Yet there was also a second, related discourse that developed in the tradition of the pre-Socratic physikoi. It differed from medical treatises insofar as these authors did not sustain their theories through a set of material and technological practices but through a set of explanatory norms. In fact, over time, the investigations "concerning nature" [περὶ φύσεως] began to address a relatively stable set of physical phenomena. This was certainly true in regard to corporeality. For example, the extant fragments of Thales, Pherecydes, Anaximander, and Anaximenes, all writing in the sixth century BCE, show no real interest in the body itself or in its behaviors as objects of inquiry.[4] Instead, they remain at the most general level of analysis, occasionally discussing, as the Hippocratics charge, the fundamental constituent element from which living things are composed, or the substance of the soul itself.[5] Greater

4. Since the extant fragments of the physikoi arrive through the filter of later authors, assessing this material requires acknowledging the complications and difficulties of historical reconstruction; see Mansfeld 1999, 2017; Mejer 2006; McKirahan 2010: 1–6; Betegh 2010.

5. Anaximander discusses the origins of humans in a cosmic sense, claiming that since they cannot take care of themselves as infants, humans must have originally been born from other animals; see DK12 A10 (D8 LM) = Ps.-Plut. *Strom.* 2; cf. DK12 A30 (D38 LM) = Aët 5.19.4, which describes men originating in bark, and DK12 A30 (D39 LM) = Cens. *Die nat.* 4.7; DK12 A30 (D40 LM) = Plut. *Quaest. Conv.* 8.8.4 730E–F, which describes how human embryos must have been carried by fish. Xenophanes, too, describes the origins of humans, placing them within the cycle of generation and destruction; see DK21 A33 (D22 LM) = (Ps.-?) Hippol. *Haer.* 1.14.3–6. Xenophanes also discusses sleep at DK21 A51 (D48 LM) =

interest in the human form "qua physical compounds"[6] starts to appear at the beginning of the fifth century BCE, with Heraclitus (ca. 535–475 BCE)[7] and—even more dramatically—Parmenides (ca. 520–450 BCE). In fact, Parmenides introduces topics into the physikoi tradition that become all but requisite parts of any future accounts, including embryology, sex difference, sleep, and sensation.[8] He refers to "passages" through which

---

Tert. *De an.* 43.2, but nowhere else does he (or Anaximander) discuss body parts or functions, although he does suggest that the soul is pneuma at DK21 A1 (D24 LM) Diog. Laert. 9.19; cf. DK21 A50 (D47 LM) = Macrob. *In Somn.* 1.14.20, which suggests he considered it a compound of earth and water. Similarly, Anaximander or Anaximenes both reportedly suggest that animals are composed of air or breath [πνεῦμα] (Anaximander DK12 A14 (R13 LM) = Anaximenes DK13 B2 (R5 LM) = Aët. 1.3.3–4). Even when Anaximenes discusses breath emanating from the mouth, he does so to assert that pneuma is cooled when condensed (i.e., when we blow through pursed lips), while warmed through rarefaction (i.e., when we emit warm air through our open mouths), not to present respiration as its own object of inquiry; see DK13 B1 (D8 LM; R4 LM) = Plut. *Prim. Frig.* 7 947F–948A; DK12 A5 (D7 LM) = Simpl. *In Phys.* 149.32–150.2.

6. Holmes 2010b: 99.

7. Heraclitus describes drunkenness as the soul becoming wet, while dry souls are the wisest; see DK22 B118 (D103 LM) = Mus. Ruf. in Stob. 3.5.8 (et al.); DK22 B117 (D104 LM) = Stob. 3.5.7. He also speaks about sleep as a disconnection from the external air (e.g., DK22 A16 (R59 LM) = Sext. Emp. *Math.* 7.127–134), although this might be a Stoicized account. Nevertheless, several anecdotes characterize him as antagonistic toward physicians and medicine, although these appear in pseudepigraphic letters dated to the first century BCE (see Smith 1990: 29); cf. DK22 B58 (D57 LM) = Hippol. *Haer.* 9.10.3. For instance, Heraclitus reportedly died by dropsy, rejecting treatment from physicians and instead employing his own (unsuccessful) therapies; see R78 LM =Tat. *Or.* 3 (absent from DK); DK22 A1 (P16 LM) Diog. Laert. 9.3–5; DK22 A1a (P17 LM) = Suidae Lexicon, *s.v.* Heraclitus 472. Heraclitus also criticizes certain religious cures and propitiations as well; see DK22 B5 (D15 LM) = Aristocr. *Theos.* 2.68 (et al.); cf. DK22 B15 (D16 LM) = Clem. Al. *Protr.* 2.34.5; DK22 B68 (D17 LM) = Iambl. *Myst.* 1.11. It is only later authors, including unnamed "Heracliteans" and the Hippocratic author of *Regimen*, who employ Heraclitus's notions of harmony and reciprocity to the body, health, and illness.

8. Parmenides DK28 A53 (D42 LM) = Aët. 5.7.2 (Ps.-Plut.); DK28 A13 (D44 LM) = Cens. *Die nat.* 5.4; DK28 A53 (D45 LM) = Cens. *Die nat.* 5.2; DK28 B17 (D46 LM) = Gal. *Hipp. Epid. VI* 2.46 = 119.12–15 Wenkebach = 17A.1001–1002K; DK28 A53 (D47 LM) = Aët. 5.7.4 (Ps.-Plut.); DK28 A54 (D48 LM) = Cens. *Die nat.* 6.5; DK28 B18 (D49 LM) = Cael. Aur. *Tard. Pass.* 4.9.134–35; DK28 A54 (D50 LM) = Aët. 5.11.2 (Ps.-Plut.).Parmenides also describes sleep as a cooling down; see DK28 A46b (D59 LM) = Tert. *De an.* 43.2. He also provides a basic account of the cosmogonic generation of humans DK28 A1 (D40 LM) = Diog. Laert. 9.22.

perceptions occurred, which depended on the mixture of fire and water.[9] It is the sensory parts that later emerge as the first corporeal "organs."

Empedocles (ca. 494–434 BCE) represents a key turning point in this tradition. His own cosmogonical poem, *On Nature*, put forward his four-element theory of matter (fire, air, water, earth) and describes the cyclical generation and destruction of the cosmos during alternating periods of domination by Love and Strife. Within this narrative, Empedocles describes how the body parts of various creatures came together during the "Reign of Strife" according to some intrinsic "likeness" between them to form living entities. Out of a primordial chaos, the parts that belonged to humans collected into humans, while the parts that belonged to bulls collected into bulls, and so on. Conversely, during the "Reign of Love," parts from different animals end up collecting to create hybrid creatures with parts from different animals.[10] He also expends considerable energy speaking about human generation and embryology, discussing the formation of the embryo, sex difference,[11] parental resemblance,[12] and the numerology of pregnancy,[13] as well as remarking upon menstruation and the uterus.[14] Empedocles also describes the composition of various animal tissues, including sinew, nails, bones, tears, blood, and flesh, describing the ratios of the four elements that comprise them.[15] When

9. For mention of "passages," see DK28 A47 (D57 LM) = Aët. 4.9.6 (Stob.). Theophrastus criticizes Parmenides for explaining little, conflating thought and sensation, and then claiming that knowledge accords with which, fire or water, prevails; DK28 A46 (D52 LM) = Theophr. *Sens.* 1, 3–4; cf. DK28 A48 (R69 LM) = Aët. 4.13.10 (Stob.). This last passage mentions unnamed authors who consider Parmenides to be a ray theorist, but there is no other indication that this is true.

10. DK31 B61 (D152 LM) = Simpl. *In Phys.* 371.33–372.9; DK31 B58 (D153 LM) = Simpl. *In Cael.* 587.18–19; DK31 B57 (D154 LM) = Simpl. *In Cael.* 586.12; DK31 B60 (D155 LM) = Plut. *Adv. Col.* 28 1123B; DK31 B61 (D156 LM) = Ael. *Nat. anim.* 16.29.

11. He claims that sex difference comes because men are born from a warmer part of the uterus, which also makes them darker in skin tone and hairier; see DK31 B67 (D158 LM) = Gal. *Hipp. Epid. VI* 2.46 = 119–120 Wenkebach = 17A.1002–1003K; DK31 B67 (D171 LM) = Arist. *Gen. an.* 1.18, 722b10–12; cf. DK31 B65 (D172 LM) = Arist. *Gen. an.* 1.18, 723a24–26; DK31 A81 (D173 LM) = Arist. *Gen. an.* 4.1, 764a1–6; DK31 A81 (D174 LM) = Aët. 5.7.1 (Ps.-Plut.); DK31 A81 (D175 LM) = Cens. *Die nat.* 6.6; DK31 A81 (D176 LM) = Cens. *Die nat.* 6.10.

12. DK31 A81 (D180 LM) = Cens. *Die nat.* 6.6

13. DK31 A75 (D178 LM) = Aët. 5.18.1.

14. DK31 B66 (D159 LM) = *Schol. in Eur.* 18 (vol. 1, p. 249.24 Schwartz); DK31 A80 (D161 LM) = Sor. *Gyn.* 1.6.40–43 Burg. = 1.21 Ilb.

15. DK31 A78 (D189 LM) = Aët. 5.22.1; DK31 B98 (D190 LM) = Simpl. *In Phys.* 32.6–10; DK31 B96 (D192 LM) = Simpl. *In Phys.* 300.21–24.

describing individual parts, which occurs only seldom in the extant fragments, Empedocles outlines the physical or material processes that produced certain features. For example, when explaining the formation of the vertebrae, he says that it breaks as the embryo twists.[16] Similarly, when describing what became the canonical corporeal *explananda*, digestion and sleep, he ascribes the former to putrefaction while claiming the latter is caused by the cooling of the blood.[17]

Because of his considerable interest in living things, most notably their generation, it is typical for commentators to assume that Empedocles recognized certain corporeal functions to exist but held that it was sufficient to explain the material circumstances of their physical formation. As Jessica Gelber puts it, "[Empedocles] acknowledges that organisms are regularly generated having parts and features that are useful for performing various functions, yet nevertheless thinks that the generative process can be fully explained in terms of the movements and interactions of the material factors involved, such as earth, air, fire, and water."[18] Even ancient commentators, such as Simplicius, assume as much, suggesting that some of the hybrid animals survived, insofar as the various parts fulfill certain physiological needs.[19] Yet how can we be so sure that Empedocles recognized the body as organized around corporeal functions? To put this in other terms, both Gelber and Simplicius acknowledge that Empedocles avoided a teleological understanding of generation and growth, conceptualizing animal features as resulting from material happenstance rather than divine design, but both assume without question that Empedocles must have ascribed to a structural teleology. Since this notion of corporeality had never before been articulated, I think it makes little sense to take it for granted. This, again, is not to assert that Empedocles could not, or did not, ascribe certain functions to body parts, but we should hesitate before assuming that he conceptualized the body as an internally coherent functional system. In fact, *pace* Simplicius, it seems from the example of the hybrid animals that he characterized body parts as basic building blocks of the body, which some intrinsic relatedness consolidates into an essential unity. This, not function or physiology, creates corporeality.

There are places, however, where functions do surface, and they do so in the same places that tools appear: the sensory parts. Like Parmenides,

16. DK31 B97 (D177 LM) = Arist. *Part. An.* 1.1, 640a19–22.

17. For digestion, see DK31 A77 (D205 LM) = Gal. [*Def. Med.*] 99. For sleep, see DK31 A85 (D206 LM) = Aët. 5.25.4 (Ps.-Plut.).

18. Gelber 2021: 99.

19. DK31 B61 (D152 LM) = Simpl. *In Phys.* 371.33–372.9.

Empedocles provided a generalized account of the senses, and as part of this account, he, too, attributed all perceptions to "effluences" fitting into the commensurate "pores" [πόροι] of each sense.[20] Yet Empedocles also dealt with some senses individually, and it is in these part-specific explanations that he tends to invoke technological analogies. For example, when explaining vision, he likens the eyes to a lamp, and when explaining hearing, he compares the ear to the bell of a wind instrument in which an echo resounds.[21] In both cases, the tool analogy conflicts with his general theory of perception: the lamp presents an extramissionist theory, not an intromissionist pore theory as his generalized account suggests, while the bell requires hearing to occur when sound strikes the "solid parts" of the inner ear, not when it fits into pores. Accordingly, it seems as though explaining each individual sense can require finding some correlate technology and using it as a model, even if it presents issues. Tools are not invoked to elucidate a pre-held theory. They operate as embodied theories themselves, setting the parameters for the phenomenon that they explain.

The same basic dynamic appears when Empedocles explicates respiration. It should be noted that he is the first physikoi to explain this process (such as their fragments can be reconstructed), and no prior account even mentions the mechanisms of breathing, even as authors discuss breath or *pneuma* as part of their investigations into nature. As Antoine Thivel has noted, however, even references in Homer to ἀναπνέω, often translated as "to breathe," seem to connote "respite" or "catching one's breath," rather than inhalation or exhalation specifically, and no specific mention of air both entering and exiting a body appears.[22] To be sure, the integration with or identification of breath and life is long evident in Egyptian, Assyro-Babylonian, Greek, and other mythic traditions, and previous physikoi had made air or breath a key component of their account of living things. For instance, Xenophanes identifies πνεῦμα with ψυχή, or Anaximander and Anaximenes suggest that animals are composed of air or breath and describe the effects of opening or closing your lips on the temperature of exhaled air.[23] Nevertheless, that they did so does not mean that these authors conceptualized breathing as the iterative exchange of internal and external air, as occurring by means of the lungs, or indeed as a physiolog-

20. DK31 A86 (D218 LM) = Theophr. *Sens.* 7–8. For his account of perception in general, including vision, see D213–D234 LM.

21. For vision, see DK31 B84 (D215 LM) = Arist. *Sens.* 2, 437b26–438a3; for hearing, see DK31 A93 (D227 LM) = Aët. 4.16.1 (Ps.-Plut.).

22. Thivel 2005: 240–41.

23. See note 5 on p. 81.

ical process housed in organs at all. Indeed, the lungs, which to us seem so obviously a part of respiration, are often seen as a recipient of liquids, especially as suggested by the etymological connection between πλεύμων and the Indo-European root **plew*, "to float," and their frequently mentioned spongy texture.[24] Thus, when Empedocles discussed the mechanism of breathing, he was articulating a process that was certainly underdetermined in previous discourses and potentially imagined in ways quite different from our own assumptions. His explanation confirms just such a divergent conception.

When explaining respiration, Empedocles employs the same explanatory mode that he used when discussing sensation: he invokes a technological analogy. In the case of breathing, he produces a lengthy simile between the body and a wine-serving device called the *clepsydra*, or "water-stealer." This is a pottery vessel with perforations on its bottom that allow liquid to enter its main cavity and a hole on top that can be plugged to prevent liquid from flowing back out. This device functions in the same way as when a child sticks one end of a straw in a can of soda, places her finger over the other opening, traps a bit of liquid in the bottom, and then transfers it to her mouth. Although devices using hydrostatic pressure to hold and transfer liquid appeared in the eastern Mediterranean from at least the Bronze Age, the clepsydra itself first appears in the archaeological record in the mid- to late sixth century BCE, at which point it obtained considerable regularity in form, featuring a circular handle with a raised button directly underneath it (see fig. 6). Eurydice Kefalidou has cataloged eighteen extant examples, the majority of which are from Athens and Boetia, although some of the oldest come from Olbia, in Sardinia, and from Sicily, which may provide a material context in the southern Italic peninsula for Empedocles's analogical use of the device around the same time.[25] The clepsydra's wide and sudden geographical distribution intimates the popularity of these devices, and since a few display images that suggest a sympotic use context, such as Dionysus, satyrs, or kithara playing, they might have been exported as part of the Athenian elite drinking culture that also spread at this same historical moment.[26] That said, many other clepsydrae host no such imagery, and the difficulty of wielding them

24. Thivel 2005: 242.

25. See Kefalidou 2003 for a survey of extant devices and a description of use; cf. M. Lewis 2000b: 343–45.

26. For examples, see: University of Mississippi, Robinson Collection, acc. No. 77.3.74 (Athens, 525–475 BCE); Copenhagen, National Museum, acc. No. 12385 (Athens, 525–475 BCE); and Würzburg Universität, Martin von Wagner Museum, acc. No. 9023821.

FIGURE. 6. *Clepsydra*, 6th c. BCE, terra-cotta, Boeotian, 15.2 × 11.1 × 11.1 cm, Metropolitan Museum of Art, New York, accession no. 57.11.1.

without spillage makes it difficult to imagine that they were used to distribute wine with any success. As such, Kefalidou has expressed doubts whether these devices were ever used as wine servers and instead argues that they were employed as "sprinklers" as part of religious sacrifice.

Regardless of the primary use context, both the sudden presence of these instruments in the material record across the wider Mediterranean and the fact that literary references to the device also start in this same period together indicate that the clepsydra developed into a newly potent cultural object in the late sixth century BCE. The device made an immediate conceptual impact on accounts of nature, and it quickly became the totemic technology for conceptualizing the power of air.

According to Aristotle, Anaximenes invoked the device to explain how the flat earth remains stable by resting on trapped air beneath it,[27] as did

27. DK13 A20 (D19 LM) = Arist. *Cael.* 2.13, 294b13–23.

Democritus a century later, who also employed the clepsydra to illustrate why the earth's breadth allowed it to remain stationary, suggesting that the earth traps air beneath it, which prevents it from falling.[28] Anaxagoras seems to have likewise been interested in this device, since he incorporated an explanation of the mechanism in his accounts,[29] and from here, the clepsydra appears in multiple explanations, appearing wherever authors need to demonstrate the power of air or the mutual displacement of air and water. Even three hundred years later, Heron (ca. 10–70 CE) still uses the clepsydra to establish whether a void can exist in nature.[30] As simple as it is, the device became crucial for articulating the power of air and theories of the void in the sixth through fourth centuries BCE. It became a crucial cipher for conceptualizing the impossibility of empty space and the reciprocal exchange of liquid and air.

Empedocles is the first to employ the clepsydra to envision the process by which we breathe in and out, perhaps because of its growing totemic status to conceptualize the behavior of air, and because the device made a characteristic sound of aspiration as water entered and exited the bottom, forcing and sucking air through the narrow top spout.[31] In fact, the Aristotelian *Problemata* notes the "breath and belching" [πνεῦμα καὶ ἐρυγμός] that the clepsydra makes when it is raised and lowered into water with its top pipe open, which sounds like a heavy inhalation and exhalation.

It is quite complicated to pin down just what is going on in Empedocles's lengthy poetic comparison and how the clepsydra relates to the anatomy of the human body, but Aristotle preserves his account in full:

> All things breathe in and breathe out in the following way: all have bloodless tubes of flesh [λίφαιμοι σαρκῶν σύριγγες] stretched out to the outermost threshold along the body, and at their mouths they are pierced with many furrows right through the furthest extremity of the skin [ῥινῶν ἔσχατα τέρθρα]; as a result, gore is concealed, but a good path is cut for the air by passageways. Whence thereupon whenever

28. DK68 A20 (D110 LM) = Arist. *Cael.* 2.13, 294b13–30; cf. DK68 A94 (D111 LM) = Aët. 3.10.4–5 (Ps.-Plut.). Diogenes also lists one of Democritus's cosmological treatises as "The Struggle of the *Clepsydra*" ["Αμιλλα κλεψύδρᾳ], although the latter word is corrupted; see DK68 A33 (D2b LM) = Diog. Laer. 9.45–49; cf. [Arist.] *Sud.* 25–26.

29. DK59 A69 (D61/R22 LM) = [Arist.] *Pr.* 16.8, 914b9–15; cf. DK59 A68 (D60 LM) = Arist. *Phys.* 4.6, 213a22–27.

30. Heron *Pneum.* 1.2. [Arist.] *Pr.* 2.1, 866b9–14 also uses the *clepsydra* to argue the (somewhat odd) idea that we do not sweat when we hold our breath; cf. p. 187–88.

31. [Arist.] *Pr.* 16.8, 915a6–16.

> smooth blood darts away, air will rush in, seething with raging surge. But when the blood leaps up, the air blows outwards, just as when a girl, playing with a clepsydra of gleaming bronze and placing the tube of the pipe against her well-shaped hand, dips it into the smooth body of silver-white water—the deluge does not go into the vessel, but the mass of air, having fallen against the many perforations, prevents it from within, until she uncovers the dense stream. Then, however, when breath is falling out, the apportioned water enters. Thus, in the same way as when water holds down in the depths of the bronze, with the passageway of the channel blocked by the mortal skin, the air outside striving eagerly inwards holds back the deluge about the gates of the harsh-sounding neck, controlling its surface, until she lets go with her hand. Then, back again, in reverse from before, with the breath falling forward, the apportioned water runs down and out. Thus, in the same way, when smooth blood surging through the limbs leaps away inwards, immediately the breath of air returns, seething with a surge; but when the blood leaps up, an equal amount breathes back out.[32]

The primary interpretative difficulty stems from the fact that Empedocles claims that the pores occur at the edge of the ῥινῶν, which can be taken as either the plural of ῥίς, "nose"—in which case Empedocles is referring to "furrows pierced at the furthest extremity of the nostrils"—or it can be taken as the plural of ῥινός, "skin"—in which case he is referring to the pores of our skin to which blood vessels lead.[33] In either case, when blood inside our internal "tubes of flesh" [σαρκῶν σύριγγες] (somehow) retreats into the interior of our body, it allows the air *outside* to push its way through certain pores or orifices and into the empty corporeal space in equal amount.[34] In turn, when blood (somehow) rushes back outwards toward the mouths of these pores, air is pushed back out into the exterior environment through these same tubes, while blood is kept inside.

32. Empedocles DK31 B100 (D201 LM) = Arist. *Resp*. 473b9–474a6. Cf. DK 31 A74 (D202 LM) = Aët. 4.22.1.

33. Aristotle himself, who preserves this passage, vacillates in his interpretation, suggesting at *Resp*. 473a15–26 that Empedocles is referring to the nostrils, but saying at 473b2–4, "[Empedocles] says that inhalation and exhalation occur through certain veins in which blood is present—although they are not full of blood; rather, they have passageways to the external air" [γίνεσθαι δέ φησι τὴν ἀναπνοὴν καὶ ἐκπνοὴν διὰ τὸ φλέβας εἶναί τινας, ἐν αἷς ἔνεστι μὲν αἷμα, οὐ μέντοι πλήρεις εἰσὶν αἵματος, ἔχουσι δὲ πόρους εἰς τὸν ἔξω ἀέρα].

34. We would generally say the air is drawn in by vacuum pressure, not *driven* in by the surrounding air, although the latter is more correct; see p. 96.

Scholars have put forward multiple interpretations to try to map this comparison onto what we now consider to be our internal anatomy, but each one presents considerable problems. Diels suggests that the simile describes only cutaneous respiration through our skin, which, while widely attested as a theory in later Greek accounts, would leave the mouth and nose out of the act of breathing entirely (see fig. 7a).[35] Furley suggests that the perforations on the bottom of the clepsydra represent our pores, while the upper perforation of the handle represents our nose (see fig. 7b). Yet this interpretation does not really correspond to Empedocles's simile at all, since water only flowed in through the bottom perforations and never came in through the top spout, while the "bloodless veins" would extend from our skin directly to our nose without the lungs again appearing in Empedocles's account).[36] Booth suggests that the clepsydra's bottom perforations correspond to hypothesized pores at the "back of the nostrils" and that the top pipe is stuck somewhere into the thorax, as though it were simply a vein (see fig. 7c). This understanding, however, would require blood to rush up through our throat, only to be stopped in our nostrils, and the lungs would again be left out.[37] Lastly, O'Brien suggests that the shape of the clepsydra represents the lungs, and thus that the pores of the ῥινῶν must refer to pores in the surface of the "skin" of the lungs, connected to veins within the body (see fig. 7d).[38] This is the conception that most comfortably aligns the simile with modern physiological assumptions,[39] but despite its strengths, it almost completely ignores what Empedocles actually says, since: (1) blood would be passing through the pores that Empedocles has said hide the gore (ln. 4–5); (2) the "furthest extremity" [ἔσχατα τέρθρα] of the skin would be inside the body (ln. 2–3); and (3) Empedocles describes the transfer of air and blood in "bloodless tubes," which may or may not be in the lung.[40] The interpretations either have scant textual support or make little anatomical

35. DK 31 B100 n16; cf. Lloyd 1966: 300–301.

36. Furley 1957.

37. Booth 1960. Booth's argument also requires that the air in the body would correspond nonintuitively to water in the simile, while air in the simile would correspond to blood. He argues that since water alone moves back and forth through the pores in the simile, it must be identified with the air in respiration, since air is the only thing that can pass through the perforations.

38. O'Brien 1970.

39. This interpretation does present the difficulty that blood would gush out of the mouth if we exhaled too strongly, since there would be no pores to prevent it from doing so.

40. It is possible that Empedocles is referring to the bronchioles in the lung when he speaks of "bloodless tubes," but then the shape of the clepsydra would

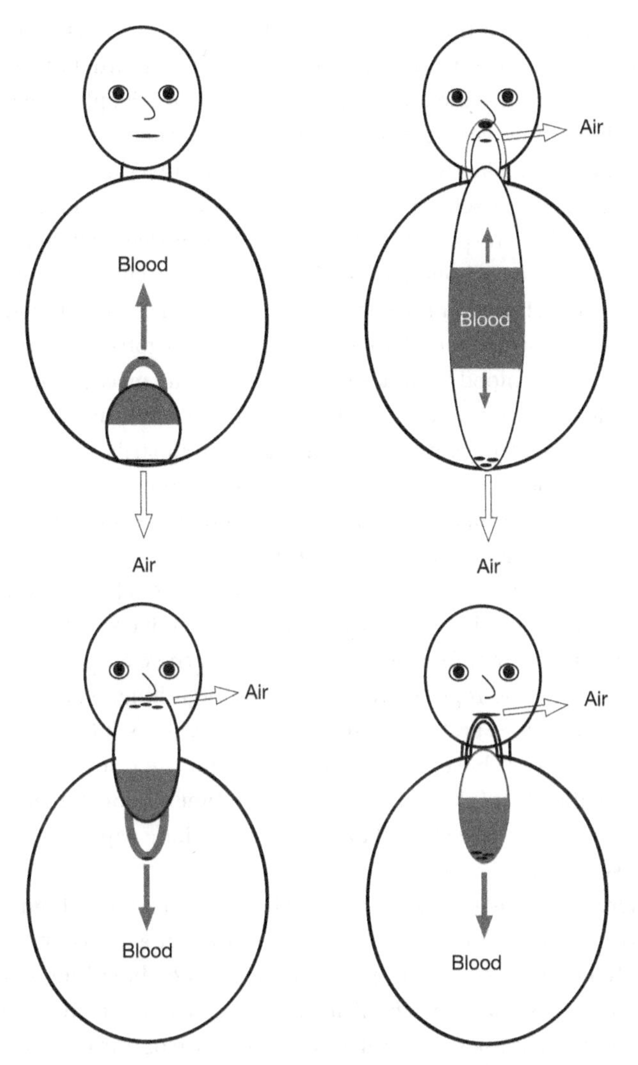

FIGURE 7A, B, C, D. From top left to bottom right, the models of Diels, Furley, Booth, and O'Brien.

sense (according to modern assumptions). The simile feels, at best, cumbersome and misleading.[41]

These difficulties led Worthen to propose that it is simply misguided to try to make the clepsydra correspond isomorphically to human physiol-

---

be less consequential, since neither the air nor the blood would flow into a common vessel through the pores.

41. To be sure, Empedocles writes poetic philosophy, which perhaps encourages a certain laxness and figuration when providing vivid descriptions of

ogy and accept that the simile does something else. It must indicate only that respiration involves the physical principle of mutual displacement, without explicating the anatomy that houses such displacement.[42] This proposition, however, does not quite seem accurate either, since Empedocles does not make the physical principles primary at all, but the technological analog itself. That is, once Empedocles has found a piece of technology that "respires," he has functionally articulated and explained respiration. The tool operates as the model to explain not only how the body breathes but what the corporeal process looks like in the first place. At that point, human physiology emerges in its negotiation with the clepsydra such that blood becomes integrated into the mechanism of breathing, as do the pores and the hypothetical vessels that must lead to them and somehow connect directly to the skin and nostrils.[43] Thus, the clepsydra simile does not produce a set of positive and negative predictions that can then be evaluated but partially embodies the corporeal processes, parts, and behaviors under investigation.

To be sure, the mode in which the clepsydra model of breathing functions relies on a unique capacity of blood to leap backward and forward without any external force applied. Therefore, the clepsydra cannot be a simple one-to-one reproduction of animal breathing. Nevertheless, the clepsydra activates and integrates certain body parts into the process of respiration (even those, such as the pores and the veins, that we might not immediately think to include). The resultant corporeal mechanism is some hybrid mix of technical apparatus and biotic capacity, interwoven and inseparable.

There are two aspects that are crucial to recognize. First, Empedocles is employing not a technology of timeless status but a device that was only recently popularized, if not invented anew. Since breathing is the only nonsensory corporeal mechanism that he discusses in the extant fragments, we might ask whether it was the presence of this tool as a totemic technology within the physikoi tradition that provided him with the impetus to address this activity, rather than his desire to explain breathing that led him to look for a heuristic analogy. If so, then, the clepsydra sup-

---

physical mechanisms. Aristotle criticizes him for as much, even as he praises his Homer-like metaphorical capacity; see DK31 A22 (R1a LM) = Arist. *Poet.* 1, 1447b17–19; DK31 A1 (R1b LM) = Diog. Laert. 8.57; DK31 A25 (R1c LM) = Arist. *Rhet.* 3.5, 1407a31–35.

42. Worthen 1970. Worthen still believes that Empedocles puts forth a model involving cutaneous breathing; cf. Harris (1973: 10–19), who is similarly agnostic.

43. We should remember that the lungs are not empty sacks, but are, in fact, composed of blood-filled vessels and air-filled bronchioles.

plied a model—but not just to explain breathing's mechanism. Rather, it supplied a model for thinking in terms of corporeal mechanisms in the first place. Second, Empedocles does not mention the lungs explicitly, which illustrates how little the idea of individuated functional parts was a necessary condition of explanation. In fact, Empedocles also puts forward an account of how embryos first start breathing, and in doing so he mentions no inner organs. Instead, respiration occurs "when the liquid in embryos retreats, into which empty space the surrounding external air enters through the openings of the vessels," while exhalation occurs when "innate heat squeezes out the air by its impulse toward the exterior."[44] His account of breathing can thus operate without organs entirely, even as the clepsydra comparison suggests that the interior of the adult body operates with just such interior tools. In other words, conceptualizing respiration as occurring within a particular functional part seems bound up with using the clepsydra as a model. It is the method of incorporating a tool analogy to explicate respiration that presents the corporeal body as potentially in need of anatomical articulation such that thinking about breathing physiologically comes from—or at the very least along with—the tool comparison. What is more, we might notice that Empedocles never discusses the purpose of breathing and makes no mention of why air should enter and exit our body without any remaining behind. As such, we might say that Empedocles's body was punctuated with technologies, but it did not yet function. In other words, his body contained tools without teleology.

It is perhaps at this point that some might object, insisting that ancient peoples, from Greeks to Egyptians to Babylonians, all breathed and knew death came with its cessation. Early philosophers like Anaximenes had made air the essential element no doubt because of its obvious importance to animal life. Moreover, ancient cultures also all discussed the viscera and knew about internal parts and divisions. They had names for the lungs and the heart, so surely they did not need the clepsydra for these parts and their vital importance to be known. Butchery separated the viscera, and practices of divination such as haruspicy examined particular parts for signs. While all this is true, adjudicating ancient physiological theories by asking whether authors "knew" about corporeal behaviors and inner anatomical features treats the body as a single, stable object of scientific inquiry that got revealed, piece by piece, as though opening windows on an Advent calendar. Considering certain behaviors relevant to the

44. DK31 A74 (D170b LM) = Aët 4.22.1; cf. DK31 A74 (D170a LM) = Aët. 5.15.3; DK31 A74 (D202 LM) = Aët. 4.22.1 (Ps.-Plut.); cf. Hipp. *Nat. Pue.* 1 = 7.12L.

scientific understanding of the body relies on a set of assumptions that get built within an explanatory apparatus. Knowing that a part exists and knowing what it does are different things, and the latter itself requires the expectation that parts "do things" in the first place. The question can thus be more productively addressed by recognizing that the body articulated and explained by scientific authors is not coterminous with the human form, but instead represents a complex of philosophical commitments, explanatory heuristics, physical principles, and material components—and that this complex changes with each explanatory apparatus. As such, it is not that either breathing or the lungs required the clepsydra in order to be known, but respiration and its mechanisms did need to be included as part of what it meant to understand the body qua physical compound. This transition occurred while—if not because—breathing was constructed around the clepsydra.

2.2 PHYSIKOI ON CORPOREAL TOOLS

After Empedocles's *On Nature*, other physikoi start to treat both perception and respiration as standard corporeal features in need of explanation. Anaxagoras (ca. 500–428 BCE), whom Aristotle says was "earlier than Empedocles in years, but later in works,"[45] describes how perception occurs as a result of the effect of opposites.[46] He also discussed breathing insofar as he asserted the universality of breathing among all animals and discussed how fish and oysters breathe, as did Diogenes of Apollonia (fl. 5th c. BCE).[47] Alcmaeon, another contemporary whose dates are

45. DK59 A43 (D18/R18 LM) = Arist. *Metaph.* 1.3, 984a11.

46. The eyes see by reflection of one color in its opposite; the tongue tastes what is sweet by the presence of the bitter, the cold by the hot, etc.; while smell and hearing somehow work in the same way. Hearing, by contrast, occurs when sound penetrates to the hollow echo chamber of the skull; see DK59 A92 (D72–74 LM) = Theophr. *Sens.* 27–28; cf. DK59 A106 (D75 LM = Aët. 4.19.5; DK59 A74 (D76 LM) = [Arist.] *Pr.* 11.33, 903a7–10; DK59 A74 (D77 LM) = Plut. *Quaest. Conv.* 8.3.3, 722A.

47. Aristotle preserves only their account of aquatic respiration, wherein Anaxagoras argues that "whenever fishes expel water through their gills, by drawing air into their stomach they breathe. For there can be no void," while Diogenes says that "whenever they expel water through their gills, they draw air from the water by the void in their stomach" (Arist. *Resp.* 2, 470b31–471a5). Although we cannot assert a great deal about their account of respiration in humans from this commentary, it is notable that both accounts rely on the mutual exchange of liquid and water, which might indicate the ways the clepsydra still structured assumptions about the lungs as involving just such an exchange—especially as both accounts locate respiration in the specific parts.

somewhat vexed,[48] explains the mechanisms of hearing, smelling, tasting, and seeing in individuated sensory parts, asserting that they are all connected, through *poroi*, with the brain (which he asserts is the location of cognition).[49] He also reportedly excised the eyes to see the channels behind them, an action that need have been little more than basic removal of the eyeballs from their sockets, rather than an interrogation of their internal anatomy.[50] Alcmaeon mentions respiration as well, but only as occurring concomitantly with smelling, which occurs through the nostrils, as pneuma rises to the brain[51] (although he also reportedly asserted that goats inhale via their ears).[52]

As the internal mechanisms of animal bodies began to be articulated, a few other authors started to address what function breathing might fulfill for the living animal. Philolaus (470–ca. 385 BCE) used respiration as evidence for asserting warmth as the primary corporeal substance. His account, as truncated as it is, suggests that respiration serves to cool the body, or as he states, "immediately after birth, an animal draws in the external air, since it's cold; then it sends the air back out again, like a debt."[53] Beyond this assertion, he offers no corporeal mechanisms that explain how we breathe, instead attributing it to the natural impulses of pneuma and the natural opposition of hot and cold. Democritus (ca. 460–370 BCE), too, seemed to have discussed respiration more extensively, including an account of its function, insofar as he suggested that respiration brings motion-producing soul atoms into the body to prevent the soul from being squeezed out. That is, inner soul atoms are constantly being pushed out of the body by the surrounding atmosphere. Breathing thus

48. Zhmud (2008: 61) suggests his birth as ca. 500–480 BCE.

49. DK24 A5 (D19a LM) = Theophr. *Sens.* 26; DK24 A8 (D19b LM) = Aët. 4.17.1 (Ps.-Plut.).

50. DK24 A10 (R6 LM) = Calcid. *In Tim.* 246; see Lloyd (1975a) and Mansfeld (1975) for doubts about whether Alcmaeon conducted dissection; cf. Democritus (DK68 A105) and Diogenes of Apollonia (DK64 A19).

51. DK24 A5 (D13a LM) = Theophr. *Sens.* 25; DK24 A8 (D13b LM) = Aët. 4.17.1 (Ps.-Plut.).

52. DK24 A7 (D14 LM) = Arist. *Hist. an.* 1.11, 492a14–15.

53. DK44 A27 (D25 LM) = Anon. Lond. 18.8–19.1 Manetti. Philolaus considered bodies to be composed solely of warmth and attributed diseases to blood thickening as a result of the compression of flesh and the excesses of warmth, phlegm (which he considered warm), and bile (which he argued that the flesh produced). In other words, Philolaus presents an emerging sense of corporeal functions, but he still adopts the stance that the author of *Ancient Medicine* attacked, producing a body consolidated around a single substance (that included component liquids for which no purpose is mentioned).

introduces more atoms of the same shape, which either replenish the inner reservoir or help keep those soul atoms that already reside in the body from being pushed out.[54] Democritus's account does not privilege—or even mention—the corporeal structures that might be involved in breathing, instead attributing the mechanism of respiration to ambient pressure and the atomic motions themselves. In general, then, respiration thus entered the physikoi tradition as a broader topic of inquiry, as the interior corporeal structures and processes gained more philosophical attention.[55] Some accounts deal with the mechanism of breathing and begin to formulate corporeal mechanics located in specific parts, while others seemingly ignore the parts to speak about the purpose of this behavior. In sum, by the beginning of the fourth century BCE, then, there was a growing interest in the interior components and processes of the body within the physikoi tradition that developed alongside and in opposition to discourses within the Hippocratic Corpus. Yet even as treatises like *On the Sacred Disease* and *Diseases* 4 described interior places and parts and used

54. Arist. *Resp*. 4, 471b30–472a26. Aristotle complains that Democritus has not explained why nature would make respiration function like this or how it would start in the first place, and that external atoms both compress and push out soul atoms while also somehow forcing new ones in and expanding the chest; cf. DK68 A28 (D136 LM) = Arist. *De an*. 1.2, 404a10–15.

55. Alcmaeon is also the first physikoi to define "health", which he ascribes to "equal rule of the powers, the moist and dry, cold and hot, bitter and sweet and the rest" [τὴν ἰσονομίαν τῶν δυνάμεων, ὑγροῦ, ξηροῦ, ψυχροῦ, θερμοῦ, πικροῦ, γλυκέος καὶ τῶν λοιπῶν] (DK24 B4 [D30 LM] =Aët. 5.30.1). MacKinney (1964) and Schubert (1996: 148–49) express reservation about whether Alcmaeon himself used the term *isonomia*. The attention that Alcmaeon pays to health and the body is likely what led Calcidius to consider him a *medicus*; see DK24 A10 (R6 LM) = Calcid. *In Tim*. 246. As Mansfeld (1975) has argued, however, this classification is likely retroactive; cf. Mansfeld 2013. Most ancient sources place him within the tradition of the physikoi, and indeed, like Parmenides, Alcmaeon discussed sleep and death (see D31–D32 LM), semen, embryology, and the production of sexual difference (D20–D29 LM). Favorinus even mistakenly claims that he was the first to write an account of nature [φυσικὸς λόγος] (DK24 A1 (D1a LM) = Diog. Laert. 8.83; DK24 A2 (D1b LM) = Clem. Al. *Strom*. 1.78.3. Galen more accurately places him in a line of those writing such treatises (*Elem*. 1.9.27 Delacy = 1.487K), but since titles could be retroactively applied, it is hard to assert who was the first to write a text with this name. Diogenes says that Alcmaeon wrote chiefly on medical subjects, which may illustrate that the types of reflection on health and disease that Alcmaeon conducted were associated with medicine more than natural philosophy by the late second and early third centuries CE; see DK24 A1 (D3 LM) = Diog. Laert. 8.83. Recently, Année (2019) has suggested that he wrote his medical poem in verse within the tradition of Spartan song culture.

tools to understand corporeal mechanisms, no author had called interior parts *organa* or corporeal tools. The body, even though it sometimes had functions, was not consolidated around this concept. This state of affairs changed with Plato (428–347 BCE), who referred to the sensory parts as "tools," before adopting the terminology to describe the "organs" of the body more broadly.

2.3 PLATO'S *TIMAEUS* AND COMPETING TECHNOLOGICAL HEURISTICS

The first explicit reference to bodily "organs" comes in Plato's *Theaetetus*, where Socrates proposes that we see through our eyes "as though through tools" [οἷον ὀργάνων].[56] Once this analogy is formally introduced, however, Socrates quickly drops the qualification "as though," thereafter referring to sensory parts simply as *organa* without any qualification.[57] Such an omission, although stylistically straightforward, occludes that this metaphor is of sparklingly recent vintage. The sensory parts were thus quickly transformed into the first naturalized organs of the body, becoming a key vector for thinking about the body parts as biotic tools. In fact, Plato expanded this vocabulary and used it to underpin some of his basic notions about corporeality within his cosmogonical dialogue, the *Timaeus*, which includes discussions of the sensory organs as well as other tool-like parts.

Technology looms large within this treatise, which describes how a craftsman god, the so-called divine demiurge, created the cosmos itself as a type of living entity and constructed the human in imitation of this rational living whole. To this end, the divine craftsman composed the human body of hierarchically arranged corporeal parts that oriented each individual toward rationality and reason as his ultimate *telos*. To describe these parts, Plato uses comparisons to mirrors, cloths, pipes, fish traps, and the clepsydra, among other tools and instruments. Paying attention to these comparisons and how they structure his explanations can reveal the traces that material technologies left on his conception of the body and its behaviors. It can show how impactful tools were for assembling Plato's somatic assumptions, even as his organism remains quite different from our own, ultimately structured to imitate the rational cosmos. Both health and life are brought under this broader rubric.

The *Timaeus* recounts the construction of the world and its components, and seemingly does so three times, once each for the so-called

56. Pl. *Tht.* 184d3–4.
57. Pl. *Tht.* 185a5, 185c7, 185d9.

accounts of Reason (29d–47e), Necessity (47e–69a), and Reason and Necessity combined (69b–92c).[58] Since each of these sections attends to different primary goals, they also each describe different corporeal parts and thus configure corporeality in slightly different ways. This difference can be quite disorienting and misleading if all sections are not seen in their proper relationship. Over the course of all three accounts, Plato puts forward a tripartite vision of the body, whereby each major division (head, chest, abdomen) supports one type of soul (rational, emotional, appetitive).[59] These, in turn, imitate the cosmos (mind, soul, matter), which imitates the rational good/God.[60] At its most essential level, then, the body is not formed around physiological function but is structured as an imitation of the living rational cosmos and, ultimately, of the demiurgic god himself. The "tools" of the body need to be understood within this context.

In the account of "Reason" or "Mind" [νοῦς], the narrator, Timaeus of

58. It was Cornford ([1937] 1997: xv–xvii) who first asserted that the creation story was broken into three sections: Reason, Necessity, and Reason and Necessity, although Timaeus only names the first two explicitly. Vlastos (1975) objected that the creation story was only divided into two sections, those of the "Triumphs of Pure Teleology" (28e–47d) and the "Compromises of Teleology with Necessity" (47e–end); cf. Harte 2002: 223. More recently, Harte (2002: 219–33) has suggested that there are only two creation stories presented, because the section after 69b2 resumes the first account of Reason. She does admit, however, that this return does begin a third, ab initio story. To my mind, the division of the creation story into three would imitate the tripartite body and the tripartite cosmos, for which reason it is strongly preferable as an interpretative structure. Moreover, the type of corporeality described in the first account deals only with the body as a means to perceiving and imitating the rational cosmos, and this paradigm differs quite substantially from that of section three, which deals with the body as a vehicle for the tripartite mortal soul, which has fears and "unavoidable emotions" [ἀναγκαῖα παθήματα] (Pl. *Ti.* 69d2), and which needs to maintain itself, grows, ages, decays, and dies. Moreover, the divine demiurge, having constructed all the divine things, turns the manufacture of the body over to his offspring in the third section, which does suggest that a different type of instantiated teleology is at work. For further reflections on the dialogue's structure, see note 59 below and, pp. 99–102.

59. This tripartite configuration is potentially prefigured by Philolaus; see DK44 B13 (D26 LM) = Ps.-Iambl. *Theol.* 25.17–26.3.

60. This iterative structure seems to unpack at least part of what it means when Timaeus names his broader account an *eikos logos*. Although the meaning of this phrase remains the subject of much scholarly controversy, it seems to resonate with how Timaeus arranges each level of existence as an *eikon* or "image" of the level above it, which it imitates.

Locri, describes how the divine demiurge, who was perfect, good, and devoid of envy, created a rational cosmos that was as like himself as possible. To do so, he imposed order on the preexisting visible disorder, thereby fabricating a single, unified cosmos composed of mind and soul united in a single body. This cosmos was an eternal living animal, one that did not decay or degrade and that contained all other mortal living animals within it, which are themselves "of like kind" [ξυγγενῆ] with the whole.[61] After the creation of the stars and planets, in which he instantiated a geometrically harmonious motion, the demiurge begins to manufacture a primordial human male, which requires constructing the body as a vehicle [ὄχημα] for the soul. In this first account, such construction largely means creating the body as it relates to the ultimate human *telos*, the perception of reason. The demiurge thus creates the eyes and ears, not so that they might spot food or hear danger, but so they might perceive the eternal cosmic motions and hear harmony, respectively, and lead them to understand these perfect geometries as intrinsic, ideal aspects of the soul.[62] This particular teleology illustrates how different Plato's view of the body is from our own, and why the introduction of tool-like parts into the body does not immediately imply the type of organ-based physiology to which we are accustomed. Within this dialogue, organs first serve as the physical tools of rationality.[63]

The notion of corporeality, or at least materiality, changes somewhat in the second creation account, that of "Necessity" [ἀνάγκη]. In it, Timaeus recounts how the divine demiurge created the four sublunary elements, fire, air, water, and earth, as well as a fifth celestial element, by structuring space itself into regular geometrical formations, which can transform into one another through their shared reconfigurable triangles (earth excluded). The four sublunary elements form the basic substances from which human tissues will be made in the next part; they therefore form another mode of looking at what comprises the human body and explains

61. Pl. *Ti.* 29e1–30d4.

62. Ultimately, the essential function of vision is not to see the stars and the heavens but to facilitate recognition of the notion of time, the nature of the universe, and the rationality inherent in it (Pl. *Ti.* 47a-b). Similarly, the function of the ears is to hear harmony so as to understand it as an intrinsic (and ideal) component of the soul (Pl. *Ti.* 47c–e). Here, Plato calls these sensory *organa* the "auxiliary causes" [ξυναίτια] that aid the soul to reach its highest good (Pl. *Ti.* 46c8–d10). In fact, Timaeus explicitly states that the god set the face on the front on which he bound *organa* (i.e., sense organs) with forethought for the soul (Pl. *Ti.* 45a-b).

63. We might compare this account to the description of the celestial bodies as the "tools of time" [ὄργανα χρόνων] (Pl. *Ti.* 42a1, 42d6).

its behavior. Since he already discussed the formation of these materials in regard to their rational harmonic ratios to build the cosmos in the first part, the account of Necessity functions as another ab initio account, rather than as a temporally subsequent event, providing a window into the physical properties of these materials. In this regard, it establishes a great interest in corporeality qua material object, although once again, this interest must be understood within the broader tripartite frame of the dialogue and the trifold nature of the human, especially since the final account, that of Reason and Necessity, is *itself* divided into three, obeying the same basic format as the whole dialogue, of which it forms an imitative part.

"Reason and Necessity" begins with the demiurge removing himself from the ongoing production of the human. Having attended to the head and the perceptual organs for the rational part of the soul, which is eternal, he instructs his divine offspring to manufacture the non-eternal aspects of the human form, as is more fitting. They take up the task and begin articulating the interior of the body to transform it into a vehicle for the two mortal types of soul—the emotive, which is situated below the neck and above the diaphragm, and the appetitive, which lives below the diaphragm but above the navel. As was the case for the cosmogonical account, Timaeus first describes the highest *telos* of these parts, which is to control and mitigate the effects of "necessary" or "unavoidable emotions" [ἀναγκαῖα παθήματα], which can disrupt adherence to reason and the mind. In other words, this part of the body is still built so as to facilitate rationality to the highest degree. To this end, the divine offspring fashioned the heart as the "guard house" [δορυφορικὴ οἴκησις] where they stationed the emotional part of the soul. This heart is a knot of veins and the fountain of the blood flowing around to the limbs via the vascular system, and it is through this system that reason can communicate with and calm all the perceptive parts of the body when anger strikes.[64] Next, they manufactured the lungs, which are spongelike and capable of receiving air and drink [τό τε πνεῦμα καὶ τὸ πῶμα] in order to cool the heart when anger makes it hot. The lungs also cushion and protect the heart from injury if rage causes it to leap up in the chest.[65] Below the diaphragm, the liver functions like a mirror [κάτοπτρον] that the rational soul can use to create imagistic impressions, insofar as thoughts can cause the liver to change taste, color, shape, and texture (or endure pain and nausea) so as

64. Pl. *Ti.* 70b6–7.
65. Pl. *Ti.* 70c–d.

to frighten or mollify the animallike, appetitive part of the soul.[66] The spleen, nestled beneath the liver, keeps the mirror clean like an accompanying towel [ἐκμαγεῖον] by collecting and absorbing impurities that can then be cleansed from the body.[67] The stomach, and potentially the upper intestines, act as a "manger" [φάτνη] for the nourishment of the entire body,[68] while the divine offspring crafted the lower intestines to wander and redouble on themselves to slow the flow of food and thus prevent excessive desire and gluttony, which would make the human unphilosophical and disobedient to reason.[69]

These portions of the *Timaeus* describe the interior parts of the body not only in terms of tool-like behaviors or mechanisms but also in terms of its tool-like *functions*. This account makes the parts not only technically produced artifacts but also technologically active objects themselves. In this regard, these passages transform the viscera into organs. Yet it is worth noting that aside from a general reference to "nourishment," Timaeus does not characterize any of the aforementioned bodily organs as serving a vital task, as though corporeal parts were in the business of keeping a person alive. Instead, these organs facilitate the soul's comprehension and imitation of the divine. Should any organ fail, the human individual would be led in negative ways by its emotions or appetites, but nothing indicates that death would result. The rationality of the soul—rather than either homeostatic balance or life—provides the goal of corporeal systems up until this point.[70] It is also crucial to recognize that these organa are only partially tool-like in operation insofar as they integrate interactions with the tripartite soul as an essential part of their function. That is, the liver communicates messages, but the animallike, appetitive part of the soul housed in the viscera still interprets and reacts to these messages as though it were a living thing. The body parts are tools, but the corporeal system still includes quasi-living components.

This configuration of corporeality changes once again in the second section of the account of Reason and Necessity (73b1–76e7), which describes the demiurge and his offspring manufacturing the tissues that compose the body and outlines the recipes that comprise these sub-

66. Pl. *Ti.* 70d6–71d5.

67. Pl. *Ti.* 72c3–d3.

68. Pl. *Ti.* 70e1–3.

69. Pl. *Ti.* 73a.

70. That said, the account of Reason includes a version of the homeostatic balance model when it describes inflows/outflows as disrupting the soul (Pl. *Ti.* 43a–44b), and the characterization of ignorance and irrationality as a type of disease in later sections makes these organs, albeit obliquely, productive of health.

stances. In this regard, the section parallels the account of Necessity describing the geometry of the primary elements, especially insofar as the god manufactures the constituent materials from which the body is made and molds them into shape. He starts with the marrow (which includes brain tissue), since it binds the soul to the body, and then proceeds to the bones, sinews, flesh, skin, hair, and nails. Moreover, just as the four geometrical elements can convert into one another through a recombination of their shared triangles, the tissues in the body also transform, although not endlessly without degradation as the primary geometrical elements do, but according to a normative, unidirectional generative process, the inversion of which causes disease.[71] Physical explanations dominate in the sections, and technical images abound. The demiurgic gods build bodies by fastening, molding, baking, gluing, and tempering various parts, and their tools and techniques derive from carpentry, pottery, breadmaking, metalworking, or other modes of manufacture.[72] He ensures that these tissues are arranged for certain purposes, sometimes with a view to necessary material ends and at other times with a view to higher, intelligent goods, such as arranging the mouth, teeth, lips, and tongue as an orifice fit to receive food but also to formulate speech.[73] Here, the connection between teleology and conceptualizing the body as a technically produced object seems clear. Yet, making bodies for certain purposes *with tools* and making bodies *of tool-like* parts are separate, albeit related, questions. Even these objectives are not quite the same as envisioning a body whose tool-like parts operate with tool-like mechanisms, absent human intervention. The powerful move toward this final schema can be seen in the account of respiration and blood distribution that Plato has Timaeus present. In these passages, Timaeus models the interior parts on a few technologies in particular: fish traps, the clepsydra, and Greek irrigation pipes. It is here where the particulars of Greek technologies in the fourth century make their mark on the imagined corporeal interior as it is getting articulated within this new "organic" frame.

2.4 RESPIRATION, THE CLEPSYDRA, AND IRRIGATION PIPES OF THE FIFTH CENTURY BCE

In the third section of the account of Reason and Necessity, Timaeus finally deals with the construction of the human with respect to its lowest,

71. Pl. *Ti.* 82d1–b7.
72. Pl. *Ti.* 73b1–76e7.
73. Pl. *Ti.* 75d5–e5.

material goals. That is, this section of the text presents some of the first explicit discussions of how the mechanisms of the physical body function to sustain both health *and* life. As normal as this notion may now seem, it represents the first explicit instance of what we currently call "physiology" in our extant Greek sources, since it presents dynamic physical processes as enacting functions that keep the animal alive. Yet even this statement must be qualified, since for Plato these dynamic corporeal processes serve to maintain the structural integrity of the body only insofar as they also imitate of the eternal motions of the cosmos. Accordingly, the functional physiology debuted in the *Timaeus* still remains rooted in his tripartite cosmic frame, even as it brings tools and corporeal functions to the forefront of its framework.

The physical processes that Timaeus describes in this section arise from a conundrum: fire and breath are materially necessary for life, and their absence causes death (just hold your breath or touch a corpse); yet, because they are so small and angular, these same two elements dissolve the material tissues of the body, a dissolution that can itself cause disease and death.[74] For the human body to imitate the cosmos to the greatest degree, it must maintain the identity and unity of its material form—at least for the greatest amount of time that accords with its nature. Accordingly, three (and only three) physiological operations become necessary—digestion, blood distribution, and respiration. The first two replenish lost tissues, while the last regulates the presence of breath and fire in the body. Through these operations, the body sustains itself for a duration of time in imitation of the eternality of the cosmos.

Plato presents these three corporeal processes as interwoven mechanisms, integrating the forces that drive each. As harmonious as this interdependency may at first sound, these different physiological functions end up placing divergent demands on the corporeal interior. These tensions can be seen by paying attention to several technological analogies that Plato uses in his explanation, including those that involve wicker fish traps, irrigation pipes, and the clepsydra. Attending to the friction caused by the competing comparisons reveals the crucial role that material tools played in Plato's conception of physiology at the inauguration of this explanatory mode.

The descriptions of both digestion and respiration are quite difficult

74. Pl. *Ti.* 77a1–4. The account does not really explain why these elements are necessary for life, especially since the dialogue presents soul alone as productive of life, and requires it to be bound to the marrow. When these bonds break in old age, natural death occurs.

to follow, especially since they does not map easily onto our own expectations of either anatomy or physiology, and as early as Aristotle, interpreters have produced multiple different versions of what the *Timaeus* is suggesting.[75] The account begins with the divine offspring, here called the "Superiors," creating the blood vessels by channeling them into our flesh as an irrigation system [ἡ ὑδραγωγία][76] that runs down along the spine:

> Having grown all these kinds [sc., plants] as nutriment for us lesser beings, the Superiors channeled through our body itself, cutting, as it were, channels in gardens, so that it might be irrigated just as though from a stream flowing in. And first they cut two veins as hidden channels along the back, under the connection of the skin and flesh, since the body happens to have a left and a right. And they sent these down along the spine, having taken the generative marrow between them, so that this might flourish as much as possible and so that the inflow arising from there might provide consistent irrigation, flowing well insofar as it flows downhill.[77]

At first, this comparison seems remarkably simple, whereby gravity flow serves as the main mechanism of propulsion for the vascular system, in which blood moves downward to reach all parts of the body, especially the sperm-generating marrow contained in the spine. Yet the Superiors also channel veins around the head, weaving them through one another in both directions so as to supplement the skin in connecting the head to the body, and so "that the experience of perceptions might become clear from each part to the whole body."[78] In other words, the vascular system operates as a pathway for both blood *and* perceptions, the latter of which typically involve the presence of air for other pre-Socratic physikoi. Although Plato is not explicit on this point, it seems probable that he, too, envisioned pneuma as a vehicle for perceptions. If so, this ambient air potentially suggests one way that breath existed within the vascular system, traveling in the same vessels as the blood.

The account of digestion and blood distribution continues. Next, the gods craft "fish weels" or "meshworks" [οἱ κύρτοι] of fire and air that serve

75. See Arist. *Resp.* 472b6–23, which presents a summary of Plato's model and Aristotle's objections.

76. Pl. *Ti.* 78a1.

77. Pl. *Ti.* 77c6–d8.

78. Pl. *Ti.* 77e4–6.

as semipermeable membranes or barriers.[79] The nature of these meshworks is difficult to determine, but Timaeus seems to mean that the membranous tissue of the esophagus, stomach, trachea, and lungs is woven together of fire and extended down into the body, thereby shutting food and drink inside these parts once they enter the stomach, while also preventing these substances from infiltrating the rest of the body without proper outlets.[80] This weaving has two "weel entrances" [ἐγκύρτια], the nose and the mouth, the former of which is bifurcated into the two nostrils.[81] Timaeus describes these as "hollow," "air-like," and woven of air.[82] Typically, commentators imagine these woven fish weels on the model of modern lobster traps or "pots," which are crates that rest while submerged in water with a single point of entry.[83] Yet Aristotle describes the technique for using these wicker traps in the fourth century BCE, which requires digging a trench as a type of artificial estuary and then covering it over. When the weather turns cold, fish swim into it, at which point one installs a meshwork basket at the base of the channel to trap and collect the fish.[84] If this is the type of trap Aristotle envisions, such a comparison supports the notion that a meshwork traps food and liquids (earth and water) inside once they pass a certain threshold.[85] These entrances allow fire and air to pass through in either direction (as in respiration) but let food and drink move only into the body (vomiting presumably notwithstanding). The nostrils and mouth connect in the throat (i.e., what we would call the pharynx)

79. Pl. *Ti.* 78a1–b2. These membranes block the passage of water and earth, insofar as all smaller particles (such as fire and air) have the capacity to bar the entrance of larger particles (such as water and earth), although the reverse is not true.

80. In this interpretation, I follow Pelavski (2014).

81. Pl. *Ti.* 78b2–7.

82. Pl. *Ti.* 78b4–c5. This description could indicate, as Solmsen ([1956] 1968) and Cornford ([1937] 1997: 308–12) argue, that actual air meshes stretch behind our nostrils and at the opening of our mouth or that these entrances are simply filled with air and shaped in such a way as to allow unidirectional flow.

83. Although Taylor (1928: 548–49) is agnostic about the nature of the trap Plato uses as comparison, both he and Cornford ([1937] 1997: 308–12) imagine that it is something like a lobster trap; cf. Reiche 1965; Pelavski 2014: 65–66.

84. Arist. *Hist. an.* 7.20, 603a2–12. He also describes a second, related method wherein one dams a river with the water flowing through a meshwork basket, which thereby collects any fish.

85. Since this meshwork is made of air, it need not be envisioned as a fleshlike anatomical feature (although we might consider whether the nasal cavity and sinuses are such weel entrances).

but then split again, with one passage (i.e., the trachea) traveling to create the lung and the other (i.e., the esophagus) running alongside it into the stomach/cavity, which operates as the other section of meshwork.[86]

As for the movements that take place in these parts, Timaeus describes how the "hollow part of the weel" (i.e., the interior of the lungs and stomach) flows gently into the weel entrances, while at other times the entrances flow back into the hollow parts. As a corollary, "the weaving, insofar as the body is porous, he made to move inwards through it [sc., the hollow of the weel] and back, and the rays of interior fire, having been enclosed, follow in each direction, when the air is moving" [τὸ δὲ πλέγμα, ὡς ὄντος τοῦ σώματος μανοῦ, δύεσθαι εἴσω δι' αὐτοῦ καὶ πάλιν ἔξω, τὰς δ' ἐντὸς τοῦ πυρὸς ἀκτῖνας διαδεδεμένας ἀκολουθεῖν ἐφ' ἑκάτερα ἰόντος τοῦ ἀέρος]).[87] It is again difficult to understand what this text means, as it could indicate either that fire enters from outside through the mouth and nose and then passes into the stomach to enact digestion, or, as Pelavski suggests, that the fire-woven membranes of the stomach lining move inwards into the organ's hollow and then back outwards, bringing the "rays" of its walls into contact with the foods.[88] Later passages that refer to the entrance of fire into the body make the former interpretation seem likelier.[89]

In any case, whether fire rays enter the body through inhalation or whether fire already composes the internal membranes, the action of breathing drives digestion by bringing fire into contact with food, which

86. Here yet another interpretative difficulty lurks, since it is unclear whether an *additional* meshwork surrounds the whole body in the form of the skin, to which the stomach and lungs are attached via veins, or whether the meshwork terminates in the lungs and stomach. The crucial passage comes when Timaeus next claims: "And from the weel entrances they [sc., the weavings] were stretched in a circle, like chords, through everything to the extremities of meshwork" [καὶ ἀπὸ τῶν ἐγκυρτίων δὴ διετείνατο οἷον σχοίνους κύκλῳ διὰ παντὸς πρὸς τὰ ἔσχατα τοῦ πλέγματος] (Pl. *Ti.* 78b8–10). Pelavksi (2014) argues that this description merely indicates that the membranes stretch down from the mouth and nose in a tube (i.e., circle) and to the extremities of the meshwork of the lung (i.e., the bronchioles). I agree that we should not consider the body as a whole as an additional fish weel, but since Plato describes veins leading from the stomach and later describes fire-woven vessels extending inwards from the skin as part of respiration, it seems more likely that this passage refers to the vascular system extending from the lungs and stomach, respectively, even if the reference to the skin as a "meshwork" is abrupt.

87. Pl. *Ti.* 72d5–8.

88. Pelavski 2014.

89. Pl. *Ti.* 82a1–7.

produces the blood that flows outwards through the veins. This outward flow nourishes, moistens, and cools the animal. Timaeus summarizes:

> For whenever fire, when respiration goes in and out, follows inside, since it is attached, and always floating to and fro, entering through the cavity, it takes the foods and drinks—in fact, it dissolves them—and dividing them up into small parts, leading them off through the exits by which it passes, just as from a fountain into channels, and drawing them off into the veins, it makes the streams of the veins flow through the body as through a pipe.[90]

From here, the vascular system acts as a set of irrigation channels by which the entire body is nourished. Inhalation and exhalation are thus tied together with digestion as two parts of the same physiological process, both of which sustain the creature and continue as long as the mortal animal maintains its structural integrity.

Up until this point, digestion and blood distribution have been presented as driven by gravity, with the movement of air supplying the motions required to push blood into the system. Breathing has simply involved the inflow and outflow of air. But how does the air flow in and out of the body? Here the forces at work change rather dramatically when Timaeus begins to explain the physical mechanism in greater depth, and vacuum pressure enters the conversation. He describes breathing as a type of "reciprocal propulsion" [περίωσις][91] wherein air is not drawn but driven into the body through the nose/mouth and the pores in succession. In this regard, he has taken up Democritus's ambient pressure as a propulsive force. For Plato, when we breathe, air moves through the nose and mouth, through the windpipe, and into the lungs (and potentially the stomach). The heat of the interior "fountain of fire," by which he seems to mean the fire-woven membranes of the lungs,[92] then warms the air until it grows hot, at which point it flows back up the windpipe so as to reach its proper place in the fiery cosmic ether according to the idea that "heat, according to its nature, moves to its own place outside, toward what is kindred" [τὸ θερμὸν δὴ κατὰ φύσιν εἰς τὴν αὑτοῦ χώραν ἔξω πρὸς

90. Pl. *Ti.* 78e5–79a5.

91. Pl. *Ti.* 79c5, 79e2, 79e6.

92. Pl. *Ti.* 79d1–5. Timaeus explicitly states that he has compared this interior fountain of fire to the weaving of a weel, which makes it clear that he is referring to the meshwork membrane of the lungs.

τὸ συγγενὲς ἰέναι].[93] When the air rises out of the body, it enters into the surrounding atmosphere—which is *itself* conceived of as an enclosed, finite space. Since a void is impossible [κενὸν οὐδέν ἐστιν], the air coming out of the lungs must somehow be accommodated in the greater atmosphere. Since this external air can neither expand nor compress, it drives a portion of the emerging air back into the body through the pores and into the veins.[94] At this point, the parts surrounding the air—or potentially the fire-woven tissue of the veins themselves—heat this new air entering through the skin, at which point the heated air, too, seeks its natural place. It then exits through the pores and thus drives more external air back into the lungs through the windpipe to start the whole process again. This completes Plato's reciprocating system and makes inhalation and respiration "a circle rolling in one direction, then the other, being performed by both" [κύκλον οὕτω σαλευόμενον ἔνθα καὶ ἔνθα ἀπειργασμένον ὑπ' ἀμφοτέρων].[95] Accordingly, respiration itself, even though bidirectional, forms a material imitation of the motion of the heavens, limited in form by material necessity.

Several questions arise from these details. For example, if the air exiting the lungs upward drives air into the vascular system, which is seemingly connected with the lungs, why does air not simply move unidirectionally without any cessation, perpetually entering through our pores and exiting through our mouth and nose? Why does it not form a perfectly circular system? Pelavski suggests that the vessels leading to the skin are a separate, terminal system, but there is no textual support for this claim, and it would be completely unique in ancient anatomical understandings.[96] Instead, I suspect that unidirectional flow is prevented because the air enters a vascular system filled not just with breath but with blood, too. Accordingly, blood implicitly prevents the unobstructed movement of air and presumably must give way in order to accommodate incoming material.

A second, more important, question also lingers. If the quantity of air exiting the body is always exactly equal to the quantity entering the body, how do you explain the expansion and contraction of the chest cavity? I suspect both questions can be answered by seeing how the clepsydra still implicitly shapes several of Plato's assumptions about respiration and the exchange of air, even though he never employs the simile explicitly. That

93. Pl. *Ti.* 79d7.
94. Pl. *Ti.* 79b1; cf. 79b10.
95. Pl. *Ti.* 79e10–11.
96. Pelavski 2014.

is, even as he cites abstract principles about the nonexistence of a void for justification, his abstract principles are already embodied in this totemic device to such a degree that when he thinks about the exchange of air and nonexistence of a void, he tacitly uses the clepsydra to do so.

Two features demonstrate this. First, in order for Plato's model to work, the lungs—indeed, the whole of the chest—would need to be a rigid body. If this were not the case, the air rushing into the body during inhalation would simply result in the expansion of the chest (not unlike an inflatable windbag), and no air would need to be pushed out of the pores. Were the chest not implicitly modeled on a rigid body, its reduced size after exhalation would offset the volume of the newly exhaled air (not unlike a deflated windbag); the surrounding atmosphere would not need to accommodate any new material, since the reduced volume of the collapsed chest would have already compensated for it. No air would need to be "driven round" into the pores. Rigidity remains requisite for all interpretations of the text, regardless of whether the stomach lining moves, fire moves in and out of the body, or the air entering through the pores actually makes its way back to the lungs. For the reciprocal exchange to work, the volume of the thorax must remain consistent.

A second feature suggests that the clepsydra lurks behind some of Plato's corporeal assumptions, since the extension of respiration into the vascular system seemingly integrates the exchange of air within a liquid-filled system. Indeed, his model functions as though the whole body itself were a large (ceramic) cavity with two sets of in/outflow points (nose/mouth and the pores), which once again looks remarkably similar to the clepsydra with its pores on the bottom and control outlet at the top. We might even consider whether his insistence that the transfer of digested foods, drinks, and air requires semipermeable meshworks reflects the implicit effects of this totemic technology, especially since no visible meshworks are present at either the back of the nose or the mouth. This absence makes it even more likely that Plato is implicitly imagining pores as an *essential* feature required in the mutual exchange of liquids and air. Such a conception suggests that the clepsydra sits somewhere behind his account, informing his understanding of the movement of air and the nature of the void. In fact, the technological model seems to have naturalized expectations of human anatomy to such a degree that the pores predicated by the technology get explained by yet *another* technical analogy (the wicker fish traps).[97] In any case, the contours of respiration look

97. Plato even follows this discussion by claiming that its basic physical principles can be used to explain a whole set of behaviors [παθήματα], including med-

dramatically different from our own expectations, since it is an entirely automatic process, and no role is given to volition at all. In other words, by outlining the physical mechanics that necessitate respiration, Plato has seemingly left no room for us to hold our breath at will.

There are, to be sure, important differences between the operation of the clepsydra and Plato's mechanics of respiration. The latter incorporates fire as a motion-inducing force, even when the clepsydra incorporates no such heating mechanism. Accordingly, this tool cannot completely dominate Plato's conception. Nevertheless, it provides the site where he activates his abstract principles about air exchange and the void, and this tool embodies his principles to such a degree that some of the clepsydra's features necessarily get implicitly integrated with those of the body, an integration that then confounds our modern expectations of human anatomy and physiology. In other words, Plato does not quite use the clepsydra as a heuristic analogy to *explain* certain previously observed corporeal behaviors or certain corporeal parts; rather, he first employs the clepsydra, which then helps him to *articulate* certain corporeal behaviors and certain corporeal parts. Acknowledging this bidirectionality reveals the ways that technologies structured the application of seemingly abstract principles well beyond the application of a formal analogy. It also shows that as the organism started to develop, even behaviors as basic as breathing emerged as an assemblage of observations from bodies and tools. Both the body's inner landscape and its behaviors depended upon the technological environment in which it was born.

2.5 IRRIGATION AND WATER DISTRIBUTION TECHNOLOGIES

The traces that material tools of the fourth century BCE left on Plato's theories are not imprints from the clepsydra alone, as important as this technology is. In fact, the impact that Plato's technological environment had on his implicit assumptions becomes visible when his account of breathing is stitched together with his discussion of blood distribution. Up to this point, Plato has presented a rather pastoral view of the blood vessels as irrigation pipes, flowing gently through the body to distribute nourishment to the spinal marrow and the rest of the body in a constant flow. Fire divides the food in the stomach as it moves back and forth, following

ical cupping vessels, the act of swallowing, and the ability of projectiles to stay aloft; see Pl. *Ti.* 79e12–80a3. That is, he points to other tropes of the power of air and the totemic tool used to understand attraction, namely, the cupping vessel.

the breath, potentially taking some nutriments with it, and "by oscillating as it follows the breath inside, it fills the veins from the cavity by pouring in the divided particles from it" [αἰωρουμένου δὲ ἐντὸς τῷ πνεύματι ξυνεπομένου, τὰς φλέβας δὲ ἐκ τῆς κοιλίας τῇ ξυναιωρήσει πληροῦντος τῷ τὰ τετμημένα αὐτόθεν ἐπαντλεῖν]. These veins and their "streams of nourishment" [τὰ τῆς τροφῆς νάματα] flow in this way for all animals.[98]

Here a tension arises. If the stomach is the distribution point for the blood, a considerable difficulty arises, since the blood needs to flow upward for all points above the diaphragm. Even the heart, which he previously called "knot of the vessels and the fountain-spring [πηγή] of the blood traveling around forcefully through all the limbs," sits above the stomach that generates the blood.[99] Plato attends to this difficulty obliquely when Timaeus describes how nourishment moves through the body according to the two principles introduced in the mechanics of respiration: the impossibility of a void and the natural tendency of everything kindred to move toward itself [τὸ ξυγγενὲς πᾶν φέρεται πρὸς ἑαυτό]. Both principles operate in the body as they do in the cosmos, and in doing so, "the movement [of the body] by necessity imitates the movement of the whole" [τὴν τοῦ παντὸς ἀναγκάζεται μιμεῖσθαι φοράν].[100] In the case of blood distribution, exterior forces damage parts of our body and break down the structures of our tissues into their component elements. These dissolved bits of fire, air, water, and earth then exit from our systems and move outwards toward their kindred substances in the broader cosmos. This movement creates spaces that must immediately be filled for the structure of the animal to remain stable and the animal to maintain its unity, and the disruption of this process causes illness.[101] Nourishment from the blood thus serves this reparatory function, insofar as the appropriate substances move from the blood toward the developing

98. Pl. *Ti.* 80d4–5.

99. Pl. *Ti.* 70a7–b2; cf. Hipp. *Oss.* 2 = 9.168L, which also calls the heart the "innate spring" [πηγὴ ξυγγενής].

100. Pl. *Ti.* 81a2–b4.

101. As Harte (2002: 212–66) has shown, this structure is what provides the identity of the whole; disease is a disruption of the living animal's oneness. Timaeus deals with corporeal diseases from Pl. *Ti.* 82a1–86a8, and these sections again break into a tripartite structure: those of the primary elements, those of the compound tissues, and those of the corporeal processes such as breathing. The text then discusses diseases of the soul, including ignorance, irrationality, and injustice (Pl. *Ti.* 86b1–88b7), before then discussing how exercising both the body and the mind imitates the motion of the cosmos and keeps the body healthy and oriented toward rationality (Pl. *Ti.* 88b8–90e5); cf. Stalley 1996; Betegh 2021.

gaps according to the movement toward what is kindred. As such, "each part, being irrigated, fills the place of what is being emptied" [ὑδρευόμενα ἕκαστα πληροῖ τὴν τοῦ κενουμένου βάσιν].[102] Although this account seemingly harmonizes all the corporeal processes as materially necessary instantiations of the same mechanical principles, the situation is slightly more complicated. The irrigation metaphors now seem incongruous with the bidirectional movement of air and blood in his account of respiration, most notably the assertion that the downward-flowing veins along the spine ensure "consistent" nourishment. Yet Plato does not give any indication that the veins used in respiration differ from those used to distribute blood. That is, when discussing respiration, he proposes that air moves in and outward within a vascular system as though in the confines of a clepsydra-like pressurized vessel, whereas the demands placed on the veins in blood distribution now make blood another moving part. This proposition, at the very least, would seem to complicate the perfectly symmetrical volumes required in air exchange. In other moments still, Plato mentions air in the veins, which could refer either to the air moving in and out through cutaneous breathing or to freer-flowing air responsible for communicating messages from the heart outwards.[103]

In sum, whereas his physiology of respiration relies on blood and air constantly displacing each other in the chest and veins, his account of nutrition constructs a more ambiguous corporeal interior in which air, fire, and blood can potentially fill the same vessels without problem, sometimes performing different tasks according to overlapping mechanics.[104] Rather than simply invent new details to harmonize these actions, I find that the tensions themselves are instructive. Indeed, I would argue that there are competing explanatory frameworks, employed in alternation to explicate related physiological aspects, even within the same system of vessels and membranes. When discussing blood distribution through the vessels, Plato uses gravity-flow irrigation pipes. When conceptualizing the transfer of air into and out of the body, he is informed by the clepsydra as the totemic technology of the void. When thinking about the replacement of dissolved flesh, he mediates between the two while remaking the internal mechanics of the body in the image of the cosmos itself.

We might step back and ask another question that arises at the intersection of these various theories: Why was Plato so comfortable with air in

102. Pl. *Ti.* 81a2–3.

103. Pl. *Ti.* 82e1.

104. Plato attributes the transmission of sense perceptions to the veins, which might indicate the presence of pneuma in them (Pl. *Ti.* 77e).

the veins in either theory? If the clepsydra model explains the exchange of air and liquid at pressure in cutaneous breathing, why did Plato seemingly also accept unpressurized breath present in the veins? To be sure, one might point to the many other contemporary theorists who also thought that pneuma moved within the vascular system, since he was certainly not alone in this notion. Alcmaeon seems to distinguish a special class of "blood-flowing veins" [αἱμόρροι φλέβες] into which blood can retreat as it exits from other vessels, which suggests that other veins are not solely for blood.[105] Diogenes of Apollonia argues that pneuma—the substance he holds responsible for thought—travels around the body through the blood vessels.[106] Philistion, a contemporary of Plato, describes health as the unobstructed flow of pneuma throughout the body, while disease is caused by its blockage.[107] Similarly, the Hippocratic author of *On the Sacred Disease* describes a corporeal system that includes the distribution of sense-causing pneuma throughout the same passageways that carry phlegm, bile, and blood.[108] In fact, this author characterizes the veins as both vents for air [ἀναπνοαί] and channels for blood [ὀχετοί] at the same time.[109] To modern minds it perhaps seems obvious that air and blood cannot flow smoothly in the same pipe—especially if that pipe moves up and down, ascending and descending across uneven terrain as it curves along the contours of the body.[110] In such a system, air pockets would form that would capture blood just like the trap in a kitchen sink. Bubbles would develop. Blood would pool at our feet. If we lifted our arm, blood

105. DK24 A18 (D32 LM) = Aët. 5.24.1 (Ps.-Plut.); cf. Harris 1973: 8. The manuscripts, however, claim that these vessels are "same-flowing" [ὁμόρρους], and Lloyd ([1975a] 1991: 177 n49) states that Reiske first proposed the emendation to "blood-flowing" [αἱμόρρους].

106. DK64 A29 (D33 LM) = Aët. 5.24.3 (Ps.-Plut.); cf. DK64 B4 (D9 LM) = Simpl. *In Phys.* 152.18–21; DK64 B5 (D10 LM) = Simpl. *In Phys.* 152.22–153.16; DK64 B6 (D27 LM) = Arist. *Hist. an.* 3.2, 511b–512b12; cf. Harris 1973: 25–27.

107. Anon. Lond. 20.43–50 Manetti; cf. Harris 1973: 19–20, 36–38.

108. See also Hipp. *Flat.*; *Carn.* 6 = 8.592L; and *Oss.* 11 = 9.182L; 13–18 = 9.184–90L. Due to the presence of a quotation from Aristotle and the inclusion of the work in the list of Hippocrates's treatises by Bachhaeus of Tanagra in 200 BCE, *Nature of Bones* must have been written either in the latter half of the fourth or in the third century BCE; cf. Harris 1973: 51.

109. Hipp. *Morb. Sacr.* 7 = 6.368L.

110. Multiple ancient authors seem to imply that water can flow upward. For example, Hippon DK38 B1 (D19 LM) Schol. Genav. In Hom. *Il.* 21.195 argues that all groundwater comes from the sea. The prevalence of porous karst within the Greek archipelago also leads to springs appearing at the top of hills, which illustrates that such upward flow can in fact happen (given the right circumstances).

would immediately drain out of it. Despite such difficulties, however, the double function of the veins was an extremely common idea, often assumed without argument.[111] Despite the fact that theorists operate with vastly different physiological ideas, almost every author of the fifth and fourth centuries BCE assumes that both blood [αἷμα] and air [πνεῦμα] flow through the same vessels at the same time. Veins almost always perform this double duty.[112]

We might also ask why the gravity-flow model of blood distribution was so attractive to Plato, especially if it conflicted with his assertions about the impossibility of a void. Once again, he is not alone in this conception. It seems implicit in the Hippocratic treatise *Nature of the Human* insofar as it suggests that the thick vessels for the transference of humors descend from the head.[113] Similarly, as we saw in the last chapter, both Diogenes of Apollonia and Syennesis of Cyprus likewise have vessels running downward from the top to the bottom of the body. In fact, when relating these details in *Historia Animalium*, Aristotle claims that "everyone identifies the source of [the veins] as in the head and the brain" [πάντες δ' ὁμοίως τὴν ἀρχὴν αὐτῶν ἐκ τῆς κεφαλῆς καὶ τοῦ ἐγκεφάλου ποιοῦσι].[114] This claim is perhaps an exaggeration—largely so as to emphasize the novelty of his

111. Certainly, ancient medical thinkers describe blockages and clogs. For instance, along with the consequences of blockage seen in *On the Sacred Disease* above, Praxagoras talks about "bubbles" rising up from the feet and causing both mania and epilepsy (Praxag. Fr. 25 Lewis = 70 Steckerl). He makes a distinction between exterior πνεῦμα and the substance that activates the body, which is "breath-like" [ἀτμῶδες]; cf. Gal. *Art. Sang.* 2.2 Furley = 4.707K. (See also Lewis 2017, esp. ch. 2). Similarly, Philotimus, Praxagoras's pupil, mentions bubbles arising from the process of digestion (Oribasius, *Coll. Med.* 5.22; Clem. Al. *CMG* 6.1.151). Nevertheless, these phenomena are pathologies. In a healthy body, air and blood would flow through the same vessels at the same time without interruption.

112. Although one could argue that ancient medical theorists believed that air and blood flowed in the same veins because they knew only one set of vessels, I would argue that these authors were not yet looking for another set of vessels to solve the problem of dual function veins—precisely because they did not see this dual function as a problem. Praxagoras (b. ca. 340 BCE) was the first theorist to argue that the arteries contained air, while the veins contained blood, although Alcmaeon may have hinted at a similar distinction (see p. 113 above; cf. Longrigg 1988: 467; von Staden 1989: 173). See also sections 4.3 on p. 172 and 4.4 on p. 176.

113. Hipp. *Nat. Hom.* 11 = 6.58–60L. Yet, like Plato, this author also adds at 11.40–46 that veins from the stomach carry nourishment all over the body. For a description of Polybus' account, see Arist. *Hist. an.* 3.3, 512b11–513a2; cf. Arist. *Hist. an.* 3.2, 511b24–30 = Hipp. *Oss.* 9 = 9.174–176L; cf. note 92 on p. 46.

114. Arist. *Hist. an.* 3.3, 513a10–12.

own theory that the heart supplies the source of the veins (see pp. 127–28 below)—but Aristotle's assertion indicates how common the idea that veins descended from above was, thus implicitly relying on gravity flow as a propulsive force.[115]

An examination of the technological environment of the fourth century BCE may reveal why the notion that air and blood flow together and the gravity-flow model are so common and how they potentially both hang together. Some potential insight comes from Plato's irrigation metaphor, especially since Plato is not alone in this comparison. *On the Sacred Disease* likewise speaks of the vessels as "channeling" [ὀχετέουσι], as does Aristotle, who likens the veins to channels [ὀχετοί] and irrigation systems [αἱ ὑδραγωγίαι] while describing how blood travels around the body in both *Parts of Animals* and *Historia Animalium* (see pp. 140–44).[116] In fact, ancient Greek pipelines—specifically, domestic water-delivery systems of the fifth century BCE—display a significant feature: the majority were composed of short sections of terra-cotta pipe about 20–25 cm in diameter and 60 cm long (presumably restricted by the length that could be handled on the potter's wheel).[117] These sections were fitted together using their male and female ends and then sealed with mortar. Most importantly, each section had a hole cut in the top, which was then covered back over with a close-fitting lid (fig. 8).

115. For the chief Hippocratic descriptions of the blood vessels, see *Epid.* 2.4.1= 5.120–124L, which corresponds with *Oss.* 10 = 9.178–180L; and *Nat. Hom.* 11 = 6.58–60L, which corresponds with *Oss.* 9 = 9.174–180L; *Morb. Sacr.* 6 = 6.366L; *Carn.* 5 = 8.590L; cf. Jouanna (1999: 311), who notes that all these accounts are slightly different.

116. Arist. *Part. An.* 3.5, 668a14–17; *Hist. an.* 3.4, 515a23–24. Hipp. *Oss.* 13 = 9.184L; 16 =9.190L; 19 = 9.194–196L and *Cord.* 7 = 9.84L also use this same metaphor, but these texts are dated to the late fourth or third century BCE and will be discussed separately in the next chapter.

117. Hodge 2000: 41; (1992) 2002: 25. Compare Thomson and Wycherley (1972: 197–98), who approximate pipe lengths at 30 cm in diameter; Wilson (2008: 293–96) gives the internal diameter as 15–25 cm. If Greek water engineers wanted to increase capacity, they would have increased the number of pipes, not their size. Greek water supply systems, however, did not rely on pipelines alone but also employed wells and bottle cisterns. For a general account of Greek water supply systems, see Thomson and Wycherley 1972; Hodge (1992) 2002: 25–31, 48–66; Crouch 1993; Jansen 2000; Wilson 2000; 2008: 285–318; Humphrey 2006: 35–51. See also Bruun (2000: 557–73), who examines the legislation relating to water distribution in the Greek world. For other aspects of ancient water technology, see Wikander 2000. Tamburrino (2010) indicates that Assyro-Babylonian canal systems also included terra-cotta pipes.

FIGURE 8. Terra-cotta pipes from the Peisistratid aqueduct, late 6th century BCE. Found during subway construction and displayed at Syntagma Station, Athens. Photo by author.

This design most likely allowed the pipefitter to reach inside to seal the joint, although it has been suggested that these holes may have instead simply allowed for cleaning.[118] Regardless of their purpose, the con-

118. Scholars disagree about the purpose of the holes; cf. Tölle-Kastenbein 1991, 1994: 71–72; Fahlbusch 1994: 109; Hodge 2000: 41; and Jansen 2000: 106.

sequence of the design remains the same: these pipes could not have run at pressure. With a potentially leaky outlet at the top of the pipe, water could have only filled some of the passageway, while air must have filled the rest.[119] The system could therefore have no real flow control and did not incorporate valves to shut off the water at any given point except at its source. Instead, these water-delivery systems relied solely on gravity flow to maintain constancy. Air and water flowed in the same pipes. They were both vents and channels.

If this is the technological context in which ancient physiologists lived in the fifth and fourth centuries BCE, when they thought about blood (or humoral) distribution and used their water-distribution systems as conceptual analogues, it would have been simple to assume that both blood and air likewise flowed in the same vessel simultaneously. In fact, since the water-distribution technology around them worked in exactly this manner, it might actually have been odd for them not to think in this way. And indeed, once the metaphor is employed, it is a convincing enough comparison to suggest that the vascular system functions in the same way despite any difficulties with two substances traveling along the same passageways, sometimes traveling upward together, as to the brain. Plato's conception of blood distribution seems to include traces of this technological reality. Even despite a broader theory whereby the internal motions of nourishment and blood operate according to vacuum pressure and the movement of like to like, he seemingly discusses the presence of fire and air in the veins, moving alongside or potentially within the blood, without difficulty. This co-presence creates additional tensions with his account of respiration, which mobilizes the vasculature as a clepsydra-like space wherein air moves into the veins as space is created for it. These modal heuristics establish two competing mechanical operations that rest on two different technologies, and the body that Plato thereby creates becomes a bricolage, assembled from different tools. He employs abstract physical principles to justify his explanations, but he activates those physical principles through models that rely on specific tools available in the fourth century BCE.

119. Hodge ([1992] 2002: 25) notes that calcium deposits occur only in the bottom half of ancient pipes, notably the Enneakrounos pipeline in Athens. This observation supports the conclusion that they did not run full; cf. Tölle-Kastenbein 1994, 1996; Wilson 2008: 294. There is some evidence that the varying levels of terrain in Athens caused a short section of pipeline to run under very slight pressure, although even this possibility is debated; cf. Hodge 2000: 42; Jansen 2000: 108.

2.6 CONCLUSION

This chapter has traced the emergence of the idea that individuated parts serve functions within the body and outlined how this basic idea is connected to tools. The investigation began with Empedocles and suggested that the incorporation of tool analogies into the explication of corporeal behaviors was crucial for assigning these processes to individuated parts. In this way, tools helped make organs and facilitated the location of bodily processes within these internal components. It then looked at Plato's extension of this concept within his teleological generation of the cosmos. Although the divergent aims within the *Timaeus* present us with a layered view of the body and the purpose of its parts, Plato's essential corporeality seems aimed at comprehending the creator god himself, such that the body's parts supported human rationality more than they sustained or produced life. Nevertheless, he was the first to explicitly refer to corporeal parts as *organa*, as he created the body as a type of functional artifact. Both authors included non-pathological somatic processes in their account of the human whole, and in explicating these processes with tools, they built different types of epistemic objects.

In addition to limning this overarching conceptual shift, this chapter also looked at how this view of corporeality created novel interactions between technologies and bodily processes, emphasizing that the inclusion of tool analogies to account for respiration and blood flow did not simply create new theories but helped construct the basic behaviors that Empedocles and Plato attributed to the body. Even something as elemental as breathing was a phenomenon that emerged from the negotiations within their explanatory apparatuses. These hybrid arrangements helped determine which parts were or were not involved in breathing, as well as the dynamics of the parts that had been included. Moreover, these explanations operated modally, such that Plato's account of respiration integrated the same vessels as his account of blood distribution, even though the former operated by clepsydra-like vacuum pressure while the latter operated by irrigation-like gravity flow. Whereas the Hippocratics created therapeutic interfaces to determine the fundamental nature of the body, the explanatory techniques used within the physikoi tradition and by Plato privileged other investigative modes, and the particular tools that were used to both structure the body and articulate its parts left their traces on the assumptions about corporeality and the activities it involves.

CHAPTER THREE

Aristotle and the Emergence of the Organism

3.0 INTRODUCTION

Whereas Plato's articulation of the human body appeared embedded within the larger cosmological account of the *Timaeus*, Aristotle (384–322 BCE) devoted a huge portion of his corpus directly to the study of living things. His treatises range from the most general investigation into the nature of the soul and its capacities to an expansive observational program that catalogs animal parts, characters, and modes of life. His sustained and systematic investigation of living things expanded well beyond any previous efforts and rightfully gained him the title of the "father of biology." Despite the different aims of his treatises, Aristotle's various texts can be arranged within a hierarchical causal science of living things, with the study of the soul qua soul at the top of his biological project. The syllogistic science of generalizable statements that he outlines in *Posterior Analytics* provides the rough methodological structure for examining animal species in *Parts of Animals* and *Historia Animalium*. In both these texts, parts and their individuated functions form the conceptual components that build bodies, and he catalogs and systematizes a teleology of differentiae to determine how each species is suited to its unique mode of life by the possession of particular parts and features, both internal and external.[1] In the so-called *Parva Naturalia*, Aristotle addresses the indi-

1. Aristotle spends considerable time outlining his methodology at the beginning of *Parts of Animals*, especially *Part. An.* 1.4, 644a13–b21. Along with establishing essential differentiae, he is also interested in finding correlations between parts—e.g., since the elephant is very large, its feet must be flat and dense to hold up the weight, which is why it must use its nose/trunk as a prehensile appendage. There are too many overviews and investigations of Aristotle's biology to summarize them concisely, but special attention has been given to how his broader project relates to a syllogistic natural philosophy; see especially Gotthelf

vidual corporeal activities more directly, explicating the senses, memory, sleep and dreams, respiration, aging, and death. Although this part of his project owes much to the physikoi tradition and Plato's *Timaeus*, Aristotle was evidently familiar with at least some of the Hippocratic treatises as well, borrowing most heavily from *Diseases* 4 and the gynecological texts, perhaps insofar as they come closest to articulating the body as a system of operational components.[2] Nevertheless, his biological project moved onto radically new ground, where bodies became hierarchically arranged systems whose functional components performed or supported vital tasks. Far from a banal title, *Parts of Animals* is almost a revolutionary cry, implicitly containing a new vision of corporeality. With Aristotle, parts and organs became the crucial divisions within the body, the primary goal of which was now the maintenance of life. By marrying form and function within parts, Aristotle *organized* the body, and in so doing, he solidified a new epistemic object.

His consolidation of what might be called the "organism" brought with it a growing commitment to what we call physiology, now understood to require both an explanation of corporeal mechanisms and their purpose for the living animal. This approach was accompanied by a corresponding set of investigative and explanatory practices in which technologies play crucial but complex roles. On the one hand, Aristotle is adamant that the categories of living and nonliving are strictly divided and that living things are distinct from artificial objects: living things alone possess intrinsic sources of motion and change, while nonliving artificial objects are created and moved by external impetus. On the other hand, tools structure his model of what a living thing is and how its parts relate to the biotic whole.[3] He extols a type of structural teleology wherein the corporeal components now completed a set of hierarchical vital functions. Pneuma is a tool-like agent of these operations, but so too are tissues, structures,

1985, 2012; Gotthelf and Lennox 1987; Devereux and Pellegrin 1990; Lloyd 1996; Lennox 2001; Ebrey 2015; and Connell 2021. For a broader overview and lively (if somewhat fanciful) biographic reconstruction, see Leroi 2014, and for a full bibliography on Aristotle's biology, see Lennox (2006) 2021.

2. For Aristotle's connection to the gynecological texts, see Byl 1980 and Oser-Grote 2004.

3. His metaphysics is deeply indebted to techne in general, insofar as his teleological ontology uses tools to structure his notion that all things have material and formal arrangements that allow them to serve certain purposes, just as an axe needs to be made of hard metal in a certain shape so as to chop, and if it cannot accomplish this purpose, it is not an axe.

and organs. To this end, he often involves comparisons with technologies in his explanations, and although this practice is certainly not novel, his analogies tend to be more robust than his predecessors', insofar as he aligns both the mechanisms and functions between his analogues and their corporeal correlates. Aristotle thereby tends to establish broad networks of correspondences between tools, their material operations, and the behaviors of the body.

This chapter examines networks and how the material specifics of fourth-century BCE tools impact his explanations, focusing on the heart as the locus of such dynamics. It illustrates that Aristotle is careful to delineate the differences between the heart and the tools that resemble it, such that any easy correspondence should be avoided. Nevertheless, his technological comparisons can sometimes bridge his detailed cardiovascular anatomy and the physiological processes housed in these parts. In fact, his comparisons often direct attention away from problematic anatomical features that he himself has described and shift cognitive focus to the operations of surrogate tools. As a result, the somatic processes often end up articulated in the space between living and technological things, with the corporeal parts themselves formulated as objects within this hybrid explanatory object. Technologies thus directly impact the features he ascribes to the body. Moreover, Aristotle's tools can sometimes predicate certain behaviors that he then attributes to corporeal parts, even though they could not have been observed acting in this way. Technologies thus support his notions of what a body is, how it works, and what properties it displays.

The last section of this chapter expands on this observation by moving beyond the heart and toward the extremities, illustrating how features of automata from the fourth century structure some of Aristotle's assumptions about the mechanisms of animal motion. It emphasizes that animal tissues, like sinews, were themselves used as the raw materials for technological devices, and that these uses in turn informed assumptions about their operations in the body. Making artifacts with a biotic substance could therefore act as a proxy for knowing about its natural operations. In sum, this chapter outlines how tools structured Aristotle's notions of corporeality, illuminated the operations of the organs, and created hybrid networks of physical assumptions that moved between living things and artificial objects. It continues to track the exchange and interaction of technologies and bodies by following how Aristotle established a new model of the body and articulated its interior using new material and cognitive tools.

3.1 THE SOUL AND THE ORGANISM

For Aristotle, the body is neither a machine that can be brought to life by the addition of a soul nor a living thing that can be reduced to mechanistic physical processes. Rather, it is a biotic whole whose first-order active function is life. Just as the activity of an ax is chopping, which serves as the purpose for which this tool was made, the body's essential activity is living. Aristotle identifies this activity as the soul itself. As clear as he makes the distinction between natural and artificial objects, Aristotle relies heavily on tools for thinking about the nature of the body so conceived, especially since he insists that life can only occur within an "organized," "tool-like," or "organic" system:

> That is why the soul is an actuality of the first kind of a natural body potentially having life. Such a thing is that which is tool-like [ὀργανικόν]. . . . If, then, it is necessary to declare some general definition applicable to every kind of soul, it would be the primary actuality of a natural tool-like body [σώματος φυσικοῦ ὀργανικοῦ]. (Arist. *De an.* 2.1, 412a27–b6)[4]

Within this organized body, the parts themselves become tools [ὄργανα], each now fulfilling some necessary role in the broader vital system.[5] As he states in *Parts of Animals*:

> Since every tool [τὸ ὄργανον] is for the sake of something, and each of the parts of the body is for the sake of something, and what they are

4. Arist. *De an.* 2.1, 412a27–b6. Cf. *De an.* 2.1, 412a20–22, which states: "Therefore it is necessary that the soul is the substance of a natural body that potentially has life, as its form. And this substance is actuality. Therefore, [the soul] is the actuality of such a body" [ἀναγκαῖον ἄρα τὴν ψυχὴν οὐσίαν εἶναι ὡς εἶδος σώματος φυσικοῦ δυνάμει ζωὴν ἔχοντος. ἡ δ' οὐσία ἐντελέχεια. Τοιούτου ἄρα σώματος ἐντελέχεια]. See also *De an.* 2.1, 412ab10–24.

5. The ellipsis in the passage above contains the assertion: "The parts of plants are *tools* [ὄργανα], although wholly simple—for example, the leaf is a covering for the pericarp, the pericarp for the fruit. And roots are analogous to mouths, both draw in nourishment." Aristotle further refines this at *Part. An.* 3.4, 665b20–26, where he calls the parts required for the maintenance of life the "necessary body" [τὸ ἀναγκαίον σώμα]. These end where the residues get expelled. Limbs, for instance, are not necessary for life. Moreover, not every body part is an organ, although Aristotle generally refers to the compound parts as tool-like [ὀργανικά] (*Part. An.* 1.7, 491a26ff).

> *for* is the sake of a certain action [πρᾶξίς τις], it is apparent that the entire body too has been constituted for the sake of a certain multipart action. For sawing is not for the sake of the saw, but the sawing; for sawing is a certain use [χρῆσις]. So the body too is in a way for the sake of the soul, and the parts are for the sake of the tasks [τῶν ἔργων] in relation to which each of them exists by nature.[6]

We should not underestimate the force of this relatively new terminology of *organa* and corporeal "tasks," which most translators simply render without reflecting on the considerable conceptual shift that such nomenclature captures. In fact, translators frequently apply these metaphors retroactively to earlier texts as though this terminology did not imply an entire concomitant theory of corporeality.

These views about living things are on full display in *Parts of Animals* and *Historia Animalium*. These texts investigate animals by lemmatizing their bodies, breaking them down into their hierarchically arranged component parts in order to map any structural differences onto the demands of each animal's lifestyle, character, and social practices. All animal bodies are composed of incomposite, "homoiomerous" [ὁμοιομερῆ] substances (e.g., blood, flesh, bone, skin, sinew) and residues [τὰ περιττώματα] (e.g., nails, hair, menstrual blood, semen). Although these parts can serve their own functions, they also supply the basic materials that compose the multiform or "composite" [σύνθετα] parts. It is these more complex parts that Aristotle formally terms *organs*.[7] These include the hands, feet, legs, face, liver, heart, and kidneys.[8] Each such organ performs an essential activity that operates within one of four systems: (1) sensation, (2) voluntary motion, (3) nutrition (to which respiration and blood distribution

6. Arist. *Part. An.* 1.5, 645b15–20.

7. Arist. *Hist. an.* 1.1, 486a6–25; cf. *Hist. an.* 3.2, 511b1–11. See also Arist. *Part. An.* 2.1, 647a3–10, which makes a further distinction between sensory [αἰσθητήρια] and tool-like [ὀργανικά] parts insofar as tool-like parts are compound, whereas perceptive parts are uniform throughout.

8. Some organs are themselves components of even more composite organs (e.g., fingers to hand; nose and mouth to face). What we call the "internal organs" (e.g., heart, liver, lungs, kidneys, spleen), Aristotle tends to call "viscera" [σπλάγχνα], although these also qualify as organs. In addition, Aristotle identifies a third category, "residues" [τὰ περιττώματα] (e.g., feces and urine, but also hair, antlers, and fingernails). These are produced by the homoiomerous substance and can serve ancillary functions but not essential ones (cf. Arist. *Part. An.* 2.2, 247b27–29).

both belong), or (4) reproduction.[9] Aristotle argues, much like Plato, that Nature is a good craftsman who has designed each part of an animal for a particular goal or function, although Aristotle gives more acknowledgment to material constraints upon these demands and is sometimes forced to accept that certain parts seemingly have no purpose in certain animals.[10] He chastises his predecessors for failing to see nature as goal oriented in this way, especially insofar as they fail to explain the purpose of each corporeal element, which he calls the "that for the sake of which" [οὗ ἕνεκα]. Some commentators treat this as though Aristotle were the first to privilege teleological *explanations* as a scientific methodology, almost assuredly since Aristotle frames it as such.[11] Yet Aristotle's biological outlook did not merely formalize certain types of explanation. It also constructed a different epistemic object. The boundaries between inside and outside, so productively malleable within Hippocratic logics, became far more rigid, as the body transformed into an ontological unity now consolidated around a primary and essential function: the maintenance of life. The eye was for seeing, the ears hearing, but so too now were all interior parts understood only relative to their activities, which operated within a system that collectively produced or—more accurately—enacted life. Whereas the Hippocratics had typically identified a body's physis with its health-balancing humors, Aristotle prioritized structures rather than the substances. The organism had arrived.[12]

Along with shifting the essential nature of a body to its corporeal structures and components, Aristotle also privileged a different investigative practice for understanding biotic things: systematic dissections.[13] Prior to Aristotle, some anatomical inquiries took place, but they seem to have

9. The parts of each species within a genus will have parts that differ from one another according to "more or less" (e.g., long legs or short legs), while those across genera will have parts that differ from one another by analogy (e.g., fins to wings, lungs to gills); see Arist. *Part. An.* 1.4, 644a13–24. This principle applies to internal parts as well as external; see *Part. An.* 3.4, 665b2–6 for the latter.

10. See Arist. *Part. An.* 1.1, 639b16–640a1; cf. p. 177.

11. There have been numerous investigations of Aristotle's teleological arguments, including the recent contributions of Gotthelf (2012) and Leunissen (2010). See Leunissen 2010: 2 n3 for a bibliography on the subject. For broader approaches to teleology in the ancient world, see Rocca 2017.

12. Similarly, Bianchi (2017: 7) states, "Aristotle's thought is without doubt the site where the organism is instituted as metaphysically hegemonic, where its healthy functioning is offered as a paradigm of being as ἐνεργεία."

13. See Aristotle's plea at *Part. An.* 1.5, 645a33–b14 for the necessity of understanding the minutiae of flesh, blood, bones, etc. in order to understand the beauty of Nature.

been piecemeal, ad hoc, and, in some cases, imagined rather than performed. The first mention of deliberate dissection occurs with Alcmaeon, who, as mentioned above, excised the eyes. Anaxagoras, Alcmaeon, and Democritus are all apocryphally described as anatomists who conducted dissections, despite the extant fragments evincing no real evidence for sustained anatomical inquiry (although some individual instances are certainly possible).[14] Similarly, explicit references to dissection are all but completely absent within the (pre-Aristotelian) Hippocratic Corpus, and authors certainly do not advocate for its use, which accords with their greater interest in humors than in structures.[15] That said, certain levels of anatomical knowledge suggest that some level of animal dissection must have taken place.[16] Indeed, in *Historia Animalium* Aristotle details how Syennesis, Diogenes, and Polybus mapped the vessels through the body, and he states that some predecessors examined dead and dismembered animals.[17] Similarly, Aristotle elsewhere asserts that his predecessors incorrectly understood the lungs because they removed them from the body during dissection rather than examining them in situ.[18] He also claims, however, that Anaxagoras's and Diogenes's theories of respiration are in-

14. See chapter 4, p. 195–98.

15. Hipp. *Morb. Sacr.* 14 =6.382L mentions cutting open goat heads to reveal that excess phlegm (and not divine agency) causes seizures. This account mirrors that of Anaxagoras cutting open the skull of a one-horned ram to illustrate the inner causes of its aberrant form (DK59 A16 [P21 LM] = Plut. *Per.* 6). Neither of these episodes, however, illustrates any interest in the functional arrangements of internal parts, and instead they both remain focused on exploratory autopsy and the proximate needs of their textual arguments. A few texts do evince some dissection practices. Of these, *On Fleshes, On the Heart*, and *On the Nature of the Bones* are dated to the third century BCE (see Abel 1958 and Lonie 1973a), whereas the *Anatomy* is either contemporary with Aristotle or slightly after (see pp. 161–64).

16. Elsewhere in the corpus, there is an interest in the arrangement of the sensory parts and the blood vessels, which may supply indirect evidence for some level of dissection. For example, Hipp. *Epid.* 2.4.1= 5.120–124L. describes the pathways of interior vessels, which is copied at Hipp. *Oss.* 10 = 9.178–180L. Similarly, *Places in Man* 2 = 6.278L talks about voids in the regions of the ears, three membranes of the eyes, and veins that nourish sight, although the author describes the information as coming from injury to the eye, not from dissection. Bubb (2022) collects and assesses the relevant Hippocratic evidence.

17. Arist. *Hist. an.* 3.2–3, 511b10–513a16. He complains that these predecessors missed seeing important vessels because they collapsed when blood flowed out of them, but he also insists that others who traced the blood vessels externally in emaciated human subjects missed crucial internal components as well.

18. Arist. *Hist. an.* 1.17, 496b5.

correct precisely because they pay insufficient attention to basic animal anatomy. In doing so, he explicitly attaches anatomical investigations to a teleological account of the body, claiming that "the chief reason that they are not correct about these things is that they are ignorant of the interior parts and they do not apprehend why nature makes all things."[19] In other words, the increased interest in figuring out how the body *worked* was paired with a dramatic increase in the epistemological weight granted to dissection as a relevant mode of knowing.[20] This shift, in turn, changed the systematicity of dissection and what structures investigators were interested in examining when they opened animal bodies.

Since each part or organ was designed to perform a certain function and completing such a task [τὸ ἔργον] was its "activity" [ἐνέργεια], Aristotle's organism came with a fuller commitment to physiology, now understood as articulating internal mechanisms *and* the purpose that they serve. Despite the fact that dissection seems particularly well suited to investigating the now-crucial parts and structures, *Parts of Animals*, *Historia Animalia*, and his lost treatise *Anatomy* are more involved in using this practice to systematize anatomical differentiae and discover correlative variables between parts. To be sure, Aristotle describes each part's goal, or "that for the sake of which" it exists, in these treatises, but a more sustained focus on the processes and mechanisms of the body occurs in the so-called *Parva Naturalia*, which, as mentioned above, explain the senses, memory, sleep and dreams, respiration, aging, and death. When explicating these behaviors and phenomena, Aristotle generally establishes the relevant "accompanying facts" [τὰ συμβαίνοντα], including any previous theories, and then supports his account with a mix of logical assertion, teleological probability, and anatomical argumentation, often including several technological analogies to illuminate different aspects of the mechanism he proposes. Lloyd has illustrated that in these physiological or functional accounts Aristotle often omits anatomical details when doing so serves his proximate argumentative interests, even as he invokes

19. Arist. *Resp.* 3, 471b25–29.

20. Indeed, at *Hist. an.* 1.16, 494b23–25 Aristotle claims that the internal parts of humans are for the most part unknown, which suggests a relatively new interest in dissection and anatomy. Moreover, the most robust anatomical knowledge prior to Aristotle concerned the sense organs and the blood vessels, both of which present the first examples of function. Even Aristotle himself does not appear to have dissected adult humans, relying instead on comparative studies of other large mammals and some anatomical investigations of human embryos (see *Part. An.* 3.4, 665b2; 3.4, 666a20–24).

his anatomical works by name especially.[21] Indeed, he can remain remarkably elliptical about connecting his physiological explanations to the anatomical specifics described with great precision elsewhere. Moreover, Aristotle sometimes places conflicting anatomical demands on the same organs, especially when he is describing different physiological behaviors and systems, so much so that Shaw insists that Aristotle did not integrate his anatomy and physiology in the way that modern science expects.[22]

Instead of characterizing these discrepancies as failures to adopt to modern scientific practices, attending to these moments of tension can reveal how, just as with Empedocles and Plato, technological intermediaries help conceptualize and construct certain corporeal behaviors and components, such that the body Aristotle articulates becomes entangled with—although never reducible to—multiple technological objects and behaviors. Tools can interact with, crowd out, and occasionally supervene upon his anatomical assertions. For this reason, understanding the impact ancient technologies had on conceptions of the body cannot remain at the highest level of analysis, since leaving it there misses the role that individual material tools had in conceptualizing the body and co-constructing its behaviors. In this regard, analogies cannot be treated simply as didactic, but must be seen as shaping the epistemic object(s) they articulate.

3.2 THE TOOLS OF THE HEART

The heart sits at the center of Aristotle's organism as the *arche* of all his vital systems, including cognition, perception, motion, nutrition, respiration, and reproduction. This plurality of operations places manifold demands on its anatomy, many of which sit in tension with one another. Often, then, conflicting needs emerge from or are smoothed by technological analogies. The heart therefore presents a useful site to witness how the generalized metaphor of "tool parts" meets with the particulars of fourth-century BCE technologies. In his anatomical treatises, Aristotle presents a description that is somewhat confounding to modern readers and has resisted satisfying interpretation, not least because of several conflicting statements and its incongruities with modern assumptions. Here Aristotle claims that the heart is "source and spring of the blood, or its first receptacle" [ἀρχὴ καὶ πηγὴ τοῦ αἵματος ἢ ὑποδοχὴ πρώτη], and he points to the anatomical obviousness of this assertion, since "[the blood

21. Lloyd 1991.
22. Shaw 1972.

vessels] manifestly come from it and do not go through it" [φαίνονται γὰρ ἐκ ταύτης οὖσαι καὶ οὐ διὰ ταύτης].[23] Elsewhere he insists that "[blood] is channeled away from the heart into the blood vessels, but it is not channeled into the heart from elsewhere."[24] As for the heart itself, its structure differs in various animals. In smaller animals (that possess a heart), the heart is only a single chamber [κοιλίη]. Intermediate-sized animals such as fish have a heart with two chambers, while the largest animals, such as humans, oxen, horses, and sheep, possess a heart with three chambers.[25] In these most developed hearts, the largest chamber sits on the left and connects to the "Great Vessel" [ἡ μεγάλη φλὲψ] (i.e., the vena cava), the middle-sized chamber sits in the middle and connects to the aorta [ἡ ἀορτή], while the smallest sits on the right.[26] All three are also said to be connected or "perforated" [τετρημένας/συντέτρηνται] to the lungs, although he states that only one of these connections is obvious. As such, it is unclear whether he here refers to the pulmonary artery extending from the right ventricle, or to some other, smaller structures.[27]

23. Arist. *Part. An.* 3.4, 666a6–10; cf. *Resp.* 8, 474b6–9. This claim seemingly conflicts with his comments at *Hist. an.* 3.3, 513b2–b6, which states that the Great Vessel attaches to the right/largest cavity on the upper right side and then "runs through the intermediary cavity and reappears as a blood vessel again, as though the cavity were part of it, like a lake formed by the blood" [Ἡ μὲν οὖν μεγάλη φλὲψ ἐκ τῆς μεγίστης ἤρτηται κοιλίας τῆς ἄνω καὶ ἐν τοῖς δεξιοῖς, εἶτα διὰ τοῦ κοίλου τοῦ μέσου τείνασα γίγνεται πάλιν φλέψ, ὡς οὔσης τῆς κοιλίας μορίου τῆς φλεβὸς ἐν ᾧ λιμνάζει τὸ αἷμα]. I am convinced by Bubb (2019), who argues that διὰ τοῦ κοίλου τοῦ μέσου should be translated as "through the intermediary hollow space" (i.e., of the right ventricle) rather than "through the middle cavity" (i.e., of the middle cavity/left ventricle). This translation eliminates any contradiction.

24. Arist. *Part. An.* 3.4, 665b16–17. He also points to the fact that it is the first body part to be formed in the embryo.

25. Arist. *Part. An.* 3.4, 666b23–26; cf. *Hist. an.* 3.3, 513a32–b6; *Hist. an.* 1.17, 496a4–37. In large mammals, the largest cavity sits on the right, the smallest on the left, and the middle cavity in between them. The largest chamber holds the hottest blood, while the smallest contains the coldest (*Part. An.* 3.4, 667a1–6).

26. Arist. *Hist. an.* 1.17, 496a5–27; cf. *Resp.* 16, 478a26–28. At *Somn.* 458a15–19, Aristotle presents a somewhat different picture of the heart, insofar as he claims that the aorta is connected to the left/smallest ventricle, not the middle cavity.

27. Arist. *Hist. an.* 1.17, 496a22–23 says that "[the heart] has all perforated/connected into the lungs, even the two small ones, but it is clear in one of the cavities" [ἁπάσας δ' ἔχει, καὶ τὰς δύο μικράς, εἰς τὸν πνεύμονα τετρημένας, κατάδηλον δὲ κατὰ μίαν τῶν κοιλιῶν]. This account suggests that the largest cavity/right ventricle has the most obvious connection, which would suggest that he is talking about the pulmonary artery, which extends from it to the lungs. Similarly, *Hist.*

In addition, Aristotle also asserts that middle chamber is a "common source" [ἀρχὴ κοινή] to the other two for their supplies of blood, although it is not clear whether this assertion indicates that he assumes direct openings lead between the ventricles (as is visible between the right atrium and right ventricle) or whether they are indirectly connected via vascular pathways.[28] Regardless, the largest/left cavity is the hottest and contains the most blood, while the smallest/right cavity is the coolest and contains the least amount of blood. The middle/intermediate cavity holds an intermediate amount of blood of intermediate heat, and its blood is also the purest. Although the specifics are difficult to pin down, the middle cavity is also the site where blood is manufactured insofar as it seems

---

an. 3.3, 513a32–b6 states that "all [the ventricles] are perforated/connected toward the lungs, but these are unclear on account of the smallness of the passageways, except in one of them" [συντέτρηνται μέντοι πᾶσαι αὗται πρὸς τὸν πλεύμονα, ἀλλ' ἄδηλοι διὰ σμικρότητα τῶν πόρων πλὴν μιᾶς] (this text is from Balme 2002). Recently, Bubb (2019) has argued that the "perforations" that Aristotle describes as connecting the heart to the lungs are in fact additional access points through the back of the heart wall and into the trachea (i.e., not vascular connections at all). She draws attention to Aristotle's comments at *Hist. an.* 1.16, 495b12–16, where Aristotle states that "the heart too is attached [συνήρτηται] to the trachea by fatty, gristly, and fibrous bounds, and where it is attached, there is a hollow. When the trachea is inflated, although it does not do it clearly in some animals, in larger ones it is clear that pneuma enters into the heart" [συνήρτηται δὲ καὶ ἡ καρδία τῇ ἀρτηρίᾳ πιμελώδεσι καὶ χονδρώδεσι καὶ ἰνώδεσι δεσμοῖς· ᾗ δὲ συνήρτηται κοῖλόν ἐστιν. Φυσωμένης δὲ τῆς ἀρτηρίας ἐν ἐνίοις μὲν οὐ κατάδηλον ποιεῖ, ἐν δὲ τοῖς μείζοσι τῶν ζῴων δῆλον ὅτι εἰσέρχεται τὸ πνεῦμα εἰς αὐτήν.]. Since Aristotle does not here mention a perforation between the trachea and the heart, but merely an attachment, I think that it is more likely that he is simply describing the fact that when the trachea is inflated, it fills both the lungs (as he has just mentioned) and the heart by means of the lungs, especially since at *Hist. an.* 1.17, 496a28–34 he mentions that pneuma enters into the heart via the lungs. In addition, *Hist. an.* 1.16, 495b7–8 mentions that the trachea is attached [συνήρτηται] to the Great Vessel and the aorta, and if this were grounds for pneuma to enter, it should reasonably inflate them too. See note 28 below for a further discussion of this issue.

28. Arist. *Part. an.* 3.4, 666b33–36. Shaw (1972) argues that Aristotle is describing the ligamentum arteriosum, a connection between the pulmonary artery leading from the right ventricle and the aorta that forms in the embryonic heart as the ductus arteriosus, which allow blood to bypass the not-yet-functioning lungs. Aristotle may envision direct connections, however, since even five hundred years later, Gal. *Nat. Fac.* 3.15 = 2.207–208K still thinks that the left and right atria can directly communicate, identifying the small pits on the septum of some animal hearts as perforations that can transmit blood; cf. Gal. *AA* 7.10 Singer = 2.623K.

to receive nourishing liquid from the stomach and concoct it further.[29] In any case, Aristotle is explicit, at least in the anatomical treatises, that no blood flows into the heart from elsewhere.[30]

Since many of these assertions do not correspond to modern notions, much scholarship has tried to make sense of Aristotle's competing and confusing claims.[31] Despite widely divergent reconstructions, most commentators agree at the very least that the three chambers of the human heart most likely correspond to the right ventricle, the left atrium, and the left ventricle, with the right atrium considered to be part of the "Great Vessel" and thus not an independent chamber.[32] Beyond that, little can satisfy all commentators. Assessing the role that individual technologies play for Aristotle's conceptualization of the mechanisms at work in corporeal functions requires examining how these anatomical details (such as they can be understood) feature in his explanations of some key physiological processes, most notably respiration, pulsation, and blood delivery. These different systems make different claims about the heart.

In *On Respiration*, Aristotle attempts to explain how and why breathing occurs. For Aristotle, animal life and the activity of their souls depend on the maintenance of heat in the body. Nourishment itself requires this heat, so without it, neither plants nor animals could grow or survive.[33] He defines birth as "first participation" [ἡ πρώτη μέθεξις] of the nutritive soul in this heat; death, as its extinction.[34] Although our chief worry

29. See Arist. *Somn.* 458a15–19. Bubb (2020) provides a detailed summary of how Aristotle thinks blood is produced. See note 67 on p. 143.

30. Arist. *Part. an.* 3.4, 666a6–8.

31. Most dramatically, Aristotle's three-chambered heart does not correspond particularly well to our depiction of the heart as four-chambered. Harris (1973: 126–33) ultimately concludes that no satisfactory answer can be discerned from the evidence (either textual or anatomical) and suggests that perhaps the mistake simply arose from Aristotle's doctrine of the mean. Dean-Jones (2017) suggests it comes from Aristotle's dissections of human embryos, not adult cadavers. More recently, Zierlein (2005) has outlined all of the controversies; see also Bubb 2019, 2020; Shaw 1972.

32. Huxley (1879) argues that the specific technique of strangulation that Aristotle prescribes can leave this part of the heart distended and full of blood, facilitating the idea that the left atrium is a continuation of the vena cava, while engorging the left ventricle and increasing its size relative to the other chambers. Shaw (1972: 375–77) accepts the identification of the three chambers as the right ventricle (biggest), left ventricle (middle), and left atrium (smallest); so too does Harris (1973: 127–29).

33. Arist. *Resp.* 8, 474a25–29.

34. Arist. *Resp.* 18, 479a30–b7.

when tending fires is that they will run out of fuel and burn out, Aristotle's assumption is that our vital inner heat, if left unchecked, will burn too hot and consume itself. Respiration, Aristotle argues, serves to cool the blood in the heart so as to preserve vital heat and prevent death.[35] After spending the majority of the treatise discussing his predecessors' views and how such refrigeration occurs in other animals, including insects and fish, he finally turns to the mechanism in blooded animals with lungs. He starts this part of his account by outlining three physical behaviors, all of which involve the heart: throbbing, beating, and breathing [πήδησις καὶ σφυγμὸς καὶ ἀναπνοή].[36] Although these phenomena may at first appear to be manifestations of the same basic behavior, he insists that they are actually separate corporeal processes. This assertion highlights that the behaviors that Aristotle attributes to his heart are not the same as modern assumptions and have been constructed differently. He consolidates two of these processes around technological analogues.

Aristotle spends little time on the first behavior, throbbing, which he claims constitutes the pounding of the heart that accompanies fear. This pounding, he asserts, results from heat retreating swiftly to the heart when residues or secretions chill it. This chilling can occur in either disease or fright, and forcing the heat so quickly into so small a space produces throbbing (almost like the reverse of the glugging that accompanies the swift exit of liquid through a narrow opening), and, he argues, this throbbing can even cause death.[37] He provides no more details about this occasional and incidental behavior.

He spends more time explicating the next two behaviors linked to the heart, beating and breathing, both of which he links to the heart's function to manufacture blood.[38] Although we might assume that the heart's beat is obviously part of its functional mechanism, Aristotle explains the pulse as another incidental feature, a behavioral by-product of the process of concoction similar to boiling, whereby the blood is heated in

35. Arist. *Resp.* 3, 471b23–29; 10, 476a7–14. He cites our need to breathe more when we are hot and notes that air is cooler when it is inhaled than when it is exhaled. As mentioned above, he may have taken this idea from Philolaus of Croton (DK44 A27 [D25 LM] = Anon. Lond. 18.8–19.1 Manetti) or from within the Hippocratic Corpus; cf. p. 95.

36. Arist. *Resp.* 20, 479b17–19. Freeland (1990) examines Aristotle's use of συμβαίνοντα as part of his general endoxic practice; cf. Owen 1986; Nussbaum 1987; Smith 1995; Frede 2004. For an account of the rise of συμβαίνοντα in the methodology of post-Aristotelian science, see Sedley 1982.

37. Arist. *Resp.* 20, 479b20–27.

38. See note 67 on p. 143 for a full discussion of blood manufacture.

the process of manufacture such that "moisture becomes pneumatized" [πνευματουμένου τοῦ ὑγροῦ] and begins to expand.[39] When this expanding air strikes the "furthest membrane," it causes a beat:

> [The beating of the heart] is similar to boiling; for boiling happens when liquid is turned into air by heat; the liquid increases in size because its bulk gets larger . . . in boiling there is an expulsion over the rim (of the container), but in the heart the liquid flowing in from nourishment expands because of the heat and causes pulsing whenever the expansion increases to the furthest membrane of the heart [πρὸς τὸν ἔσχατον χιτῶνα τῆς καρδίας]. And this is always continuous, since liquid is always continuously flowing in, from which the nature of the blood is generated. For blood is manufactured in the heart first.[40]

There are a few ways to interpret how the boiling blood produces a "beat." The interior liquid could boil upward, somehow causing a pulsation when the entire mass reaches the outer membrane and strikes it, or this mass could (somehow) lift and resettle the membrane like the rattling lid of a pot. Another option appears in a lemma from the pseudo-Aristotelian *Problemata*, which insists that water does not sputter when boiled, whereas both gruels and silver will produce a "blow" [πληγή] when heated.[41] We thus might also consider the pulse emerging like bubbles that sputter and strike the inside of the ventricle at regular intervals like a low-boiling porridge.[42]

Although this analogy is not particularly technologically complex, we should note that Aristotle has selected an implement and process that correspond both functionally and physically. Since blood is concocted during the process of manufacture, selecting a boiling pot to understand the pulse links the form of the heart as a vessel, the function of concoction that both perform, and the behaviors associated with this activity. Many of Plato's analogies only loosely connected corporeal mechanisms and corporeal functions, either leaving the inner processes underdeter-

39. Cf. Arist. *Pr.* 24.7, 396b10–11, which claims that fire produces pneuma. See also Hipp. *Flat.* 8 = 6.100–102L, which advances the notion that the blood gives off vapor.

40. *Resp.* 20, 479b30–480a8.

41. Arist. *Pr.* 24.9, 936b39.

42. Arist. *Resp.* 20, 480a12–16 goes on to mention that all veins pulse because they are connected to the heart; cf. Arist. *Hist. an.* 3.19, 521a7–9.

mined (e.g., the liver as mirror, spleen as towel) or providing only an indirect relationship between inner and outer functions (e.g., comparing the semipermeable membranes of the stomach and lungs to fish weels). By contrast, Aristotle's comparisons establish a deeper reciprocity between internal and external activities, and as a result, his tool analogies produce more robust networks of mutually reinforcing physical associations, pulling technological practices deeper into the body. That is, trying to establish the boundaries of the tool and the body becomes more difficult when corporeal behaviors are built around technologies that complete similar tasks utilizing shared physical properties. To use the vocabulary of the introduction, didactic comparisons slide into heuristic comparisons, while these in turn entertain a type of metaphorical slippage, transforming into the explanatory object itself.

At the same time, the boiling pot is not the heart, since it radically simplifies the anatomical picture of the three-chambered organ. Does only the middle chamber bubble? If there are three chambers with blood of three different thermic levels, are there three separate bubbles? If not, why does not the hotter, left cavity (left atrium) boil and pulse more, or at all? These difficulties help illustrate that the analogy facilitates a physiological explanation at the expense of anatomical rigor. Even the somewhat imprecise reference to the "furthest membrane" underlines this blurriness. Of course, Aristotle could potentially have answers to all these questions, but the boiling pot comparison activates certain corporeal features, while drawing attention away from others, shifting the reader's cognitive focus. In so doing, it alters the site where the heartbeat takes place, now somewhere between the organ and the tool. If "pulsation" were fully imagined as taking place within the heart that Aristotle described elsewhere, it would open up a whole host of difficulties. If pulsation occurs somewhere between the heart and the pot, the theory appears almost self-evident. Cognitive focus shifts to this new heuristic plane.

After explicating both throbbing and the pulse, Aristotle moves on to an activity that we generally do not ascribe to the heart directly: breathing. Once again, the demands that this activity places on this organ are varied, and to explain the heart's role in respiration, he constructs another technological analogy, this time with the furnace and its bellows. Respiration, he argues, occurs when the natural heat of the heart's nutritive principle starts to increase (insofar as the heat's tendency is to grow, especially with the influx of new nutrition), although now, instead of the heat causing blood to aerate and expand, it supposedly causes the entire heart to expand (a potential tension that is not addressed). This expansion, in turn, causes the "part encompassing it" (i.e., the lungs) to rise as well.[43] As the lungs thereby increase in volume, they draw air from the mouth and nose, through the windpipe and into their "tubes" [σύριγγες].[44] These air-filled vessels run alongside an accompanying set of blood-filled vessels that extend from the heart (most notably from the pulmonary artery from the largest/right ventricle).[45] These systems "do not share a common passageway," but they are in close contact with one another and share a "synapsis" [σύναψις] which may be either perforations or simple points of contact.[46] Because of their close proximity, the external air cools the blood through heat transfer. This refrigeration shrinks the lungs and heart back down, and as these organs collapse to their original size, the warm air is pushed back out into the atmosphere until the inner heat grows once again, and the process repeats.[47]

As obvious, even *natural*, as the comparison now seems, Aristotle is seemingly the first extant source to explicitly make this analogy between the lungs and bellows:

> It is necessary to understand the structure of the organ as similar to the bellows in forges [ταῖς φύσαις ταῖς ἐν τοῖς χαλκείοις] (for neither the lung nor the heart are far from taking on such a shape), but such

43. How this transfer of expansion takes place is left unarticulated. It can be construed as the heart's expansion causing either the thorax as a whole to expand, or just the blood in the lungs, which envelop the heart, thereby causing the organ as a whole to rise; see Arist. *Resp.* 21, 480a16–b20.

44. The word σῦριγξ originally refers to a shepherd's pipe, presumably made out of a reed; see Hom. *Il.* 10.13, 18.536. Since this reed can be sealed by the fingers or used to draw up liquid as through a straw, it takes on the secondary meaning of "tube," a meaning first seen in Empedocles's clepsydra comparison, where the word σῦριγξ is used to describe a tube that transports liquid (DK31 B100 [D201 LM] = Arist. *Resp.* 7, 473b9–474a6); cf. entry in *LSJ*. See also Pl. *Ti.* 70c1–d6, which describes the lung as being spongy and perforated with cavities.

45. At *Resp.* 16, 478a26–29, Aristotle directs his reader to his *Anatomy* and *Historia Animalium* to learn "the manner by which the heart has a connection to the lungs" [Ὃν δὲ τρόπον ἡ καρδία τὴν σύντρησιν ἔχει πρὸς τὸν πλεύμονα]. As discussed above, he describes all three chambers as connected to the lungs, although the pulmonary artery seems to feature as the main and most notable connection, which Aristotle says splits around the base of the windpipe and then enters into each lung; cf. Arist. *Hist. an.* 1.17, 496a22–34, which claims that the middle chamber (right ventricle) connects to the lungs via the aorta.

46. Arist. *Hist. an.* 1.17, 496a27–33; cf. Arist. *Part. an.* 3.4, 668b33–669b13 for another discussion and description of the lung.

47. Arist. *Resp.* 21, 480a16–b20.

> a thing is double [διπλοῦν δ' εἶναι τὸ τοιοῦτον]. For it must have the nutritive part of the capacity for life in the middle. And so, as the heat increases, it expands, and as it expands, the part encompassing it must also expand. This is what men seem to do when they breathe; they expand their chest because the principle of the part described above resides in the chest and causes this same expansion [αἴρουσι γὰρ τὸν θώρακα διὰ τὸ τὴν ἀρχὴν τὴν ἐνοῦσαν αὐτῷ τοῦ τοιούτου μορίου ταὐτὸ τοῦτο ποιεῖν]; for as the chest rises, the air from outside must flow in, as it does into the bellows, and being cold and refrigerative, quench the excess of fire. Just as increase makes this part rise, so decrease must make it subside, and as it subsides, the air which has entered must pass out again. It enters in cold and passes out hot, because of its contact with the heat which resides in this part, especially in those whose lung contains blood. For each of the many canal-like tubes in the lung, into which the air passes, has a blood vessel alongside it, so that the whole lung seems to be full of blood. The entry of the air is called inhalation, its exit exhalation. And this occurs continuously as long as the creature lives and keeps this part continuously moving. This is why life depends upon inhalation and exhalation.[48]

The reference to "the part" in this paragraph leaves it hard to determine whether Aristotle means that the lungs and heart together, considered as a compound tool, resemble the bellows and furnace, or whether the lungs and heart both individually resemble this technological apparatus. Turning to what fourth-century BCE furnaces looked like can once again help us to interpret this passage to see how certain material realities shaped Aristotle's views on corporeal parts, as well as establishing new behaviors as essential to the act of respiration.

Although most of the archaeological evidence has been destroyed, images of stack furnaces from the fifth century BCE show a tall central chamber (fig. 9) with a single bellow built into the base, next to the forge's mouth, where the craftsman inserts metal to be heated. Along with this basic structure, these furnaces also depict a small pot or crucible at the top of the stack. This detail helps explicate why Aristotle's account of the heartbeat and respiration can operate broadly under the same analogical scheme, and it aligns with (although need not be the source of) the idea that the nutritive principle sits in a central chamber.[49] Saying "such a thing is double" might indicate that furnaces had two bellows to allow for

48. Arist. *Resp.* 21, 480a17–b13.

49. It may come from his philosophical commitment to the mean.

FIGURE 9. Oinochoe (Athenian), 500–475 BCE, pottery, 26.67 cm high, British Museum, London, museum no. 1846,0629.45; cf. Berlin Foundry Cup, 490–480 BCE, pottery, found in Vulci, Italy, Antikensammlung Berlin, no. F 2294.

a constant inflow of air rather than the halting blasts from a single bellow, which would need to pause so as to take in air before expelling; alternately, it might simply allude to the duality between the central nutritive chamber and the mechanism needed to cool it.

How important are these bellows analogies for Aristotle's account? Are they, as Shaw puts it, "purely metaphoric"?[50] Do they serve a didactic function to make comprehensible what Aristotle has independently come to believe? Do they operate heuristically? Determinatively? To be sure, Aristotle is careful to delineate multiple aspects of the bellows that do not correspond to the form of the lungs and heart. Most notably, he points out that breath comes in and exits through the same passage in the case of the lung (i.e., through the windpipe), but this is not the case in the bellows (in which air enters through a flap valve in the rear and is pushed out through the nozzle in the front).[51] As such, the bellows cannot simply replace the body parts that he is explicating through a type of analogical drift. That said, to envision breathing in the body, he has incorporated a technology that replaces human breath to control and manipulate the heat used in the manufacture of metal.[52] This analogy joins together form and function, as the human technological process sets up crucial expectations for the process of blood production in the body. Moreover, by foregrounding the bellows, Aristotle privileges a corporeal behavior that neither Empedocles nor Plato highlighted: that the chest and torso expand upon inhalation. Although this aspect of breathing now seems incredibly obvious, a phenomenon that the physikoi must have witnessed, neither Empedocles nor Plato privileged the expansion and contraction of the chest, since they viewed respiration as the transfer of external air and internal blood on the model of the clepsydra. This device was the totemic technology of the impossibility of a void [κενὸν οὐδέν ἐστιν] and relied on a rigid cavity. Accordingly, for Aristotle, the comparison does real work, helping to cement the expansion and collapse of the chest as an *essential*, rather than *incidental*, feature of respiration. Bellows consolidate the phenomenon.

If perhaps it still seems that these tools cannot be that important, since Aristotle simply used the comparisons to explicate what is obvious (that the lungs expand and take in air like bellows), we should remember that lungs are not empty sacks but spongy, blood-filled tissue, so as naturalized as the bellows comparison has become, it is not an immediately obvious physical correlate. Moreover, because of their spongy texture, mul-

50. Shaw 1972: 385.

51. Arist. *Resp.* 474a12–17. The lungs produce bidirectional airflow through a single entrance/exit passageway, not unidirectional flow through separate intake and outtake passages like the bellows, and human force, rather than heat, drives the bellows. That said, Aristotle does state in this passage that "people breathe out by collapsing and pushing down [the lungs]."

52. Although we think that the bellows *increase* heat by adding oxygen and kindling the fire, Aristotle thinks that the cool air compresses the heat and sustains it from burning out; see pp. 130–31.

tiple ancient authors, argued that we took in liquids via our lungs, as was mentioned above. This makes the bellows comparison far less natural.[53] In addition, there are other ways to see the conceptual work that his hypothetical mechanism completes, namely by constructing physiological behaviors in ways that seem strange to modern readers. For example, the bellows analogy connects blood production to heat regulation, such that the heart becomes its own homeostatic control mechanism. As it gets too hot, its expansion produces the actions in the lungs that cool it. Two corollaries emerge. First, since this process is driven by the vital heat's natural tendency to increase and the blood's natural tendency to expand when heated, it operates more or less automatically—as a self-regulating physiological mechanism articulated according to normal, even if hypothetically engineered, physical behaviors.[54] This model leaves the role of human volition in breathing underdetermined and undiscussed, just as it was for Plato. Second, and more importantly, the rate at which the heart expands and contracts is linked directly to the rate of respiration. That is, the heart's expansion and contraction are *not* what constitute its beat. Instead, the heart rate and breathing rate correspond at a one-to-one ratio, such that a single inhale and exhale would produce a single cycle in the heart.[55] As discussed above, this one-to-one correspondence proves no true difficulty for Aristotle's theory of pulsation, which he considers an accidental behavior resulting from the heart's concoction of the blood, and which he implies occurs separately from coronary expansion and contrac-

53. For example, Plato argues that the lungs take in liquid at *Ti.* 70c–d;cf. 91a, and Craik (2006: 160) suggests that Hipp. *Morb.* 3.16 = 7.142L implies that drink moistens the lung and removes pus; cf. Hipp. *Morb.* 1.28 = 6.196L; Hipp. *Loc. Hom.* 26 = 7.606L. Aristotle argues that lungs do not receive liquids at both *Resp.* 11, 476a31 and *Hist. an.* 1.16, 495b17, but the idea perseveres. Hipp. *Cord.* 2 = 9.80L, which postdates Aristotle, argues this point explicitly and conducts a dissection of a pig to demonstrate it; cf. Hipp. *Oss.* 1 = 9.168L, which postulates something similar and suggests that air, blood, and liquid all move through the lungs. Hipp. *Anat.* also seems to believe that fluid enters the body via the lung, insofar as he describes the trachea, lungs, kidneys, and bladder in succession; see Craik 2006: 160 and chapter 4.1 on p. 161. We should also note other ways the bellows analogy is not seamless, especially insofar as the lungs do not get larger and contract through their own expansion but because the volume of the thorax changes as a whole.

54. Arist. *De motu an.* 11, 703b3–10 refers to sleep and respiration as "not volitional" [οὐχ ἑκουσίους] as opposed to "involuntary" [ἀκουσίους], such as the heart movements and erections are, but nothing in *On Respiration* mentions the quasi-volitional nature of breathing.

55. Harris (1973: 164) and von Staden (1989: 260) both put forward this same interpretation.

tion. Nevertheless, it does illustrate how the bellows-forge complex constructs a suite of hypothetical corporeal behaviors as much as it explains predetermined bodily phenomena. The synchronized expansions of the heart and lungs are not something that Aristotle could have felt, let alone observed in any straightforward sense. Thus, his bellows-and-forge theory cannot simply explain a preexisting set of "accompanying facts," since the analogy complex is itself responsible for predicating unseen phenomena and incorporating them into a network of interwoven assertions, some anatomical, some physiological, some physical, and some technical. The bellows comparison essentializes the thorax's expansion, which we take to be obvious, while also creating other behaviors that seem all but risible to us.

In addition to the conceptual work that the bellows analogy does for Aristotle's physiological account, the comparison also occludes certain anatomical difficulties, especially since Aristotle places competing demands on the vessels leading from the lungs to the heart. As mentioned above, he claims that the vessels that connect these two organs (pulmonary arteries) bring warm blood from the heart but share no common passageway with the air-filled tubes reaching into the lungs from the trachea (bronchioles). Yet at the same time, *Historia Animalium* claims that "through connection [of these two sets of vessels] they receive air/pneuma and send it along to the heart. For one of the passageways [of the pulmonary arteries] travels to the right hollow [of the lungs], the other to the left."[56] In other words, despite his physiological insistence that the bellows analogy is distinct from the situation of the lungs because the bellows have separate entrance and exit points for air, while the lungs do not, his anatomical descriptions insist that the lungs actually have two entrance/exit points for air, a major one leading upward via the trachea, and a secondary one into the heart via the pulmonary vessels. In fact, in *Historia Animalium*, Aristotle declares that if you blow down the windpipe of a large animal, you will inflate the heart.[57] These anatomical details disrupt the normal functioning of the bellows, since the expansion of the lungs should therefore draw air from both the outside and the inside of the body. Of course, Aristotle could

56. Arist. *Hist. an.* 1.17, 496a31–34; cf. *Hist. an.* 3.3, 513b13–24. I am persuaded by Bubb (2019) that φέρει γὰρ ὁ μὲν εἰς τὸ δεξιὸν κοῖλον τῶν πόρων, ὁ δ' εἰς τὸ ἀριστερόν is describing the passages of the heart entering into the "hollow" of the lungs, not passages from the lungs entering into the cavities of the heart.

57. *Hist. an.* 1.16, 495b8–14; cf. *Hist. an.* 3.3, 513b17–23. See note 27 on p. 128 for a discussion of this passage. Even if Bubb (2019) is correct, and this pneuma enters the heart via direct connection with the trachea, then the bellows comparison would still be directing attention away from these complicating anatomical details.

have supplied some additional arguments to accommodate and explain away these troublesome details, perhaps adding that only a small amount of pneuma transfers from the lungs into the heart via the intervening vessels and therefore does not affect the basic mechanics of respiration. Nevertheless, these anatomical issues do reveal how the bellows analogy helps simplify the exchange mechanism within the lung, establishing that it works on the principle of mutual exchange and vacuum pressure, even when Aristotle's own anatomical assertions complicate this picture. As such, both his comparison of the lungs/heart to the bellows and his careful delineation between the two direct our cognitive focus to certain anatomical features necessary for the mechanism at hand and away from other problematic parts. The technological comparison elides disruptive details and privileges a set of corroborating physiological behaviors, some of which are bound up with the tools that he uses as proxies for internal corporeal processes.

3.3 THE JOURNEY OF THE BLOOD

Another surprising feature of Aristotle's account of the heart—especially for a modern reader familiar with steam generators and pumps—is that when discussing how blood gets distributed as nutriment to the entirety of the body, he does not incorporate the bellows or the boiling pot as mechanisms of propulsion.[58] Instead, when Aristotle describes how blood travels around the body in *Parts of Animals*, he turns to the familiar comparison of irrigation, whereby blood flows through the body via simple gravity flow:

> [The systems of vessels] are similar to irrigation systems that are built in gardens to transmit [water] everywhere, [leading it] from one source and spring into numerous other channels; and just like in building, when stones are set along the entire outline of the foundation, nature has also channeled blood through the whole body in the same way, since this is the constituent matter of it all. This is clear in people who are especially thin. For in these cases, nothing other than the blood vessels are visible, just as in vines and fig leaves and all other such things—for when these things wither, their veins alone remain. The cause of this is that the blood (and its analogue) is potentially the body and the flesh (or its analogue). And so, just as in channels, the biggest

58. Cf. Hipp. *Flat.* 8.18–40, where the boiling-pot analogue explains how fevers heat the body's air, which forces open the jaw.

> of the trenches persist, while the smallest disappear first and fastest under a coat of muck, only to reappear again when the muck recedes; in the same way the biggest veins persist, while the smallest become flesh in actuality, but potentially they are nothing less than veins. For this reason, whenever the flesh is healthy, blood will flow wherever the flesh is cut. However, without veins there is no blood, even if the smallest vein is not actually visible, just as in the channels the trenches are not visible until the mud has been cleared.[59]

Instead of employing models of propulsion in the above passage, Aristotle leaves the flow of blood basically unexplained, allowing the analogue of irrigation to do most of the work for him. He even allows this comparison to dictate other aspects of physiology, including hypothetical veins that cannot actually be seen because they are too small and only appear once the "muck" of the flesh recedes. Moreover, this irrigation model of blood distribution implies roughly uniform, unidirectional flow away from the heart. Yet this placid picture is substantially complicated with assertions he makes elsewhere about the processes of nutrition and blood production.

Although the treatise in which Aristotle must have dealt most directly with blood production, *Nutrition*, has been lost, he speaks about the subject elsewhere. In *Parts of Animals*, he discusses how blood is concocted and transformed through the power of heat. Aristotle names several parts that act on the nourishment entering the body. The first work occurs in the mouth, without concoction, where heat is applied in the "upper and lower cavity" [ἡ δὲ τῆς ἄνω καὶ τῆς κάτω κοιλίας], by which he means the stomach and intestines, when the food reaches them. From these cavities, the body siphons off its basic nourishment into the blood vessels.[60] In *On Sleep and Dreams*, Aristotle claims that when this incoming nutrition enters the blood vessels via the stomach and mesentery, it starts to emit a type of vapor because of the heat.[61] The natural tendency of such a sub-

59. Arist. *Part. an.* 3.5, 668a11–32. Cf. *Part. an.* 2.1, 647b1–8, which compares the viscera to mud-like deposits left by the blood running through the irrigation channels of the veins. It is possible that Aristotle was describing a particular set of irrigation channels when thinking about the interior vessels of the body, since Theophr. *Hist. pl.* 1.7.1 mentions the plane tree growing in the Lyceum "along the watercourse."

60. Arist. *Part. an.* 2.3, 650a3–b14. Aristotle uses the same comparison that Plato did, likening this reservoir of nutriment to a manger, but then adds that the siphoning occurs in the same way as plants draw off nourishment from the earth.

61. Arist. *Somn.* 3, 456b18–19, calls it "the exhalation concerning nutriment" [ἐκ τῆς περὶ τὴν τροφὴν ἀναθυμιάσεως].

stance to rise drives it and the accompanying blood upward through the vessels *through* the heart and on to the upper parts of the body. When this collective material reaches the head, it cools and descends in a mass back down to the heart, which causes the head to droop.[62] This downward-moving, cooled vaporous mass drives the heat into itself by means of "surrounding" or "reciprocal pressure" [ἀντιπερίστασις].[63] Aristotle identifies sleep as the consequent "natural collection" or "inward compression" of heat in the heart (and feet) resulting from the general cooling of the upper parts.[64]

As comprehensible as this account is, the upward and downward flow of blood, nourishment, and vapors creates a set of behaviors that sit uncomfortably with the uniform flow implied by the irrigation model of blood distribution. The vessels through which blood travels to the rest of the body in blood distribution (e.g., aorta, Great Vessel) are the *same vessels* though which the blood, incoming nutriment, and vapors supposedly move up and down, and the same passages through which respiration takes place.[65] Blood, nourishment, and vapors travel in both directions through the same veins.[66]

62. Arist. *Somn.* 3, 456a18–b26.

63. By this term, Aristotle means a supposed phenomenon that occurs when cold air does not cool a fire but condenses and intensifies its heat, a concept he elsewhere uses to explain why hot summer days can condense clouds and produce hail; cf. Arist. *Mete.* 1.12, 348b3–8; see also [Arist.] *Pr.* 2.16, 867b31–33, which asks, "Why do those who are sleeping sweat more? Or is because of *antiperistasis*? For the heat having been collected inside drives out the moisture." Similarly, *Pr.* 23.6 describes how pouring cold water on a bleeding nose might drive hot blood inside.

64. Arist. *Somn.* 3, 457b1–7; cf. 3, 457b20–458a10, which offers several potential mechanisms for the cooling, including "thinness and narrowness of the vessels around the brain" and the natural frigidity of the brain. See also Arist. *Hist. an.* 3.19, 521a15–18.; *Part. an.* 652b36–653a16. Aristotle's account draws upon and refines the ideas of several predecessors. Parmenides identified sleep as a "cooling down" (DK28 A46b [D59 LM] = Tert. *De an.* 43.2, as did Empedocles, who claimed that sleep was a moderate cooling, death a complete one (DK31 A85 [D206b LM] = Aët. 5.24.2). Alcmaeon's explanation of sleep was that it occurs when the blood retreats to the blood-flowing veins (DK24 A18 [D32 LM] = Aët. 5.24.1 [Ps.-Plut.]), which seems similar to Hippocratic accounts at *Epid.* 6.5.15 = 5.320L; *Epid.* 6.4.12 = 5.310L.

65. Arist. *Somn.* 3, 457a13–14.

66. Cf. Shaw (1972: 388), who makes this observation. Boylan (1982: 116 n4) comes up with two "solutions" to this issue, namely that there are perhaps two channels in the Great Vessel, or perhaps blood and nourishment are like oil and water and can (somehow) flow in different directions at the same time; cf. Bubb 2020: 149.

Bubb has recently tried to smooth the seams where Aristotle's underdetermined and conflicting theories of blood distribution, respiration, and sleep meet, but even she acknowledges that friction remains between accounts.[67] Accordingly, I argue that it is more useful to acknowledge how much work his analogical models do to direct attention away from these seams and toward the network of corporeal and technological overlaps. At the very least, his assertion that the heart "channels" blood out of its

67. Bubb (2020) argues that the standard model of Aristotle's vascular system as involving only unidirectional flow (e.g., as put forward by Boylan [1982]) cannot be correct and has pushed back against the claims of Harris (1973: 135), who argues that Aristotle provides no clear account of how the heart produces the blood, which supplies both nourishment for the body and material for its constituent parts. She puts forward an excellent reconstruction of Aristotle's account of blood production that illustrates the complexity and robustness of his account. That said, I am far less optimistic about the coherence of his theory than she is, for a number of reasons. First, Bubb acknowledges that Aristotle is unequivocal in multiple passages that blood does not enter the heart, but she suggests that insofar as the fluid that comes from the stomach and passes through the heart is not technically blood, but *chyle*, it does not violate his repeated assertions (cf. Arist. *Resp.* 480a16–20, which implies as much). Yet she also acknowledges that accompanying blood does enter the heart along with the *chyle*, and to avoid outright contradiction she accepts Aristotle's reconstruction of the heart at *Somn.* 458a15–19, which there sees the Great Vessel and aorta connect to the leftmost and rightmost ventricles, rather than the aorta connecting to the middle chamber. Bubb argues that with this configuration, Aristotle is suggesting that the blood moves so swiftly through these outside chambers that it does not really enter into the heart (now identified only as the middle chamber). This interpretation may indeed reflect the theory of blood production in *On Sleep*, which confidently makes these assertions without acknowledgement of any potential contradiction (*Somn.* 3, 456b2–4 even repeats the assertion that all blood comes from the heart). Yet these two versions of the heart do conflict, as does the idea that the two outer ventricles are only heart-adjacent. Moreover, when asserting at *Part. an.* 3.4, 666a6–10 that no blood enters the heart, Aristotle points to the anatomical self-evidence of the heart being the source of the blood vessels and the blood. If either nutriment or blood enters the heart, the anatomical obviousness of his claims is rendered false (not to mention that such a case does not account for the one- and two-chambered hearts of smaller animals). It would be like insisting that a reservoir is obviously a source for all the channels that touch it, since no water flows into the reservoir from the channels and the channels all proceed from it, only then to assert that mud flows into the reservoir via the channels, and, in fact, so too does water. The obviousness of the first fact would be negated by the possibility of the second. That this contradiction exists does not mean that Aristotle himself saw a problem (or at least not an unresolvable one), but his theory does have some friction points.

chambers radically simplifies a set of physiological demands that sit uncomfortably with one another. Indeed, in his explanation of the pulse, the heart was an inelastic pot; in respiration, it forms part of an expanding and contracting mechanism.[68] In blood distribution, the heart is an irrigation source, but also a potential thruway and the site where vaporous nutriment is concocted into its final form. If nothing else, Aristotle's various physiological systems and their associated technological proxies shift cognitive focus away from one set of anatomical features and direct it to others. His explanatory apparatus thereby creates a hybrid object composed of biotic bits and technological processes. As a result, it is unclear where one looks to determine what final form his theories take. Do we privilege his anatomical assertions, which would make his physiology impossible? Do we infer his true anatomical beliefs from his technologically infused physiology? The role that these technologies play is therefore never determinative but interwoven within a complex network of composite parts and competing claims.

3.4 AUTOMATA AND ANIMAL MOTION

Having spent time with the heart as the *arche* of multiple physiological systems and the site of throbbing, pulsation, respiration, blood production, and blood distribution, the last section of this chapter addresses another system of which the heart sits at the center: voluntary motion. Although Aristotle's account of animal movement places even further demands on the heart, rather than simply adding additional complications to his coronary physiology, we can turn to how other corporeal activities are bound up with tools and technologies, notably the movement of joints and limbs. When describing the mechanics of animal motion, Aristotle outlines how all voluntary movements arise from an original impulse in the heart. This impulse then leads to a series of sequential steps transferring motion to the extremities, and he likens this process to the actions of an automaton. This comparison has invited much commentary, in terms of both its psychological and physiological consequences, but almost all scholars have accepted that Aristotle was not suggesting that voluntary motion occurs automatically, nor was he implying any substan-

68. This expansion also complicates Aristotle's account of animal motion in *On the Motion of Animals*, which he claims originates in the heart expanding and contracting (because of our volitional states) and thereby moving sinews in the ventricles, which in turn set off a sequential set of motions through the other motion-generating sinews in the body.

tial isomorphism between the mechanisms of an automaton and those of animal bodies.[69] Although the former conclusion is certainly true, the latter needs clarification, especially since Aristotle insists that sinews [τὰ νεῦρα] were responsible for generating force at each joint, and an examination of automata and other motion-generating devices invented in the fourth century BCE shows how sinews were used to generate movements artificially, whether to drive wondrous devices, move carts, or launch projectiles. This final section thus emphasizes that animal parts and tissues were themselves used as tools and technologies. Their functional uses transformed them into what Sophia Roosth calls "persuasive objects" that both shaped and guaranteed theorists' assumptions about how they operated in the body.[70] This argument highlights another aspect of the recursive relationship between technological devices and living things, and it reveals how technological transformation and manipulation can itself become a type of theoretical act.

As was discussed above, Aristotle carefully delineates between natural and artificial objects, insisting that the ontological distinction between the two rests on the former having an intrinsic source of movement, whereas the latter must always be set in motion by some external force. This distinction has occupied a dominant node for modern philosophies of biology, and for some thinkers, this question serves as the only crucial—and sometimes only relevant—one to ask when evaluating the overlap of biological and technological entities. That is, these thinkers ask whether the movements occurring with living entities can be explained by mechanical forces alone, and whether a living thing can, even theoretically, be manufactured with nonliving materials. Because of the preoccupation with this inflection point, a great deal of interest flows toward automata, since they are machines that perform complex, multistep movements often designed to imitate living beings. Automaticity therefore becomes deemed the essential feature of technical objects, and the automaton, the technical surrogate for a life par excellence. It is worth noting, however, that automata are relatively absent from ancient philosophical debates concerning the nature of living things and certainly do not supply a particu-

69. Motion originates from a goal-oriented desiderative process that involves "intellection, imagination, choice, consideration, and desire" [διάνοιαν καὶ φαντασίαν καὶ προαίρεσιν καὶ βούλησιν καὶ ἐπιθυμίαν]. We see or imagine an object or action, decide whether it is something to flee or pursue, and then move our body to achieve that goal. Because this process involves intellection, it is not automatic. For further details, see Corcilius and Gregoric 2013; Corcilius 2020; Cooper 2020.

70. Roosth 2017: 5.

larly large philosophical anxiety for Aristotle, largely because he believed that living things simply had different types of change available to them.

Automata appear twice in Aristotle's biological explanation. In *Generation of Animals*, he outlines the processes of animal generation. He argues that viviparous female animals produce menstrual blood,[71] and they do this because they possess insufficient heat and are thus, as Aristotle infamously declares, "like a deficient male" [ὥσπερ ἄρρεν ἐστὶ πεπηρωμένον]. In this scenario, "menstrual blood is semen, but not purified" [καὶ τὰ καταμήνια σπέρμα, οὐ καθαρὸν δέ],[72] and for this reason it possesses all animal parts in potentiality but is not generative. Even if menstrual blood is produced from material necessity and a supposed deficiency, nature still uses it to serve a biological purpose by using it as the matter for the embryo. The introduction of semen into the womb therefore instills a type of soul principle insofar as it activates growth through the addition of pneuma and its heat.[73] This instillation starts the process by which a living thing develops, much like fig juice or rennet curdles milk.[74] At this point, the embryo develops in a determinate order. First, it sends out the umbilical cord to receive nutriment, then the heart grows as the source of the blood vessels, the internal source of nourishment, and the goal for which nourishment exists.[75] After the heart, the rest of the "tool-like parts that are generative by nature" develop, then the parts that exist for

71. Arist. *Gen. an.* 2.5, 738a10–18.

72. Arist. *Gen. an.* 2.5, 737a28–29.

73. As Aristotle states, "Heat exists in the seminal residue that possesses the degree and quality of motion and activity that is commensurate for each part" [Ἡ δὲ θερμότης ἐνυπάρχει ἐν τῷ σπερματικῷ περιττώματι τοσαύτην καὶ τοιαύτην ἔχουσα τὴν κίνησιν καὶ τὴν ἐνέργειαν, ὅση σύμμετρος εἰς ἕκαστον τῶν μορίων] (*Gen. an.* 2.5, 743a28–30). He also outlines the material processes involved in the formation of the homoiomerous tissues that occur first, including blood vessels, sinews, bones, brain, and eyes, followed by the body parts, such as nails, hair, and teeth, formed from residues. Like Plato before him, he draws on comparisons with baking, pottery making, carpentry, and other technical processes for insight into the behavior of physical substances; see esp. *Gen. an.* 2.5, 743a2–b34.

74. Arist. *Gen. an.* 2.5, 736b34–737a17; cf. 2.5, 739b22–25. Semen thus imparts nutritive movement as well as the "sensate soul," which guides growth. As Aristotle states, "The female animal always provides the matter, the male the builder" [ἀεὶ δὲ παρέχει τὸ μὲν θῆλυ τὴν ὕλην, τὸ δ' ἄρρεν τὸ δημιουργοῦν] (*Gen. an.* 2.5, 738b20–21). He also expresses this notion as the female supplying the body, the male the soul, but we should remember that souls, for Aristotle, are essentially the activity of a living body and not some life-giving substance; see *Gen. an.* 2.5, 738b26–27; cf. note 166 on p. 65.

75. Arist. *Gen. an.* 2.5, 740a17–24.

the sake of something else.[76] Aristotle likens this process to the sequential movements of automata, almost as an aside:

> Since the parts are present in the matter potentially, whenever the initiation of movement occurs, a succession of events follows of itself, just as in those marvelous automata. And one ought not say what some of the physikoi claim, that "like moves to like," as though the parts move spatially, but [that they change] by remaining in the same place and altering in softness, hardness, color, and the other differences found in homoiomerous tissues, becoming in actuality the things that they were potentially before.[77]

Here Aristotle is clear to denote that the mechanics of the automata, which involve the spatial movement of internal elements from one place to another—spools unwinding, cogs spinning—do not correspond to the mechanisms by which an embryo grows. As such, the invocation seems merely to be an illustrative example of the transitive causality embodied in these devices, and a good example of a didactic, rather than heuristic, analogy.[78] He is more than willing to acknowledge that a certain amount of material necessity drives physical processes without intervention, but since he sees the embryo as imbued with soul, this predetermined physical process of growth is self-directing, even if proceeding in a determined sequence. This sequence is not some order imposed on the embryo externally but belongs to the intrinsic alterations already belonging to the matter potentially, activated by the presence of the soul principle introduced via the semen. As a corollary, the crucial distinction between automata and living bodies is not, for Aristotle, a question of automaticity, but rather one of what type of change the automaton can enact by itself.

Although automata do not supply a model for the physical mechanisms available to the embryo, this separation remains less clear for animal movements. In *On the Motion of Animals*, Aristotle explains that when we see an object that we want or something that we fear (or simply imagine either of these things), the associated pleasure and pain warms or cools the pneuma around the heart. Heat causes pneuma to expand, and pain-produced cold causes it to contract. These small initial motions of

76. Arist. *Gen. an.* 2.5, 742b5–6. This assertion means that the vital or necessary organs of the upper cavity develop first, with the other parts that support them following.

77. Arist. *Gen. an.* 2.5, 741b3–16.

78. See Berryman 2009: 71–74 for this point; cf. Berryman 2003: 359.

"pushing and pulling" [ὦσις καὶ ἕλξις] then move the sinews [νεῦρα] in the heart's interior, likely to be identified with the chordae tendineae, which themselves "pull and release" [ἕλκειν καὶ ἀνιέναι] to produce a cascading series of sequential motions that move through the limbs, with the sinews driving the motions.[79] The small initiate force is amplified by the principle of the lever, just like the small motions of a rudder can lead to large directional shifts for boats.[80] Changes in the configurations of the tissues around the joints, likely conceived of as the muscles, allow the limbs to move in different directions,[81] and in this way a small qualitative change in connate pneuma thus translates into the mechanical impulse necessary to produce all animal motions, whether coughing, shuddering, running, jumping, or waving. Aristotle includes few details about the physical mechanics involved in any of the activities, beyond the original impulse, instead letting a suggestive analogy do much of this work:

> Just as automata are moved when a small motion happens, when the twisted cords are released and knock one another, and the little wagon, as though being ridden, moves itself in a straight line and is moved back again in a circle by having unequal wheels (for the smaller acts as a center, just as in cylinders), so too the animal is moved. For such instrumental parts have the nature of sinew and bones, the [bones] corresponding to the wood and iron there, and the sinews to the twisted cords, which, when they are let free and released, are moved.

> Ὥσπερ δὲ τὰ αὐτόματα κινεῖται μικρᾶς κινήσεως γινομένης, λυομένων τῶν στρεβλῶν καὶ κρουόντων ἀλλήλας τὰς στρέβλας καὶ τὸ ἁμάξιον, ὥσπερ ὀχούμενον, αὐτὸ κινεῖ εἰς εὐθύ, καὶ πάλιν κύκλῳ κινεῖται τῷ ἀνίσους ἔχειν τοὺς τροχούς (ὁ γὰρ ἐλάττων ὥσπερ κέντρον γίνεται, καθάπερ ἐν τοῖς κυλίνδροις), οὕτω καὶ τὰ ζῷα κινεῖται. ἔχει γὰρ ὄργανα τοιαῦτα τήν τε τῶν

79. Arist. *De motu an.* 10, 703a18–21. Corcilius and Gregoric (2013) propose that the thermic alterations of the pneuma cannot be directly identified as the hylomorphic consequences of desire. Instead, desire is the warming of the blood around the heart, which then warms the pneuma; cf. Corcilius 2008: 332–43; Gregoric 2020: 433–34; Rapp 2020: 57. By contrast, Nussbaum ([1976] 1985: 156–57) presumes that desire directly warms the pneuma. Berryman (2002a); Buddensiek (2009); Gregoric and Kuhar (2014: esp. 96); and Gregoric (2020) all provide broader accounts of the role connate pneuma plays in voluntary motion. See also Freudenthal 1995: 107 nn2–3 for additional bibliography.

80. Arist. *Part. an.* 3.4, 666b16. For rudder comparison, Arist. *De motu an.* 7, 701b26–27.

81. Arist. *De motu an.* 8, 702a7–10; cf. Gregoric and Kuhar 2014.

> νεύρων φύσιν καὶ τὴν τῶν ὀστῶν, τὰ μὲν ὡς ἐκεῖ τὰ ξύλα καὶ ὁ σίδηρος, τὰ δὲ νεῦρα ὡς αἱ στρέβλαι· ὧν λυομένων καὶ ἀνιεμένων κινοῦνται.[82]

Textual issues have prevented clear consensus on what devices Aristotle here invokes and what degree of physical isomorphism he means to imply (for which reason I have supplied the Greek in full). The most influential interpretation has come from Martha Nussbaum, who argued that Aristotle must be talking about a type of marionette-like automaton; she turned to Heron's first-century CE text *Automata* to elucidate what he must mean. In this text, Heron provides instructions for how to build two separate devices, a multipart theatrical automaton that enacts the story of Nauplius and a moving temple of Dionysus at which the god performs a ritual libation.[83] Both devices rely on a central drive-chain mechanism whereby a cord is wrapped around a central bobbin and extended upward to loop around a pulley, then attached to a counterweight set in a tube filled with sand or mustard seed (see fig. 10). When a plug on the bottom of the tube is pulled, the seeds pour out, the weight slowly lowers, and the automaton starts to move. The "twisted cords" [στρέβλαι], Nussbaum thus argued, must refer to the wound drive cords of such a device. As for the "little wagon" [τὸ ἁμάξιον], she suggested that it is simply a child's cart that turns in a circle because the wheels on one side are smaller. Even though Nussbaum's readings required emending the text, most scholars took up her assumptions.[84] As a result, most have accepted that the comparison implies no extensive isomorphism between the cords of the

82. Arist. *De motu an.* 7, 701b2–15 Louis, with emendations. This is Louis's 2015 text and not Primavesi's from 2020. The latter includes unsupported conjectures and liberally adds to the passage; see note 84 below. In addition, I, with Ruffell, have accepted the emendation of αὐτὸ κινεῖ εἰς εὐθύ to αὑτὸ κινεῖ εἰς εὐθύ. See Webster, forthcoming, for these arguments.

83. For a recent edition and commentary of Heron's *Automata*, see Grillo 2019.

84. Nussbaum ([1976] 1985) emended the text to "For the one riding it moves it in a straight line" [ὁ γὰρ ὀχούμενος αὐτὸ κινεῖ εἰς εὐθύ]. Primavesi and Morison (2020) accepted this emendation, but rewrote the passage as follows: "Just as the automatic theatres are set in motion when only a small movement has happened—when the cables are released and the <figures immediately> strike their <sabres> against one another—and as the little wagon which <the> driver himself tries to move in a straight line <again> and again but which is moved in a circle because it has unequal wheels" [ὥσπερ δὲ τὰ αὐτόματα κινεῖται μικρᾶς κινήσεως γινομένης—λυομένων τῶν στρεβλῶν καὶ κρουόντων πρὸς ἀλλήλας τὰς στρέβλας <εὐθὺς τῶν ζωιδίων τὰς μαχαίρας>, καὶ τὸ ἁμάξιον, ὅπερ <ὁ> ὀχούμενος αὐτὸς κινεῖ εἰς εὐθύ <πάλιν> καὶ πάλιν, κύκλῳ δὲ κινεῖται τῷ ἀνίσους ἔχειν τοὺς τροχούς].

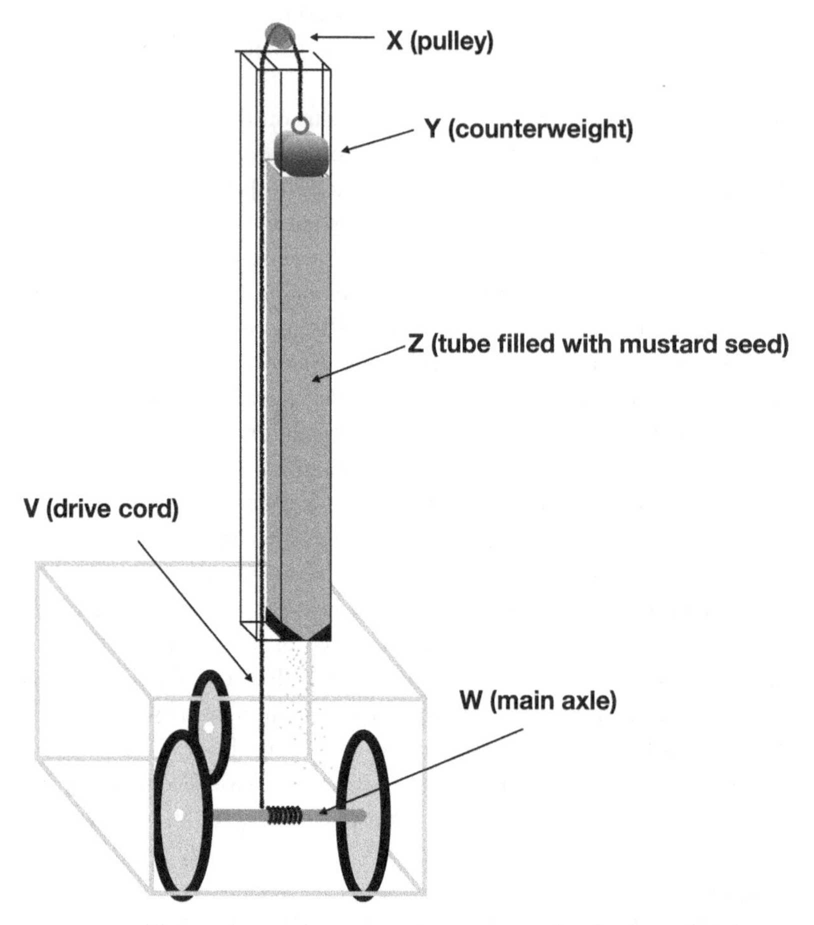

FIGURE 10. Heron, *Automata* 1.2.7–10, 1.5.3–5, 1.9.4–6, drawing by author.

automata and human physiology, despite the fact that Aristotle himself draws explicit parallels between the *neura* (sinews) and bones in the body and the motion-generating "twisted cords" and wood in these devices.

As I have argued elsewhere in far greater detail, I think that Aristotle is referring not to the wound cords of a drive-chain mechanism but to a torsion spring, and this interpretation helps us understand why he assumes that the sinews [νεῦρα] move our bones, which they do by "pulling and releasing" [ἕλκειν καὶ ἀνιέναι], even though they are a discontinuous system with its origins in the heart (i.e., in the chordae tendineae).[85] Modern com-

85. Arist. *Part. an.* 3.4, 666b16; cf. *Hist. an.* 3.5, 515a27–b7.

mentators, Nussbaum included, have presumed that Aristotle must believe that the limbs move by contraction of muscles attached to the bones with tendons, with their relaxation simply facilitating the contraction of some other muscle.[86] After all, this fact, to us, feels obvious, especially in light of Aristotle's description. Yet as Gregoric and Kuhar have correctly noted, Aristotle does not attribute any motive capacity to the muscles. In fact, Aristotle does not use the term "muscle" [μῦς] at all; the word only appears in his corpus twice in this way, both times in the pseudo-Aristotelian *Problemata*.[87] Rather, Aristotle speaks about "flesh" [σάρξ], and he does not make this tissue responsible for generating motion but for perceiving touch.[88] Sinews, not muscles, move the bones. Yet even Gregoric and Kuhar, who spend great efforts detailing this fact, still presume that by "pulling and releasing" Aristotle must mean that sinews exert force by pulling, with their release facilitating the contraction of other sinews. Therefore, they argue, Aristotle must mistakenly think that sinews run through the muscles and function as though they were themselves muscles.[89] Yet Aristotle describes the sinews as clustering around each joint, where they bind the bones together.[90] To be sure, he does not seem to distinguish between tendons (which bind muscles to bones) and ligaments/articular capsules (which bind bones to bones), and sometimes he lists certain neura by name, such as the *epitonos* and *omiaia*, that are difficult to identify,[91] but imagining marionette-like pull cords leading back to the heart seems to sit uncomfortably with these statements. Therefore, rather than presume that Aristotle must see the sinews as quasi-muscles, operating like marionettes, we can examine how animal sinews were used in technologies of the fourth century BCE to get an alternative sense of what he might mean by sinews generating force by pulling and letting go.

Sinews were known for their high tensile strength, which Aristotle himself comments upon.[92] This elasticity allowed for multiple uses, notably in bowstrings, which are pulled, then released to generate force. More

86. For scholarly assumption that the *neura* refer to muscles, see Gregoric and Kuhar 2014: 96 n3, who cite Preus 1981: 81; Nussbaum (1976) 1985: 281 and 284; Bos 2003: 37; and De Groot 2008: 62.

87. Arist. *Pr.* 5.40, 885a37–38; cf. Gregoric and Kuhar 2014: 95 for this observation.

88. See Arist. *De an.* 2.11, 422b17–423b2.

89. Gregoric and Kuhar 2014: 106–7.

90. Arist. *Hist. an.* 3.5, 515b10–13.

91. Arist. *Hist. an.* 3.5, 515b8–10. See Gregoric and Kuhar 2014: 100–106 for some of the confusions; cf. Webster, forthcoming, for my responses.

92. Arist. *Hist. an.* 3.5, 515b16.

importantly, in the generation prior to Aristotle they were used in a new type of device, a torsion spring called the *hysplenx*, which was first used to create swift-moving starting gates and then adapted to power a new type of weapon: the catapult. Onomarchus of Phocis seems to have deployed torsion-spring-driven catapults called "one-armers" [μονάγκωνες] in the middle of the fourth century BCE for use against Philip of Macedon's army (fig. 11).[93] Philip of Macedon's engineers then adapted the device to create the two-armed torsion catapult by at least 340 BCE.[94] This latter weapon employs two springs (or "half-springs," as they become known) turned vertically, into which the catapult's arms are inserted horizontally. A winch then draws back the arms, stretching and tightening the sinew spring until a trigger mechanism releases and the arm springs forward with explosive force (fig. 12). Aristotle surely witnessed these devices at some point, insofar as he was a Macedonian and had been summoned by Philip to tutor his son Alexander in 338 BCE, two years after their deployment in the field.

In addition to weaponry, the hysplenx spring was also used to power early automata. Heron describes how to build and employ the counterweight mechanism but mentions that you can also use a hysplenx, although it is an inferior power source.[95] This torsion spring works just like a catapult, wherein an arm or "axle" [ἄξων] (A) is inserted into a skein of sinew cords (B), which are then twisted tight. The arm is then pulled backward, which stretches the spring and creates stored energy. You can then attach a drive cord that has been wound around a bobbin or main drive axle, so that when the arm is released, the torsion spring pulls the cord, rotates the axle, and drives any attached mechanisms (fig. 13). The power of the spring needs to be carefully balanced against the friction inherent in the automata as a whole, which makes adding elements and programming quite difficult and unreliable. For these reasons, the counterweight is much preferable, which suggests that the hysplenx was an earlier power source, potentially employed during the same decades in which torsion springs were being adapted to multiple other technological uses. As such, it seems likely that this is the type of mechanism that Aristotle envisions when he compares the sinews in the body to the "twisted cords" of an automaton, and bones to the wooden parts, especially since in the same passage where Heron describes the hysplenx he directs, "Do not use any sinew (for the drive cords), since it stretches or contracts depending on the

93. Rihll 2007: 60–75.

94. Rihll (2007: 78–79) claims that concrete archaeological evidence for torsion springs dates to 326 BCE at the latest.

95. Heron *Aut.* 1.2.6 Grillo; cf. 1.2.8 Grillo.

FIGURE 11. Modern reconstruction of an onager. Room XVI, Museo della Civiltà Romana, Rome.

FIGURE 12. Drawing of reconstructed first-century Roman catapult at Maiden Castle, Dorset, England. Obtained from Art Resource.

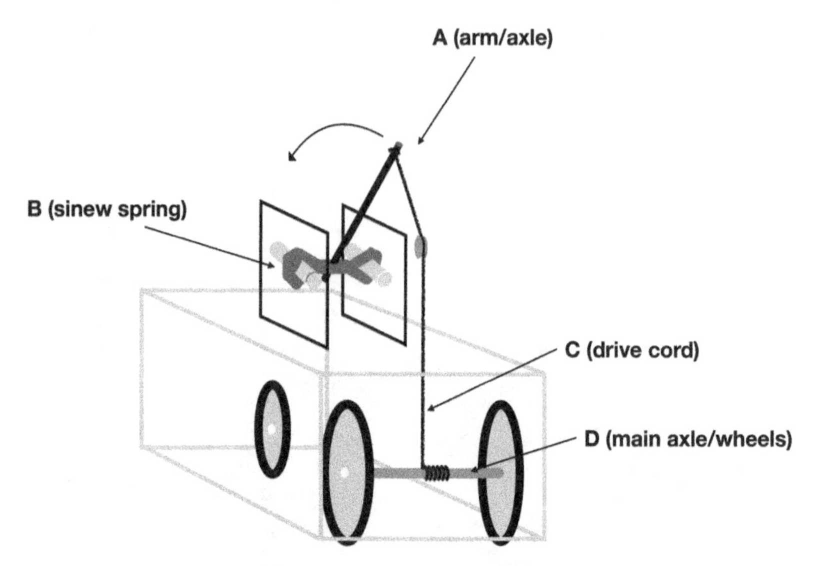

FIGURE 13. *Hysplenx*-driven automaton cart, after Heron, *Automata* 1.2.6–8, drawing by author.

conditions of the air" [νευρίνῳ δὲ οὐδενὶ δεῖ χρῆσθαι, ἐπειδὴ παρεκτείνεται ἢ συστέλλεται κατὰ τὴν τοῦ ἀέρος περίστασιν].[96] If Nussbaum were right, Aristotle would be comparing the sinews in the body to cords that should not be made from this very tissue in an automaton.

As for the second part of the comparison, later evidence suggests that the "little cart" is not a child's toy but a specific type of automaton invented in the fourth century BCE, potentially also powered by a hysplenx. The pseudo-Socratic *Epistle* 35 mentions how the Syracusan tyrant Dionysius I dedicated "clever things" [σοφά] at Delphi, and seemingly alludes to one such object when it describes how Apollo, the patron deity of Delphi, "heard and saw *the little wagon* that ran around in the hippodrome automatically" [ἀκούσας καὶ τὸ ἁμάξιον ἰδὼν τὸ ἐν τῷ ἱπποδρόμῳ περιτρέχον αὐτόματον]."[97] Both Ruffell and Grillo have suggested that this "little wagon" refers to the same type of cart that Aristotle describes.[98] Indeed, since a hippodrome is a racetrack where competitors travel in a straight line before turning around a post and returning, this interpretation fits Aristotle's assertion that the little wagon moves in a straight line as though being ridden, before it turns back again in a circle. Identifying

96. Heron *Aut.* 1.2.6 Grillo; cf. 1.2.8 Grillo.
97. [Soc.] *Ep.* 35.154 Hercher.
98. Grillo 2019: lxxxiv–lxxxv.

this device as another type of automaton becomes even more likely, since Aristotle says the same thing about the "little wagon" that he did about the automata in embryonic growth: whereas automata can create preprogrammed sequential motions, the alterations that cause them to make directional shifts are spatial, whereas in biotic things, such shifts arise by qualitative alterations.[99]

If these are the devices that Aristotle uses to imagine animal motion, what we are left with is a potential biotic mechanism for the movement of the limbs, one wherein the bones are bound together by sinews, which can create force by being stretched and released. The pneuma around the heart expands or contracts, which causes the sinews in the heart to release, which leads to a cascade of explosive force, sequentially moving to the extremities from the heart as the center of this discontinuous system. Whereas we tend to think of animal motions implicitly through marionettes, such that contraction is the sole generator of power, and lifting something with the bicep can become the paradigmatic human movement, Aristotle's model seems to involve more propulsive force, perhaps such that jumping with the legs might be more exemplary. This paradigm does not preclude the possibility that he thought sinews could contract to pull the bones, but it foregrounds more explosive movement. To be sure, since emotions ultimately cause the pneuma to expand, which supplies the initial impetus, the human body cannot be reduced to an automaton. Nevertheless, the question of technological influence does not get answered there. Instead, his explicit comparison between the motion-generating sinews of the body and the "twisted cords" of automata does seem to imply some degree of isomorphism, since they are potentially made by the same materials.

It should be noted that even if this interpretation is incorrect and Aristotle is imagining a different type of automata and the little wagon is not a hysplenx-driven cart, Aristotle still identifies the sinews, not the muscles, as the force-producing tissues in the body. This notion is shared by other ancient authors, such as Plato.[100] At the very least, this long-standing idea must come from the use of sinews in bowstrings and other similar devices, where their tensile strength is used to generate force, even before the hysplenx or the "one-armer" supplied an explicit facsimile of a human limb and a closer model for the bones being bound by sinews in the

99. Arist. *De motu an.* 7, 701b11–16.

100. Pl. *Phdr.* 98c5–d6 claims that the bones and neura are responsible for movement. He, too, states that neura both bind the bones and move them by "stretching and releasing" [ἐπιτείνεσθαι καὶ ἀνίεσθαι]; cf. Pl. *Ti.* 74a7–d8.

joints. Accordingly, we end here with an example of the same recursive relationship, where the ideas about the operation of parts inside the body is determined by the technical uses for these parts outside the body. Since biotic tissues can themselves be parts of implements, tools, and complex machines, their imagined corporeal role is entangled with their use as a type of technology in their own right.

Roosth has examined modern synthetic biologists, who seek to understand life by artificially recreating it in a laboratory. These scientists genetically modify bacteria and other life-forms so that they, in the biologists' minds, "function better," as though the scientists themselves were engineering evolutionary selection. They then see the resultant organisms as demonstrations that their theories regarding natural life-forms are correct. In these moments, their synthetic living things function as what Roosth calls "persuasive objects" and quasi-guarantors of their theoretical accounts.[101] Even as I write these words, scientists have just used human heart cells to generate a self-propelling artificial fish, which will help us "to better understand the design principles and coordination mechanisms of biological systems."[102] Fish get understood as underwater pumps, and hearts get understood as biotic tools. Although the technologies at play in Aristotle's account are less complex, it helps to think about bowstrings and torsion springs as similar types of "persuasive objects." Insofar as they employ sinews to generate motion, they inform and guarantee theories about sinew-driven motion in the body. Making things with animal tissues thus operates as a surrogate for knowing, as the distinction between biotic and abiotic things gets blurred, even before the moments where Aristotle confidently draws a line between them.

Even if automata triangulate a particular problem concerning the boundaries of life, they were not, for Aristotle, ontology-destabilizing machines. Their turning parts and shifts in direction could mirror living motions, but the types of self-directed qualitative change available to living things did not operate within artificial objects. Nevertheless, there are other consequential ways that tools can structure and influence conceptions of corporeality beyond whether some vitalist force flicks on the lights or pulls the trigger to launch a series of subsequent actions. The body can be structured as a tool-like entity, and discrete aspects of the human apparatus can mirror technological operations, even if the entire system cannot be reduced to cogs and cords. Moreover, technologies can confuse natural and artificial categories before the theoretical impulse

101. Roosth 2017: 5.
102. Lee et al. 2022.

even starts to formulate explanations, insofar as biotic tissues perform tasks for technological instruments. In the case of ancient automata and the hysplenx, these devices transformed sinews into material theories about propulsive force in the body. Even animal motions are not sitting there, waiting to be explained; rather, they need to be articulated as objects of explanation, and Aristotle did so in a moment when the hysplenx was transforming in a flourish of technological development, as it began to produce multi-set movements and deadly force. These devices did not explode his carefully articulated organism, but they influenced his biotic assumptions nevertheless.

3.5 CONCLUSION

Aristotle fully organized living things, construing them as hierarchical arrangements of instrumental parts that collectively enacted and maintained life. Health became a bioproduct of this organic system functioning properly, rather than the homeostatic corporeal state that patients and physicians tried to preserve through balancing inputs and exertions. The organism had arrived. Armed with this teleological theory of corporeality, Aristotle launched an ambitious and incredibly detailed observational program to understand all life-forms. His research into animals led him to investigate and prioritize corporeal parts, determining what function they supplied for the body and how the variations between animal components were suited for the lifestyles of each individual species. This investigation involved cutting animals open and interrogating their messy interiors. Yet even exposed in this way, their bodies did not simply disclose themselves, and even major organs left few clues as to their operation other than their macroscopic structural elements and their interconnections. Technologies thus served as ready-at-hand heuristics to decode these inner structures, which were themselves already conceptualized as somatic tools. Yet even this account presents a somewhat idealist model of how theory formation works, since it presumes that the observations come first, and that theorists, Aristotle included, then employ technological heuristics to explain what they see. In practice, observable features and behaviors are not merely sitting there, awaiting explanation, but must themselves be delineated, conceptualized, and constructed as objects and aspects of scientific explanation. Technological comparisons can therefore play a more important role in predicating what anatomical structures authors situate in the body, what functional role they ascribe to them, and the basic behaviors that the body itself displays. The resulting organism available for scientific explanation is therefore some hybrid of all these vectors, overlapping, tangling, and binding the body together into an organic whole.

The lesson drawn from this hybridity should not be that Aristotle was a bad scientific observer or that he confuses his analogies with the reality he describes. He sometimes draws strict limits on his comparisons, and he subjects his physiological theories to robust observational and analytic tests. Yet it would be incorrect to insist that either observations or theoretical structures can themselves be fully decoupled from their technical contexts or analogic supports. Indeed, Aristotle articulated his organism and filled out its contours, although not all the gears of this task-fulfilling body meshed. The heart boiled like a pot, expanded like the bellows, overflowed into one-directional unpressurized pipes, and served as a passageway for rising vaporous blood moving in the other direction. Insofar as Aristotle found tools that paralleled organic parts in both form and function, his technological analogies smoothed the seams where these varying demands met, directing cognitive focus away from tensions or difficulties. As a result, the anatomical structures that Aristotle articulates come into focus, and then recede in importance, as physiological demands move to the foreground. The body that he constructs therefore remains a type of hybrid epistemic object, wherein an analytic framework built around the concept of tools scaffolds its structural ontology, and his in-depth observational program fills out the anatomical and behavioral details, but these details can themselves be entangled with the individual technologies used to conceptualize and explain them. At the very least, this chapter has outlined the continued development of the organism and illustrated the powerful conceptual pull that material implements exerted on Aristotle to comprehend and structure the phenomena that he attributed to this new corporeal object, even in those moments where he was keen to sharpen the lines dividing bodies, organs, tools, and machines.

CHAPTER FOUR
The Rise of the Organism in the Hellenistic Period

4.0 INTRODUCTION

The last two chapters examined first how the physikoi became interested in corporeal mechanisms and how tool analogies facilitated and encouraged conceptualizing the body's interior in this way. It then tracked how these ideas met with the broader teleological views of Plato and Aristotle, who both began to privilege functional interpretations of the body and its parts. With this development came a new epistemic object: the organism. This particular discourse emerged not in medical treatises themselves but in natural philosophical texts. Nevertheless, the impact of the organization of the body on medical conceptions and practices was manifold. At a basic level, multiple post-Aristotelian authors adopted the vocabulary of *organa* and began to articulate a body containing parts that explicitly operated like tools. Moreover, whereas the Hippocratic treatises of the fifth and fourth centuries BCE had leveraged pathologies and therapeutic interfaces to reveal the nature of the body, by the third century BCE, explaining how the body "worked" absent these states and interventions became a far greater part of the medical endeavor. This shift dramatically altered the type of corporeality that fell under the gaze of physicians, generated new definitions of disease, and created different ways of making health an object of concern. An accompanying set of epistemological and investigative practices attended this new epistemic object, as the Hellenistic era saw a dramatically increased interest in anatomical investigations as a way of knowing, with functional teleology as a core rationale supporting this investigative mode. Although there were disagreements about the limits of teleology in corporeal explanations, the body that function had assembled reoriented medical discourses, practices, and training.[1]

1. Theophrastus and Strato both seem to have questioned the scope of cosmic teleological arguments (see Sharples 2017), although this questioning does not mean that they rejected function as consolidating living creatures.

Two figures, Herophilus of Chalcedon (ca. 330/320–260/250 BCE) and Erasistratus of Ceos (ca. 315–240 BCE), were crucial in elevating the status of dissection as an investigative practice, and they cemented the organism as a viable model of the body in the process. Both physicians conducted extensive dissections and identified corporeal features at a far more detailed level than previous theorists. Both also incorporated both function and operation into the heart of their epistemic object, leveraging the heuristic power of tools in novel ways. They took different stances than Aristotle and presented their own versions of an organic body, but in so doing, they developed, as Polybius states, the "theoretical" [λογικός] part of medicine tied more closely to the philosophical tradition.[2] These two medical theorists engaged in these activities in a city that became the main site for anatomical instruction in the Mediterranean for the next five hundred years and beyond: Alexandria.

Only a few decades after its founding in 331 BCE by Alexander the Great, Alexandria had become a new cultural and intellectual capital of the Greek-speaking world. Thinkers from around the Mediterranean descended upon the Mouseion and Library established by Ptolemy I Soter, who sought to leverage this cultural capital to legitimize his city as the de facto seat of power after the fragmentation of Alexander's empire. The city hosted some of the major intellectual forces of the era, including Euclid, Callimachus, Apollonius, and Eratosthenes, and this thriving climate led to the growth of innovation across multiple fields, including science, medicine, engineering, literature, and mathematics, as Greek thinkers came in closer contact with Egyptian, Persian, and Assyro-Babylonian theoretical and technical systems. Peripatetic natural philosophers formed an integral part of this intellectual fabric. Theophrastus, Aristotle's successor at his Lyceum, was reportedly invited there to teach, while Strato, Theophrastus's successor, reportedly tutored Ptolemy II Philadelphus, the second of the Ptolemaic monarchs ruling over Egypt.[3] This vibrant intellectual environment also facilitated technological shifts and saw the birth of both mechanics and pneumatics as theoretical disciplines. Inventors such as Ctesibius and Philo created new devices including water clocks, wine dispensers, and pneumatic automata.

This chapter examines how these forces—the organism, mechanics, and pneumatics—altered views of physiology developing in the third century BCE. It notes how Herophilus drew from the rich artifactual culture of Alexandria to name the parts he had identified, which suggests a con-

2. Polyb. *Hist.* 12.25d2–6 = T56 von Staden. For Herophilus and Erasistratus's Peripatetic connections, see note 59 on p. 176 below.

3. See Diog. Laert. 5.37; 5.58; cf. von Staden 1989: 39, 97.

ceptual overlap between mechanics and medicine in the period. Moreover, Herophilus adopted features of Aristotle's bellows model of the lungs but increased the number of places that these devices appeared in the vascular and pulmonary systems. Just as new technologies can be modified to suit multiple different purposes in the period of their adolescence, as engineers attempt to find problems to fix with the solutions their new tools provide, Herophilus and other medical theorists likewise deployed tools like the bellows in multiple places in the body once they entered physiological discourses. In this way, heuristic tools and physiological problems maintain a dialectical relationship, where the behaviors of the body in need of explanation change with in the introduction of a new totemic technology into the discourse. Theories invoking technologies can, in this way, mirror the dynamics of technological developments themselves.

The second part of the chapter turns to Erasistratus, who also used the bellows as an analogue to explain coronary operations, although he seems to have been influenced more deeply by contemporary developments in pneumatics. Much scholarly attention has focused on assessing whether his model of the heart as a propulsive engine relied upon a powerful contemporary invention, Ctesibius's force pump. This chapter evaluates the influence that this pump had on Erasistratus but expands beyond a single analogic relationship to demonstrate how pneumatic devices influenced Erasistratus's theories more broadly. He argued that the body was composed of separate and carefully calibrated systems of liquid and air, where disease could result from the "infiltration" of venal blood into the air-filled arteries, or vice versa. Such a pathology diverges quite substantially from the imbalance, accumulation, or blockage models that the Hippocratics more frequently relied on, and closely reflects the operation of pneumatic devices, which rely upon the delicate interplay of water and air, often separated into different chambers. The infiltration of one substance into the wrong spaces can cause these instruments to malfunction and ruin their desired effects. While tracking the development of the organism at a conceptual level, this chapter thus reveals the impact that technological developments had on a shifting terrain of medical theories in the third century BCE. It shows the interplay of technologies and theories of corporeality in the early Hellenistic period, as the body became tool-like in new, complex ways.

4.1 THE RISE OF ANATOMY

The difference between pre- and post-Aristotelian anatomy is glaring. For example, a Hippocratic treatise simply called *Anatomy*, which seems to come from the fourth century BCE,[4] consists of only two short pages and provides very brief descriptions of the trachea, lungs, heart, liver, kidneys, and bladder in succession, with passageways linking the last four and presumably extending from the first two:

> An air pipe growing out of the throat on each side ends at the apex of the lung; it is composed of symmetrical rings, which in their circular course meet one another in a plane. The lung itself occupies the chest, facing toward the left, and possesses five prominences called lobes; it is of ashen color, marked with spots, and in structure like honeycomb. In the center of the lung the heart is set, being more spherical than in all other animals. From the heart a large tube (bronchia) descends to the liver; running with this tube is what is called the Great Vessel, through which the whole frame is nourished. The liver has a symmetry with all others, but more blood flow than the others; it possesses two prominences called "gates" which lie on the right side. From the liver a vessel slanting downward reaches the kidney. The kidneys are symmetrical, and in color are like apples. From the kidneys two oblique ducts

4. Dating this treatise is very difficult, since it includes recondite vocabulary, and its truncated entries provide little context. Jouanna (1999: 530) considers it a late Hellenistic or early Roman compilation, since it does not seem to have been known before Galen, but nothing otherwise suggests so late a date. In fact, Craik (2006: 155) suggests that the text is a "guileless abridgement" of a Democritean anatomy from the fifth century BCE, an argument that she makes based on connections between this text and the pseudo-Hippocratic *Epistles* 17–23, but the latter are only apocryphal depictions arising after 200 BCE and seem to retroject assumptions about anatomy back onto Democritus (Smith [1990: 28] dates these letters to 100 BCE–100 CE; see note 13 on p. 197 below). Harris (1973: 83) suggests that *Anatomy* dates to after Praxagoras, sometime in the third century BCE, especially since it makes distinctions between human and animal anatomy, and human dissection did not begin in earnest until the late fourth century. That said, the text reflects no real post-Aristotelian notions otherwise. Lastly, we might consider the fact that the reference to the cavity as "putrefying" [σηπτικὴ κοιλίη] aligns with Diocles of Carystus's conception of digestion as a type of putrefaction, an idea also held by Plistonicus (fl. 3rd c. BCE), a pupil of Praxagoras (Celsus *Med.* 1 *pref.* 20). This shared theory may intimate contemporaneity, although not with any degree of certainty. For Diocles's views on digestion, see fr. 35 van der Eijk; cf. fr. 109 van der Eijk, which also speaks about the role of heat in the veins surrounding the stomach in digestion; and fr. 134 van der Eijk, which mentions that Diocles wrote a treatise on digestion. The preserved fragments of Diocles that deal with anatomy are found at Diocles, fr. 23–29 van der Eijk. For Diocles's dates, see note 8 on p. 163.

> reach the topmost apex of the bladder. The bladder is large and entirely sinewy, and out of it grows a channel. Nature has ordered [ἐκόσμει] these six parts in the interior around the mid-line.[5]

After this brief description, which composes literally half the text, the author provides brief descriptions of the esophagus, belly, diaphragm, spleen, cavity, intestine, colon, and rectum, with the implication that they, too, are linked in a chain. In so doing, the treatise seemingly creates two separate linear systems, one for the fluids moving through the first set of viscera and one for solid nourishment moving through the second. Even within this overarching division, the viscera are not given any individuated role, except for the description of the "putrefying cavity" [σηπτικὴ κοιλίη], which may suggest a certain concept of digestion as a type of rot.[6] Otherwise, the text assigns no individuated function to the heart or lungs and implies no mechanisms to drive fluid motion except downward flow. In other words, in the late fourth century BCE, the viscera were still stations. Even the final lines of the treatise, "the rest, nature has arranged" [τὰ δὲ ἄλλα ἡ φύσις διετάξατο], evince a loose, global teleology whereby nature situates parts in their proper order, rather than a functional teleology in which nature manufactures individuated organs with specialized tasks as a part of a broader corporeal system.[7] These parts were not tools performing unique tasks.

The difficulty in dating *Anatomy* makes the treatise difficult to rely on when constructing a chronology of anatomical approaches. Nevertheless, other evidence from this period shows the status of anatomy in the mid-fourth century. For example, Diocles of Carystus, a likely contemporary of Aristotle,[8] wrote the first anatomical treatise of which Galen was aware, and when referring to it, Galen laments that Diocles does not make clear the function or "use" [ἡ χρεία] of what he describes.[9] That said, Diocles does start to include some notions of corporeal functions into his medical theories, insofar as he regards breathing as necessary to cool the innate

5. Hipp. *Anat.* 1 = 8.538L, trans. Craik.

6. Hipp. *Anat.* 1 = 8.540L.

7. Hipp. *Anat.* 1 = 8.540L.

8. Diocles's dates are contested; Jaeger suggests 340–260 BCE, making him younger than Aristotle (see Jaeger 1938 and 1940), but this conjecture does not hold much certainty. Instead, the doxographical tradition makes him a contemporary of Aristotle (384–322 BCE). Manetti (2008a: 255–57) simply lists him as living sometime 400–300 BCE; see also von Staden 1989: 44–48 and 1992b; cf. Diocles fr. 1–12 van der Eijk.

9. Gal. *AA* 2.1 Singer = 2.281–282K = Diocles, fr. 17 van der Eijk.

heat,[10] but it seems safe to say that the earliest Greek anatomical texts did not privilege individuated function, but sequences and connections. By the late fourth century BCE, less than a generation after Diocles and Aristotle, this state of affairs swiftly changed, and medical discourses show a dramatic increase in both the interest in anatomy and the use of function to consolidate the body.

Clearchus of Soli, a Peripatetic of the late fourth and early third centuries BCE, wrote a treatise called *On Skeletons* [Περὶ σκελετῶν] that reportedly described and named the human bones and muscles.[11] Praxagoras of Cos (ca. 340–ca. 275 BCE), a physician with Peripatetic connections, wrote multiple medical treatises, including a work called *Anatomy*.[12] As Orly Lewis has argued, Praxagoras relied on minute anatomical observations when he distinguished the arteries from the veins, claiming that the former carried pneuma alone, whereas the latter transported blood.[13] Moreover, he insisted that the arteries *pulsate* and draw substance inside them through their expansions and contractions, and in so doing transformed these vessels, formerly passive channels, into bellows-like func-

10. Diocles fr. 31 van der Eijk. It is difficult to assess whether Diocles held any broader teleological commitments, especially since his theories are preserved by later authors who can project their own ideas back onto him. For instance, Galen asserts that Diocles, like the other ancients, held opinions about the formation of blood and bile in the liver and the veins, and he refers to these parts as "organs" (Diocles fr. 27 van der Eijk = Gal. *Nat. Fac.* 2.8–9 = 2.110–140K), but we have seen that this vocabulary did not feature in earlier medical accounts. Moreover, Galen also claims that neither Hippocrates nor Diocles nor Erasistratus nor Praxagoras thought that the kidneys were organs [ὄργανα] secreting urine (Diocles fr. 36 van der Eijk = Gal. *Nat. Fac.* 1.13 = 2.30–31K), which perhaps reveals the absence of this notion.

11. Clearchus fr. 111–114 Mayhew and Mirhady; he also wrote a commentary on Plato's *Timaeus*.

12. Praxagoras also wrote *On Diseases*, *On Foreign Diseases*, *On Attendant Symptoms*, *Incidental Diseases*, and *On Cures* (see Steckerl 1958; cf. von Staden 1989: 41 n41). Praxagoras's dates are difficult to pin down, but Gal. *MM* 1.3 = 10.28K (= fr. 6 Lewis = fr. 45 Steckerl) refers to Praxagoras as Herophilus's teacher. This reference, and his seeming acquaintance with Aristotle's biological works, places Praxagoras's floruit around 300 BCE. For these dates and other biographical details, see Lewis 2017: 1–6.

13. Lewis 2017. Since the work of Steckerl (1958), it had long been dogma that this division was based solely on theoretical needs rather than anatomical observation (see Longrigg 1988: 467; von Staden 1989: 173). Lewis argued that Praxagoras's claims that the arteries turn into, or are, "tendon-like," at their termini, must be rooted in anatomical investigations.

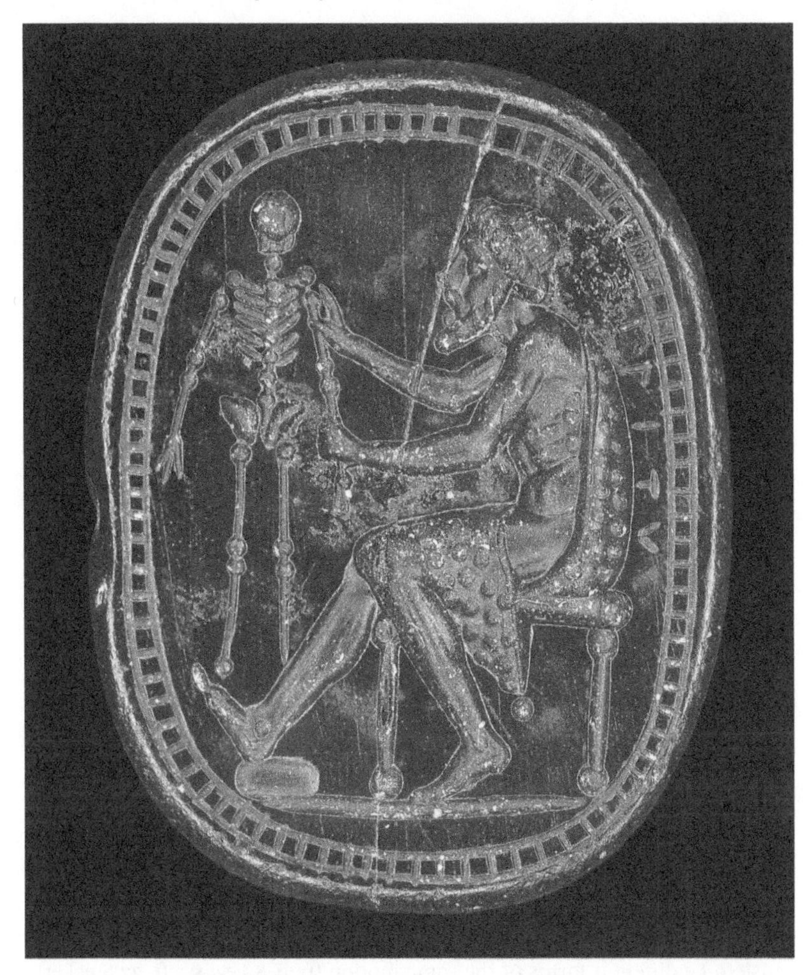

FIGURE 14. Engraved scarab of Prometheus fashioning a human, ca. 3rd–2nd c. BCE, sard, 1.3 × 1 cm, British Museum, London, museum no. 1814,0704.1312. The iconography is reminiscent of a doctor treating a patient, as seen in figs. 2 and 3.

tional parts.[14] Even amulets of the period seem to reflect a growing cultural knowledge of skeletal anatomy in contrast to earlier depictions of the same scenes (fig. 14). Galen nevertheless complains that early anatom-

14. Whereas Aristotle had considered the pulse an accidental by-product of blood concoction in the heart, Praxagoras asserted that it was functional, responsible for drawing air into the arteries, which contained only pneuma (fr. 6–10 Lewis = fr. 27–28 Steckerl). Moreover, he considered pulsation to be an action both of the heart and of the arteries, although the latter did so independently

ical authors such as Praxagoras and his pupil Philotimus (ca. 330–ca. 270 BCE) practiced anatomy "rather generally and inaccurately,"[15] and that all such early investigators "have not made clear the usefulness of the things written," although here he seems to be referring to the practical import of anatomical knowledge rather than the purpose of individual parts.[16] In sum, by the early third century BCE, physiology and anatomy were emerging as prominent parts of medicine, and functional teleology rose in prominence as an explanatory mode. Two medical practitioners helped solidify all these intellectual trends as key components of medical theorizing: Herophilus of Chalcedon and Erasistratus of Ceos. Both had Peripatetic connections and privileged dissection as a way of knowing the body.

4.2 HEROPHILUS OF CHALCEDON AND DISSECTION PRACTICES

Herophilus was born in Chalcedon and immigrated to Alexandria sometime after 315 BCE, where he conducted medical investigations and, potentially, therapeutic practice.[17] He wrote a treatise called *On Anatomy*, but his manifold discoveries and multiple discussions make his investigations seem spread wider than a single text.[18] As mentioned above, Praxagoras of Cos had made a distinction between the veins and the arteries that included a division of labor, whereby the veins delivered blood and the arteries delivered pneuma.[19] Herophilus further codified this distinction, noting that the membranes of the arteries are six times as thick as

(see fr. 9–11 Lewis = fr. 28 Steckerl), even as he considers spasm, palpitation, and pulse to be different forms of the same motion (fr. 6–8 Lewis = fr. 27 Steckerl). This account leaves open the possibility that he, too, exploited the heuristic possibilities of the bellows as a mechanism to move air, which Herophilus then used to explain how the arteries transfer both pneuma and blood, not pneuma alone (T146, 262–267 von Staden; cf. Anon. Lond. 28.47–49); see below.

15. Gal. *Ut. Diss.* 9.4–5 = 2.900–901K = Diocles, fr. 18 van der Eijk.

16. Gal. *AA* 2.1 Singer = 2.282K.

17. Some scholars have him studying in Athens first, which would push back his arrival in Alexandria by a few years. For a full account of Herophilus's life and extant evidence for chronology, see von Staden 1989: 35–66.

18. T17–19 von Staden; cf. fr. 60–62 von Staden.

19. Steckerl 1958: fr. 7, 8, 9, 11, 13b, 26–31 von Staden. Lewis (2017) has provided a crucial re-evaluation of these ossified claims about Praxagoras's doctrines and insists that his was an anatomical as well as physiological distinction; see note 13 on p. 164. Praxagoras also proposed that the body contained blood plus ten additional humors (see Gal. *Nat. Fac.* 2.9 = 2.142K).

those of the veins.[20] He also identified an entirely new set of vessels in the body, which he referred to as *neura* because of their resemblance to tendons and ligaments. He tracked these "nerves" around the head and spine and differentiated them into two types: motor [προαιρετικά] and sensory [αἰσθητικά].[21] Moreover, he distinguished the ventricles of the brain[22] and identified both the rete mirabile and choroid plexuses.[23] He discerned four different membranes of the eye and was responsible for naming both the cornea, or "horn-like," membrane [κερατοειδής] and the retina, or "net-like" membrane [ἀμφιβληστροειδής].[24] In addition, Herophilus offered the first accurate description of the liver[25] and named the duodenum, or the "twelve-finger" [δωδεκαδάκτυλος].[26]

Diocles had previously identified the "cotyledons" and "horns" of the uterus through a dissection of a female mule.[27] Herophilus paid even greater attention to female reproductive anatomy, identifying the epididymis, the ovaries, and at least part of the fallopian tubes.[28] In his ex-

20. T116, 118 von Staden. It is the anatomical distinction that leads to a difference in behavior: the thin veins *collapse* when emptied of blood, whereas the thick arteries do not; cf. Gal. *Diff. Puls.* 4.10 = 8.747K.

21. T80–85, 140a, 141, 143a–b von Staden. For a full discussion of the discovery of the nerves and how Herophilus thought them to function, see Solmsen 1961: 184–97; cf. von Staden 1989: 247–59. There is, however, debate as to whether the nerves all carried pneuma, or only certain sets of them; cf. von Staden 1989: 241–59. Harris (1973: 231–32) questions the degree to which Herophilus truly separated out these new "nerves" from νεῦρα (ligaments), since Rufus reports that Herophilus identified multiple types of neura: motor and sensory nerves, as well as those that bound bone to bone, muscle to muscle, and those that bound the joints (see T81 von Staden).

22. See T76–78 von Staden. Herophilus's distinction of the cerebrum [ἐγκέφαλος] and cerebellum [παρεγκεφαλίς], as well as the calamus scriptorius contrasts with Hipp. *Morb. Sacr.* 6, which describes the brain as "double," with a membrane separating the sections.

23. T121 von Staden and T124 von Staden. The rete mirabile is a network of veins and arteries located in the brain of many vertebrates. Both Herophilus and Galen will assume it to be a human anatomical feature as well, but no such structure exists in human brains.

24. T86, 88–89, 104a von Staden. Herophilus is reported to have written a treatise called *On Eyes* (see T20 von Staden); cf. von Staden 1989: 72–74, 159–60).

25. Fr. 6ªa–b von Staden; von Staden 1989: 161–65.

26. T96–100 von Staden.

27. Diocles fr. 22–24c van der Eijk. Humans do not have these sucker-like "cotyledons" in the womb.

28. For epididymis, see T101–105 von Staden; for ovaries and fallopian tubes, see fr. 61, T110–112 von Staden; see also von Staden 1989: 165–69.

amination of these reproductive organs, he also rejected the possibility of a bicameral womb in humans, dismissing a common notion held by his predecessors that sex difference arose from a hot and cold uterine chamber for boys and girls, respectively.[29] In addition, he demonstrated the anatomical impossibility of the wandering womb, identifying the ligaments that kept the uterus securely in place.[30] For Plato, the womb was like a separate animal living inside the female body, which made it one of the few internal parts not treated as a tool of one sort or another (the penis is another such part).[31] For Aristotle, the female animal presented a disruption to his teleology and a challenge for his vision of nature as doing nothing without purpose, since, for him, female bodies were like imperfect manifestations of male specimens. Some intrinsic thermal deficiency damaged women's capacity to generate sperm, which meant that menstrual blood was the result of material necessity rather than some ideal corporeal goal. Nevertheless, nature had used this material "for the better" to generate new creatures, and in this way, Aristotle fit women into his teleology at the level of the species, but not the individual. Emanuela Bianchi calls this disruption the "feminine symptom," which she expands to a broader destabilizing trend within Aristotle's accounts.[32] Despite Aristotle's teleological demotion of women and other female animals, the uterus surfaced as a key organ of interest for Herophilus and other anatomists interested in functional teleology. The uterus's nonvital function to house and develop embryos seemed clear enough that it fit easily within this methodology. Anatomical investigations thus reinserted female parts into the structural teleology that Aristotle's theories had kept at a slight distance.

Much has been made of the fact that Herophilus and Erasistratus not only engaged in extensive animal dissections but also started to cut apart and systematically examine human bodies, perhaps for the first time. If Celsus is to be believed, they even practiced vivisection on condemned criminals, opening them alive to examine their innards.[33] Most scholars argue that religious and cultural prohibition against cutting human

29. For a full catalogue of Herophilus's discoveries, see Lloyd 1983: 157–58; Longrigg 1988: 462–71; von Staden 1996: ch. 6.

30. T101–114 von Staden.

31. Pl. *Ti.* 91b4–d5.

32. Bianchi 2017. Mayhew (2004) finds women less disruptive to Aristotle's biological accounts; see also Connell 2016 for more recent work.

33. Celsus *Med.* 1 *pref.* 23–26 = T63a von Staden; cf. T63b–c von Staden.

flesh had previously made human dissection functionally illicit, but that these attitudes (briefly) changed in Alexandria during the third century. Many suggest that the Egyptian practice of mummification might have influenced this shift, as well as the creative plurality of this new cultural and religious geography, which might have loosened the hold of previous norms.[34] We should also consider the ethnic identity of those subject to capital punishment and used as anatomical subjects, such that the legal and racial status of Egyptian criminals could have allowed Greek physicians to treat them as non- or at least not-fully-human bodies rather than people.[35] Regardless of what sanctioned these new dissection practices, however, commentators frequently imply that it was social and religious prohibitions that prevented Hippocratic authors from recognizing intricate corporeal structures and suggest that it was dissecting humans rather than animals that produced advancements in anatomical knowledge. Yet the structures that Herophilus and Erasistratus describe (nerves, ocular membranes, duodenum, etc.) are just as apparent in animals as they are in humans and are generally magnified in larger creatures.[36] Therefore, it seems more likely that possessing this new object of inquiry—the "organism"—must have been one of the main driving forces of this new mode of investigation, insofar as it made structures far more crucial

34. Various suggestions have been made to explain this development, including Plato's and Aristotle's secularization of the corpse and the influence of Egyptian mummification practices; cf. Sarton 1959: 129–30; Lloyd 1973: 77; Longrigg 1988: 45; von Staden 1996: 85–86, 139–53; Nutton 2004: 128–39.

35. Several details point to this likelihood. First, Greeks tortured the enslaved to verify testimony, which reveals that their corporeal integrity was not protected by cultural or legal norms; see duBois 1991, although Gagarin (1996) thinks that this torture was limited, if not outright metaphorical. Second, Rufus of Ephesus (ca. 70–ca. 110 CE) later uses an enslaved person as anatomical model (see *The Names of the Parts of the Human*). In addition, Diodorus describes the vilification of the *paraschistēs* (those who cut open corpses prior to embalming), explaining that it was because the Egyptians thought that anyone who applied force to someone of the same race [ὁμόφυλος] was polluted (Diod. Sic. 1.91.1–4; see von Staden 1989: 149 n26). No evidence suggests that Egyptians did, in fact, vilify these embalmers, and so Diodorus's explanation more likely reflects his own cultural logic, which may in turn support the claim that Herophilus and Erasistratus dissected native Egyptians, since they are not *homophylos*.

36. Moreover, the Hippocratic text *Anatomy*, discussed above, compares human and animal anatomy yet fails to demonstrate any anatomical acumen. Whether the treatise pre- or postdates Herophilus, it shows that human dissection alone was not sufficient for advancement.

to the medical construction of the body.[37] Vivisection, all but epistemically null within the framework of most Hippocratic medicine, extended the logic of this new corporeal configuration to its horrific extreme: the functional body in action, displayed with its skin peeled back and its living organs visible. This practice created new types of epistemological vulnerability, since creating this type of visibility required wounding the body in extreme ways, perhaps distorting the function of the very organs that vivisection sought to expose. Nevertheless, seeking to understand the living body by revealing its biotic activities rested on conceptualizing the body as an organic apparatus.

In fact, Galen mentions Herophilus as the first medical theorist after Aristotle who wrote competently and extensively about the use [ἡ χρεία] of body parts.[38] Moreover, he alludes to the fact that Herophilus adopted a humoral theory similar to the Hippocratics[39] but shifted his therapeutic frame so that only the tool-like parts [ὀργανικά] indicated beneficial treatments.[40] Moreover, Galen reports that Herophilus (and Erasistratus) "attempt to treat rationally only those diseases peculiar to *organic* parts [ὀργανικῶν μορίων]."[41] This account suggests a shift toward conceptualizing disease as malfunction rather than imbalance and evinces a viewing of the structures of the body as performing certain functions and

37. Von Staden (1997) likewise attributes Alexandrian dissection to the legacy of Aristotle, including his "heuristically and teleologically productive use of dissection and vivisection of various animals, his abandonment of the theory of transmigration and reincarnation of the soul, and his demystifying secularization of the corpse" (von Staden 1997: 185). Von Staden (1996: 86) also tries to link the "miniaturization" of anatomy to Callimachus's aesthetic embrace of diminutive poetic scale. See also Vegetti (1998: 78–81), who also places the inspiration for Hellenistic anatomical investigation on Aristotelian biology and its functionalist approach.

38. Heroph. T136 von Staden = Gal. *UP* 1.8 = 1.15 Helmreich = 3.335K.

39. For Herophilus's humoral commitments, see von Staden 1989: 242–47; cf. T130, 132–134, 205b von Staden.

40. Heroph. T232 von Staden = Gal. *MM* 5.2 = 10.309–310K. In so doing, they do not treat ailments of the uniform parts [τὰ δὲ τῶν ὁμοιομερῶν].

41. Heroph. T233 von Staden = Gal. *MM* 3.3 = 10.184–185K. Even within his broader humoral commitments, Herophilus develops an etiology of disease that deals with organs: cardiac disorders (T214 von Staden); paralysis of the heart that causes death (T212 von Staden); and pneumonia as inflammation of the entire lung (T222 von Staden). Nevertheless, humors can still initiate disease (T205b von Staden). We might also link this shift toward organa as leading to and mutually reinforced by the expanded use of soft-tissue surgery during this time period; see von Staden 1989: 405, 453.

completing discrete corporeal tasks.[42] Indeed, Herophilus adopts the Aristotelian language of "capacities" [δυνάμεις] and "natural activities" [ἐνέργειαι φυσικαί] when describing the functional actions of body parts. As the last chapter illustrated, these Aristotelian notions were deeply wedded to the organic model of the body as essentially tool-like in nature.[43]

Herophilus's naming practices, which drew from the material world of Alexandria, evince the importance of individual tools for articulating this new corporeality. As Heinrich von Staden states, "[Herophilus] created a new nomenclature for the body—a detailed new language of the body that extensively deploys vivid metaphors drawn from Alexandrian artifactual culture."[44] Examples include naming the "styloid process" at the base of the temporal bone after a type of "reed pen" [κάλαμος] made in Egypt, and likening this same structure to the famous Pharos lighthouse in Alexandria as well.[45] Other examples are less culturally specific. Beyond the abovementioned "net-like" retina,[46] he called the tibia the "weaver's shuttle" [κερκίς] instead of the more common *knēmē* [κνήμη],[47] and he named the confluence of the sinuses at the back of the skull the "wine vat" [ληνός].[48] He labeled the passage through which nerves exit the bottom of the skull the "funnel" [χώνη] because of its shape and supposed function [ἀπὸ τῆς χρείας].[49] Herophilus's student, Andreas of Carystus (ca. 250–217 BCE), subsequently employed the vocabulary of mechanics when discuss-

42. Herophilus divides medicine into three parts, defining it as "knowledge of things pertaining to health, those pertaining to disease and those pertaining to neither of these" (T42 von Staden; cf. T43–48 von Staden). This approach contrasts quite dramatically with that of the Hippocratics, who primarily focus on pathological states, as well as with both Celsus's later division of medicine into its therapeutic categories—dietetics, pharmacology, and surgery (Celsus *Med.* 1 *pref.* 9). Herophilus's division was no doubt constructed to accommodate his reframing of medicine around physiology and anatomy (cf. von Staden 1989: 100–101). That said, the description of health attached to this definition by the pseudo-Galenic author of *Introductio sive Medicus* focuses on harmonious arrangement and fit, not function (T42 von Staden).

43. See, for example, the discussion of respiration, Heroph. fr. 143a–162 von Staden.

44. Von Staden 1996: 86.

45. T79, 90–92 von Staden.

46. T89 von Staden. He also likened the rete mirable to a net [τὸ δικτυοειδὲς πλέγμα] (T121 von Staden). Von Staden (1989: 179) claims the name originated with Herophilus.

47. T129 von Staden.

48. T122–123 von Staden.

49. T76 von Staden.

ing the body, while also employing the vocabulary of the body to name the parts of his fracture-reducing machine.[50] These moments illustrate, as von Staden argues, the multi-vector exchange of concepts, technologies, theories, language, and materials between medical and mechanical practitioners in the Hellenistic period.

4.3 HEROPHILUS'S BELLOWS

In addition to simple nomenclature, Herophilus also leveraged the conceptual power of tools to explain how the organs function. His account of respiration provides a useful example. Just like Aristotle, Herophilus seemingly models the action of the lungs on the bellows, and he, too, seems to grant the lung an explanatorily basic natural capacity to expand and contract. Unlike Aristotle, however, Herophilus suggests that the process of respiration involves four movements and ultimately serves to replenish psychic pneuma, not to regulate heat.[51] Instead, the lung alone "naturally" [φυσικῶς] dilates and contracts, and by expanding,[52] it draws in external air [ἐφέλκεται]. Second, the thorax channels pneuma into itself, insofar as it, too, dilates, pulling air from the lung. Third, when the thorax is full and unable to receive any more air, it contracts and the excess air "flows back" [ἀντιμεταρρεῖ] into the lung, which then contracts in turn and expels the air. Here is the account in full:

> Sometimes dilation, and other times contraction of the lung occurs, since repletion and evacuation occur by mutual exchanges with one another [ταῖς ἀλλήλων ἀντιμεταλήψεσι πληρώσεώς τε καὶ κενώσεως γινομένης], so that four motions occur in the lung: the first when it receives the air from outside; the second when that which it has received externally flows in turn to the thorax inside; the third when that which is contracted from the thorax is again received back into it; and the

50. Von Staden 1998; cf. von Staden 1989: 453.

51. Herophilus's arteries contract and then expand, drawing air into them; see fr. 143a–162 von Staden. The pneuma that enters might be used to fill the neura, since a fetus needs to breathe to gain "pneumatic motion" (T202a–b von Staden). For discussions of those who connect respiration and the vascular systems, see Wellman 1901: 71, 82–85, 100; Rüsche 1930: 115–26, 208–39; Furley and Wilkie 1984: 3–39; and von Staden 1989: 239.

52. Respiration, like other nonvoluntary or quasi-voluntary motions of the body, did not fall neatly into the distinction suggested by motor and sensory nerves, which is why Herophilus applies this terminology of "natural" expansion and contraction (see T143b von Staden).

> fourth when it expels to the outside that which has arrived in it from the reverse movement. Two of these motions are dilations (i.e., the one from outside and the one from the thorax), and two are contractions (i.e., the motion when the thorax draws the pneumatic substance into itself, and when the lung itself separates [this substance] into the external air). For two [motions] alone concern the thorax: dilation, whenever it draws air into itself from the lung, and contraction, whenever it returns it back again to the lung.[53]

Just as the clepsydra lurked behind Plato's account, the bellows implicitly support Herophilus's physiology, despite not being mentioned. He here makes both the lung and thorax active organs, performing certain corporeal tasks by means of "mutual exchanges" [ἀντιμεταλήψεις]. The air is moved by a series of mutual expansions and contractions from the external air to the lung to the thorax, such that unidirectional flow is maintained for each stage in the process. It is as though once Aristotle incorporated the bellows into his account of respiration, medical theorists began seeing this action operate in more places within the body. Yet "seeing" is here the operative word, since the doubled set of motions that Herophilus describes is not straightforwardly observable. At least to our eyes, breathing requires a single inhale and single exhale, with a single corresponding expansion and contract of the chest. No dual-motion sequential transfer can be easily observed. Accordingly, as Herophilus employs an implicit technological heuristic to explain the basic behaviors of the body, he partially constructs those same behaviors. In fact, Herophilus claims that certain faculties and motions in the body can only be known "by reason."[54] In this case, however, the motions are not the inferred activities of minute parts but the movements of the lungs and thorax—motions that should

53. Heroph. T143b von Staden = Aët. 4.22.3. Cf. T143c von Staden; cf. also von Staden (1989: 259–62), whose interpretation of this passage largely aligns with my own. If the vocabulary above does indeed reflect Herophilus's own terms (although this is far from certain), the older conceptual division between two different types of physiological systems remains, insofar as pneuma is drawn into the body via some type of vacuum pressure but is "channeled" [μετοχετεύει] to the thorax when it needs to be distributed.

54. Herophilus himself accepts the idea that there are "primary" things only perceptible by reason (see fr. 50a–b von Staden). He also holds that the faculties governing the arteries cannot be discovered from simple dissection and observation (T57 von Staden) and that some arterial motions are perceptible [αἰσθητή], others only known by reason (T142a–c von Staden). For a discussion of this epistemology, see von Staden 1989: 116–18.

be readily observable even without vivisection. He answers the question "why and how do we breathe?" by imagining a set of behaviors he calls "breathing" that are themselves derived from his implicit technological heuristic.

More than adopting and adapting Aristotle's bellows model for the lungs, Herophilus employed it to explain the movement of blood. Like Praxagoras before him, Herophilus divided the blood vessels into two anatomical types, arteries and veins, but argued that the arteries held a mixture of blood and pneuma, not pneuma alone. Yet he, like Praxagoras, believed that the arteries draw blood into them by expanding and contracting and may have coined the terms "dilation" [διαστολή] and "contraction" [συστολή] to describe these activities.[55] These movements rely on some innate faculty, which the arteries obtain through their connection to the heart.[56] Regardless of how they obtain this faculty, the bellows seem to underpin Herophilus's account of the arteries' actions, since if they did not, it would make little sense that the pneuma/blood mixture moves unidirectionally. Without internal valves, the arteries would push and pull the blood and pneuma back and forth. It would never reach the areas of the body in need of nourishment. In other words, Herophilus reads the functional operation of the bellows, which drive the pneuma/blood mixture forward into the arteries, even without articulating all the requisite anatomy required for such a heuristic to function (e.g., is there a valve-correlate in the arteries to prevent backflow?). In the process, following Praxagoras, he makes these vessels—formerly passive passageways—into operative organs. That he does so shows a growing interest in explicating the mechanisms of fluid and air transfer within the body.

Fluids in the Hippocratic texts move mostly of their own accord, or, if authors do explicate their motions, a few common assumptions prevail. Heat and hollow spaces both attract, while gravity flow generally moves liquids through vessels, even if fluids move in unexplained ways at other times, whether retreating into the body or flowing upward. Even Aristotle resorted to gravity flow when describing the transfer of blood as nourishment for the body. Praxagoras and Herophilus, by contrast, present a new view of blood and air distribution based on the active movements of the arteries. The ducts and channels become important structures, not sim-

55. T144, 145a, 146, 149, 154, 155, 157–160 von Staden. Von Staden (1989: 273) suggests that Herophilus named these actions.

56. T144, 145a, 155 von Staden.

ply for the inner territory they map but for their functional activities.[57] Organic logic was being expanded throughout the body.

We might consider the sudden appearance of bellows in multiple places within the body as akin to the appearance of the clepsydra a century prior. That is, whereas the clepsydra became the totemic technology of air, explicitly or implicitly structuring conceptions of pneumatic transfer, now the bellows had become a dominant tool used to explain such "mutual exchanges." It should be noted that the propulsion of the blood had not been seen as a problem before, so much so that irrigation metaphors and fourth-century pipes allowed the flow of inner liquids to appear natural. The bellows therefore provided a solution that retroactively created its own problem, first making the propulsion of air and then the mechanics of blood distribution a larger physiological question than they previously had been. Moreover, the bellows' appearance in multiple explanations, extending from the lungs to the thorax to the arteries, resembles how the hysplenx appeared as a force mechanism in multiple devices, ranging from starting gates to catapults and automata, in the decades

57. This treatment of the arteries as organa no doubt facilitated their increased diagnostic and physiological significance, which led to new technological interfaces, especially in regard to one of Herophilus's major new developments, often referred to as "pulse lore." This new diagnostic practice proposed an intricate system that attributed pathological significance to the different rhythms of the pulse. Herophilus interprets the pulse with musical terminology apparently derived from Aristoxenus of Tarentum, the primary differentiae of pulse types being rhythm, speed, size, and vehemence or strength. In fact, different ages of life should have different pulse types based on different poetic meters. Healthy youths should display a pyrrhic pulse rhythm (- -), adolescents should have a trochaic pulse (— -), grown men a spondaic pulse (— —), and old men an iambic (- —). Patients could display *heterorythmia*, when the pulse did not match their age, or *ekrhythmia*, if no characterizable pulse could be discerned. Herophilus also used the pulse rate as a way to measure fever (T178, 182 von Staden). For an overview of pulse lore, see von Staden (1989: 262–880; T159–188b von Staden. Berrey (2017a: 191–209) presents an interpretation keeping the primary diagnostic categories at five, suggesting that "size" or "mass" refers to the number of beats that occur during the time it takes for a given amount of water to flow out of a portable water clock (also incidentally referred to as the clepsydra). Following suggestions by von Staden (1989: 10), Berrey argues that this technology is an essentially Egyptian one and represents a type of cultural hybridity that forms part of emerging Alexandrian "court aesthetics." We might thus note that the new importance granted to the heart within the organa model of body parts facilitated its rising import as a diagnostic tool and opened up new technological interfaces.

after its original appearance. Technological developments often work in this way, with an innovation being deployed in multiple use cases before its potential is more fully realized and the most useful applications are found. Technological analogies can display similar dynamics, whereby a tool becomes a totem for a certain type of behavior, and the two move together to new applications until the most successful stabilize and they form a type of trope. As the case of the bellows reveals, the tools themselves might not even be new, since these devices certainly predated Aristotle's consolidation of breathing around their correspondence with the lungs. Nevertheless, the application of tool analogies to biotic circumstances naturalizes their appearance and makes it more likely for them to appear in other physiological accounts. This naturalization can further alter assumptions about the behaviors of the body. In fact, the bellows turn up again in the explanations of Herophilus's younger contemporary, Erasistratus, where they participate in another set of explanations, this time about the propulsive power of the heart.

4.4 ERASISTRATUS OF CEOS AND PNEUMATIC PATHOLOGIES

Erasistratus of Ceos arrived in Alexandria in the first half of the third century BCE. Although ancient sources frequently refer to him as Herophilus's "student," the small age gap between them suggests that Erasistratus was not quite a pupil.[58] Both the doxographical tradition and Erasistratus's own followers associated him with Aristotle and the Peripatetics,[59]

58. Gal. *PHP* 7.3.67 De Lacy = 5.602K claims that Erasistratus gave up medical practice "in his old age" so as to conduct dissections in peace (see note 73 on p. 180). This claim might bring him even closer in age to Herophilus. Nevertheless, biographical details are sparse, such that Fraser (1969) could argue that Erasistratus did not in fact practice in Alexandria at all but conducted his medical research in Antioch; both Harris (1973: 177–233) and Lloyd (1975c) have rebutted this claim; cf. Longrigg 1988: 472–73. Nevertheless, it is unclear whether Herophilus and Erasistratus researched at the Mouseion under the financial support of the Ptolemies; cf. von Staden 1989: 26–31; 1992a; Vegetti 1995.

59. Plin. *HN* 29.3 claims Erasistratus was Aristotle's grandson through Pythia, Aristotle's daughter. Although this claim is likely untrue, it certainly reflects the assumption of a Peripatetic connection. Similarly, Gal. *Art. Sang.* 7.2 Furley = 4.729K refers to Erasistratus as "associating" [συνεγένετο] with Theophrastus, and Galen mentions that Erasistratus's followers claim that Erasistratus consorted with Peripatetic philosophers; cf. *Nat. Fac.* 2.4 = 2.88K; 2.8 = 2.116K. Other sources make him a pupil of Chrysippus of Cnidos; see Diog. Laert. 7.7.186; Plin. *HN* 29.3; Gal. *Ven. Sect. Er.* 7 = 11.171K.

and there is much evidence of a philosophical indebtedness. To begin with, Erasistratus wrote a treatise called *General Principles* [οἱ καθ' ὅλου λόγοι] that began by outlining all of the "natural activities" [περὶ πασῶν τῶν φυσικῶν ἐνεργειῶν] and explained both "how they occur and through which parts of the animal."[60] This framework suggests that Erasistratus made systematically articulating the function and physiology of the body's parts the bedrock of his explanatory apparatus. He even adopted the Aristotelian language of "natural activities," which, as the last chapter emphasized, was tied directly to tool-like corporeal functions.

In addition to a functional teleology of the body that was built upon understanding physiological mechanisms, Erasistratus also construed nature as a technician [ἡ τεχνικὴ φύσις] who "did everything for some purpose and did nothing in vain."[61] It seems, however, that Erasistratus placed even greater limits on nature's commitment to these principles than Aristotle had, noting several corporeal features that displayed no obvious function, including bile, the spleen, and the omentum.[62] Galen will later argue that if someone truly believes nature to be a technician, they should accept that everything it creates has a use and is perfectly designed for that function (otherwise nature is not a particularly good technician). If we frame our investigation entirely around teleology, Erasistratus thus represents a step away from seeing natural objects as purpose oriented. Yet if we orient our discourses around the epistemic object created by thinking about corporeal parts as tool-like, Erasistratus's position looks like far less of a departure. That is, even thinking about body parts in terms of their use or uselessness requires adopting function as a heuristic through which to understand the body and the relationship between its constituent parts. Even without committing to a perfect structural teleology, then, Erasistratus articulated his own version of the organism as the epistemic object of medicine.

60. Gal. *Nat. Fac.* 2.8 = 2.112K; cf. 1.1 = 2.63K, where Galen only provides a single example, noting that Erasistratus stated that the separation of urine [from the blood] occurred in the kidneys, although Galen complains that in this instance Erasistratus did not explain how.

61. Gal. *Nat. Fac.* 2.4 = 2.91K; cf. 2.2 = 2.78K. For Erasistratus's references to ἡ τεχνικὴ φύσις, see Gal. *Nat. Fac.* 2.4 = 2.88K; cf. 2.8 = 2.116K; 2.9 = 2.131K.

62. Gal. *Nat. Fac.* 2.2 = 2.78K; 2.4 = 2.91–92K; 2.9 = 2.131–132K. It is not entirely clear whether Erasistratus made all these assertions, or whether Galen is extrapolating them from Erasistratus's basic assertion that bile was useless and thus, by extension, that both the spleen and the arteries inserted into the kidneys must also be, since, according to Galen, the former functions to remove black bile and the latter yellow bile from the blood.

Despite his adoption of a conceptual framework so clearly indebted to Aristotle's, Erasistratus differs from his predecessor's physiological theories in substantial ways, most notably in his rejection of the notion of innate heat, which was so essential for Aristotle's account of life. Moreover, Erasistratus disengaged with humoral theories, and Galen chastises him for this, complaining that Erasistratus dismissed Hippocratic ideas without discussing them at all. In fact, Erasistratus seemingly suggested that speaking about the generation of humors was simply not useful.[63] His physiological model was quite different and relied on a version of the *horror vacui* principle, which he calls "the filling toward what is being emptied" [ἡ πρὸς τὸ κενούμενον ἀκολουθία].[64] This formulation resembles that of the Peripatetic Strato of Lampsacus (335–c. 269 BCE), successor to Theophrastus at the Lyceum, who also theorized about the void.[65] Although Strato's own writings are not preserved, the first-century CE mechanical author Heron preserves Strato's theory of a noncontinuous void in the preface to his *Pneumatica*, which states that Strato held that a substance "fills up a place being emptied according to an amount correlate to the other substance" [κατὰ τοσοῦτον ἕτερον ἐπακολουθοῦν τὸν κενούμενον ἀναπληροῖ].[66] On the one hand, this account places Erasistratus squarely within a medico-philosophical tradition of arguments about empty space. On the other hand, as the discourses surrounding the clepsydra demon-

63. Gal. *Nat. Fac.* 2.8 = 2.110–113K. The rejection of innate heat relates to his assertion that digestion cannot be a process akin to cooking, since the stomach possesses barely any heat (*Nat. Fac.* 3.7 = 2.166K). Instead, he argues that digestion occurs because the stomach contracts and crushes [περιστέλλεσθαι καὶ τρίβειν] the incoming food (*Nat. Fac.* 2.8 = 2.119K). It is worth noting that he has located digestion as an activity of the stomach as an organ, rather than the more distributed approach to digestion as appears even in a Hippocratic treatise such as *Diseases* 4 or in Aristotle himself (see chapters 1 and 3).

64. See this formulation at Gal. *Nat. Fac.* 1.16 = 2.63K and passim.

65. Since as early as Diels (1893), scholars have noted the connection of Erasistratus's ideas to Strato's conception of the interstitial void; cf. Harris 1973: 200–33; Longrigg 1988: 474; von Staden 1996; although Berryman (1997 and 2009: 197–200) has refined some of these claims. For a discussion of Strato and his pneumatic theories, see Wehrli 1850; Gottschalk 1964; Fraser 1972, vol. 1: 427–28; Lehoux 1999.

66. Heron *Pneum.* 1 *pref.* 339–340; cf. Simpl. *In Phys.* 693, who suggests that Strato denied the possibility of a continuous or contiguous vacuum in nature but held that smaller vacua exist in little microvoids within all bodies. Strato argues that if there were no such microvoids, then when sunrays fall upon a glass filled to its maximum capacity, the water would overflow; cf. Heron *Pneum.* 6, who compares the idea to grains of sand on the beach. Galen indicates that Erasistratus accepts the notion of interstitial voids, while placing the mechanism of *akolouthia* at the level of perceptible voids (*Nat. Fac.* 2.6 = 2.99K).

strate, abstract principles are often embodied in material technologies, which implicitly steer the application of these principles. Hellenistic Alexandria saw a considerable development of such technologies, most notably in water distribution and pneumatics.

As part of the flourishing intellectual culture of the Hellenistic era, craftsmen turned their attention to constructing devices designed to promote wonder and amazement, notably in the emerging technical fields of mechanics and pneumatics. Ctesibius of Alexandria (290–250 BCE/285–222 BCE) and Philo of Byzantium (ca. 240–200 BCE) both developed and wrote about such devices, including pneumatic automata, birds that drank water, steam-driven temple doors, a musical horn of plenty, a water organ,[67] and a water clock with automated moving parts.[68] These machines sometimes incorporated valves to ensure unidirectional flow of water, employed smaller siphon mechanisms to drink and dispense liquid, and harnessed pneumatic pressure to make music. They utilized watertight vessels and pipes and would not only have functioned while under pressure but also by virtue of it. Erasistratus's debt to and interaction with pneumatics has attracted considerable scholarly attention, with the debate largely focusing on the effect of a single invention, Ctesibius's force pump, and the effect that this device had on Erasistratus's model of the heart and its valves. Yet pneumatics had a broader impact on his views of the body as a closed, pressurized system, as well as on how illness arises in a body so conceived.

Ctesibius was an Alexandrian inventor whom Vitruvius credits with discovering the principles of pneumatics.[69] Although he wrote on mechanics, none of his writings survive. Instead, later authors mention devices that he made, including a pneumatic drinking horn dedicated to Ptolemy II's wife Arsinoë,[70] spring- and air-pressure catapults, a seesaw tube to scale walls,[71] water clocks, the water organ, and the aforementioned force pump.[72] Galen mentions that Erasistratus gave up practicing medicine to conduct his dissections "in his old age," and if this can be approximated

67. See Vitr. *De arch.* 10.8 for a description.

68. Most of these can be found in Philo *Pneum.*, Heron *Pneum.*, and Vitr. *De arch.* 10.7.4–5. For an investigation of the hydraulic inventions of this period, see Lewis 2000b: 343–69; for a general survey of technology in this period, see Sarton 1959: 117–28, 343–78; Wilson 2008: 337–66.

69. Vitr. *De arch.* 9.8.2; 10.7.4.

70. See the epigram of Hedylus preserved at Ath. *Deipnosophistae* 11.497d–e.

71. Athen. Mech. 29.

72. Vitr. *De arch.* 9.8.2; 10.7.4; cf. Plin. *HN* 7.125, who also states that Ctesibius invented the force pump, as well as other "hydraulic implements" [*hydraulica organa*], by which he likely means the water organ. See also Tybjerg 2008.

at fifty years of age at the earliest (i.e., around 264 BCE), Ctesibius would have been about thirty at the time.[73] This possibility certainly presents enough of a window for influence, especially if we consider the dynamics of intellectual life within Alexandria.

Ctesibius left no instructions on how to build his pump, nor do any of the earliest examples survive. Nevertheless, in the first century BCE, Vitruvius—enamored as he was with Alexandrian inventions—took great pains to describe the "Ctesibian machine" and its parts. He describes a force pump with two cylinders and two well-fitted pistons that run up and down inside them.[74] Archaeological evidence suggests that later models joined the operation of both pistons by a single pivoting handle, but Vitruvius's seem to be driven up and down separately (fig. 15).[75] When the piston is drawn upward, water is drawn in through the intake valve located on or near the pipe's bottom, while the outflow valve is pulled closed by negative pressure. When the piston is forced downward, water presses the intake valve shut and forces the outflow valve open, through which the liquid exits into a shared intermediate reservoir before shooting upward out of the pump's nozzle.[76]

Although it was Iain Lonie who first suggested that this force pump might have influenced Erasistratus's conception of the heart, Heinrich von Staden has provided the fullest exposition of this argument.[77] Heroph-

73. Gal. *PHP* 7.3.67 De Lacy = 5.602K. The Suidae Lexicon (*s.v.* Erasistratos) mentions that Erasistratus was buried by Mount Mycale in Ionia, so he would have left Alexandria at the end of his life, which perhaps shortens the window.

74. Vitr. *De arch.* 10.7.1–3. Each piston was probably oiled to ease motion and to help with a seal.

75. For a full investigation of the force pump, including both literary descriptions and archaeological evidence, see Oleson 1984: 301–35; cf. Stein 2004 and Wilson 2008: 353–54.

76. Oleson (1984: 306–7) argues the water was more likely pushed in by ambient pressure; however, there would have to be enough internal pressure to pull closed the outflow valve as well. Oleson also argues that the original valves were likely made of leather flaps modeled on those of the air bellows, rather than spindle valves, which later became popular; cf. Heron *Pneum.* 10 for a description of a flap valve; cf. Heron *Pneum.* 1.28.

77. Lonie 1973a and 1973b; von Staden 1996; cf. von Staden 1997, 1998. Both Majno (1975: 332) and Vallance (1990: 70) accept the suggestion, but Longrigg (1993: 208–9) and Berryman (2009: 200) are more skeptical; Russo (2004: 147) suggests, by contrast, that the discoveries of the valves in the heart may have led to the invention of the pump, but this scenario seems unlikely given that the heart was not previously seen to be a mechanism of propulsion, and the bellows seem to be a more likely precursor to the force pump.

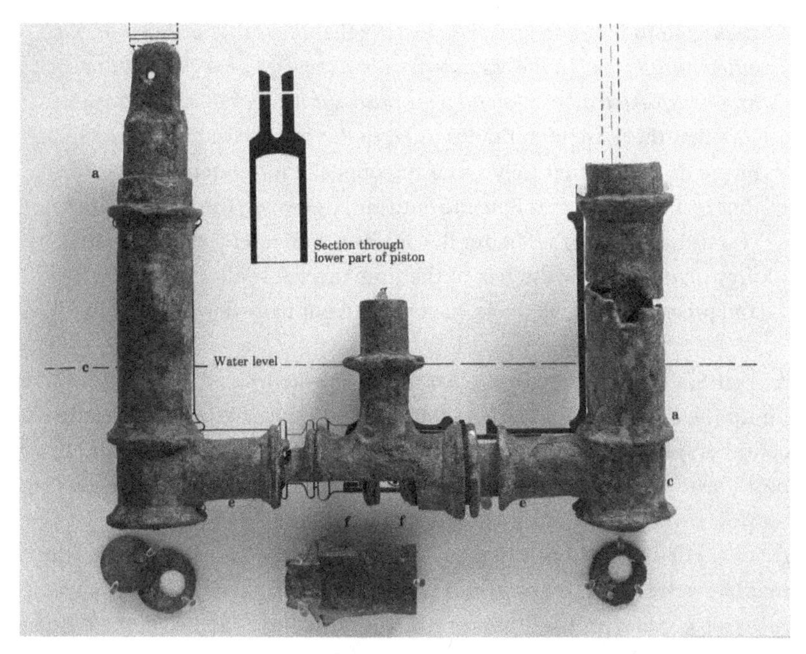

FIGURE 15. Bronze double-action water pump and parts, Roman, 3rd c. CE, 25.4 × 33 cm, found in Bolsena, Italy, British Museum, London, museum no. 1892,0517.1.

ilus seems to have discovered the heart valves first, describing them (as all subsequent authors do) as membranes.[78] Yet Erasistratus argued that they served a particular function: to ensure unidirectional movement by preventing backflow. Unlike any of his predecessors, Erasistratus also proposed for the first time that the heart serves as a mechanism of propulsion. Galen preserves his account in detail:

> At the mouth of the vena cava, there are three membranes, similar in shape to the barbs of arrows—for which reason I think some of the Erasistrateans have called them the "tricuspids." . . . As Erasistratus says when he is describing the phenomenon, one of the mouths guides blood to the lungs, and the other guides pneuma to the whole animal. The use of these membranes in the heart, as it seems to him, is toward opposite purposes, alternating at different times. The ones attached to the vessels leading matter from the outside and into the heart are turned around by the entrance of matter, and by falling inward and opening toward the cavity of the heart, they provide an unhindered

78. T119 von Staden = Gal. *PHP* 1.10.3–4 De Lacy = 5.206K.

> passage into the ventricles. For, he says that *the matter does not flow in automatically, as if a lifeless cistern were receiving it, but the heart itself, expanding like the bellows of a forge, draws it in through its dilation.* He says that there are membranes fitted to the vessels leading matter out of the heart and that they suffer the opposite movement. For he says that by turning from the inside outward, they open the mouths of the vessels because of the matter flowing out, whenever the heart distributes matter, but for the rest of the time they close the mouths tightly and prevent any of what has been pushed out from flowing back in.[79]

As von Staden has suggested, Erasistratus's heart and Ctesbius's force pump (as described by Vitruvius) have many similarities: both have two chambers; both are equipped with valves to ensure unidirectional flow; both have four sets of valves, two controlling intake and two regulating outflow from the two chambers; both function on "forked pipes" [*fistulae furcillae*] or vessels; both depend centrally on the principle of an intermediate valved chamber [*medius catinus*]; both are twin-cylinder apparatuses sitting in a round chamber (the thorax is the chamber in the case of the heart); and both utilize increasing and decreasing volume as a driving force.[80]

At the same time, Erasistratus's heart also differs from a force pump in a number of ways. Erasistratus's heart works by the expansion and contraction of its ventricles, which draw in and expel their respective substances in alternation. By contrast, the pump works on pistons within rigid chambers, such that the volume of the interior chamber changes, but not by elastic expansion. Indeed, Galen's report does not invoke an analogy with a pump, but a comparison to the bellows, which is more apt in this regard.[81] What is more, the pump conducts only water, whereas Erasistratus's heart feeds two independent systems, one with pneuma flowing through the left ventricle and into the arteries, and a second with blood flowing through the right ventricle and into the veins.[82] That is,

79. Gal. *PHP* 6.6.6–12 De Lacy = 5.539–541K, emphasis mine, trans. De Lacy.

80. Von Staden 1975; 1995; 1996: 94.

81. Lonie (1973a and 1973b) points out the discrepancy between Erasistratus's two-substance heart and Vitruvius's one-substance pump, adding that whereas the heart has separate outflows for its two substances, the pump delivers water from both pistons into a single collection reservoir. No extant archaeological examples have this reservoir but instead have pistons whose outlets flow into a shared a T-joint; see Oleson 1984: 309–10.

82. For Erasistratus, both air and blood are crucial substances in the body; blood provides nourishment for the body, while pneuma "energizes" such activ-

Erasistratus adopted Praxagoras's assertions that the veins contain solely blood, while the arteries conveyed pneuma alone, even as he rejected his predecessor's notion that the arteries were active organs, operating by dilation and contraction.[83]

These discrepancies between the early force pump and Erasistratus's heart, the tight timeline, and a lack of early archaeological evidence have led scholars such as Matteo Valleriani to reject any potential influence.[84] Yet an Arabic version of Philo of Byzantium's *Pneumatica* details multiple pneumatic devices, including several pumps, and provides further evidence.[85] Two pumps in particular might help smooth some of the discrepancies, even if they do not eliminate doubt. The first describes a dual-chambered force pump with a slightly different arrangement, whereby the two pistons work independently. In fact, they are even placed in separate leather "pots," which could at least theoretically contain two different liquids.[86] Doing so would allow you to mix the two fluids together in the same outflow reservoir. This description of the double-chambered pump, which would be older than Vitruvius's, fits Erasistratus's two independent systems of vessels more closely, albeit not precisely, insofar as

ities as digestion, movement, sensation, action, etc.; cf. Anon. Lond. 22.49–23.2 Manetti; Gal. *Us. Resp.* 5.1 = 4.502K; cf. Harris 1973: 225. Galen likewise claims that Erasistratus held that the primary function of breathing was to fill the arteries; cf. Gal. *Diff. Puls.* 4.2 = 8.714K; 4.16 = 8.760K.

83. Erasistratus, Galen claims, says that air enters the lungs, then the attached artery, then the heart, and then onward through the aorta:

The one school, following Erasistratus, assumes that the arteries in the lungs [pulmonary veins] are empty of blood like the other arteries. They hold that at each diastole of the heart the pneuma is drawn through them out of the lungs [into the left ventricle] and by its passage the pulse is produced in all the arteries throughout the body. They are persuaded that the pulse is not produced in these [arteries] by their own action, as is that of the heart, but by their being filled with the pneuma passing through them. They say, too, that the heart, when it contracts, sends forth the pneuma to the arteries. (*AA* 7.7.4 Singer = 2.597K; cf. *AA* 2.2 Singer = 2.706–707K)

Erasistratus also seems to argue that the pneuma gets pushed out of the arteries and out of the body as they collapse, as the heart is itself expanding and drawing in pneuma from the lungs; cf. Gal. *Us. Resp.* 2.9–10 = 4.481–482K.

84. Valleriani (2023) argues that Ctesibius may have invented a type of force pump, but not the dual-chambered version that Vitruvius recounts, which he sees as reflecting the material realities of the first century BCE, not the particulars of Ctesibius's design. If Ctesibius's pump is not dual chambered, he does not think it could have influenced Erasistratus.

85. Ayasofya, Istanbul, manuscript no. A.S. 3713, 14th c.; see Prager 1974.

86. Philo, *Pneum. appendix* 1.2, 192–194, Carra de Vaux.; cf. Oleson 1984: 307–9.

blood and pneuma never (ideally) mix in Erasistratus's body. Moreover, directly before describing the dual-piston force pump, Philo depicts a single-chambered pump that works by dilation and contraction and is itself even explained by analogy to the bellows.[87] If Philo's treatise is authentic and does reflect devices invented by Ctesibius, two force pumps existed that functioned in ways quite similar to, although not identical to, Erasistratus's heart.[88] It is hard not to conclude that one—if not both—of the pumps invented in his generation played an important, if not crucial role in this reconceptualization of the bellows as a mechanism that not only expelled air but could also be used to conceptualize how the heart propels blood. Erasistratus establishes the heart as an organ of both suction and propulsion in distinction to a "lifeless cistern."[89] That is, Erasis-

87. Philo, *Pneum. appendix* 1.1, 191–193, Carra de Vaux. Drachmann (1948: 4–6) holds that because this entry precedes the force pump, it likely describes another of Ctesibius's designs.

88. Since Philo's text comes via an Arabic translation, it is hard to tell what has been interpolated. Prager (1974: 126) suggests that the Arabic corresponds to the original Greek. To be sure, the traditional smith's bellows have become those of the "goldsmith," and it is clear that some author has added their Arabic name, *zaaqi*. This slight discrepancy, to my mind, suggests a gloss from the Greek rather than a full-scale interpolation. In addition, this passage suggests using leather pipes, which Carra de Vaux (1902: 191) suggests are not seen in the Greek world, but Drachmann (1948: 50–67) notes that the Arabic text of Philo seems to contain many implements that have been reinterpreted to reflect Middle Eastern vessels, and this adaptation might be similar. Valleriani dismisses the almost contemporary testimony in Philo's treatise, relying on Prager's (1974: 234–37) arguments that the latter portion of the Arabic translation (which contains sections absent in the Latin manuscript), the so-called "Extracts of Irun," reflects a later compendium of Hellenistic inventions, not Philo's own. Yet Prager's rejection relies on the supposed incongruity of focus between the chapters preserved in Latin (chs. 1–20) and those only preserved in Arabic (ch. 21 and following), since the former deals with theory while the latter deals with particular devices. Prager relied on differences of style and "quality" for his rejection of authenticity, but as Schomberg (2008) has pointed out, "since Prager had no knowledge of Arabic his work is based on the English translation of a French translation of an Arabic translation of a lost Greek original." She presents a more thorough case for authenticity of the full Arabic manuscript. Even if Philo's dual-chambered pump has pistons that sit in two separate basins, they still transfer water alone, whereas Erasistratus's heart moves both blood and air. Moreover, the single-chamber pump does not seemingly employ valves.

89. Cf. Hipp. *Cord.* 8 = 9.84–86L, which describes the heart's "ears" [τὰ οὔατα] breathing in and collapsing, and also compares the lungs to the bellows. Littman (2008) dates this treatise to ca. 350–250 BCE, perhaps contemporary to the discoveries of Erasistratus; cf. Jouanna (1999: 394), who dates it simply to the

tratus sets up a dichotomy between the static, gravity-fed spring described by Plato and his own pneumatically driven heart. With the force pump, Erasistratus now had a machine that functioned like the bellows but delivered water.[90]

While direct equivalency between tools and organs can provide dramatic instances where individual technologies appear to have been transported into the body, there are other paths of influence beyond direct identification and analogical correspondence. In fact, the pneumatic devices developed in third-century BCE Alexandria provided a far broader physiological framework. This new technological deployment of vacuum pressure created physical heuristics that provided new ways to think about the corporeal interior as composed of closed systems—as well as new assumptions about the behaviors that such systems display. As described above, Erasistratus proposes two segregated vascular systems, wherein the arteries contain pneuma while the veins contain only blood (and to these two systems he adds the nerves, thereby forming the "tri-weave" [τριπλοκία] from which all tissues are made).[91] And, like Plato and Aristotle before him, Erasistratus believes that the flesh uses up the blood distributed through the veins, as well as the pneuma that likewise enters. The veins and arteries thus represent two separate vascular networks, connected only at their endpoints, which he calls "outlets" [ἀναστομώσεις]. By the time they reach this junction, the respective substances in each vessel—blood and air—have both been used up and therefore do not enter the other passageways. Plato had applied a version of the mutual exchange of substances/impossibility of a void [κενὸν οὐδέν ἐστιν] to respiration and had leveraged the same principle to describe how the distribution of nutrition preserved a human's unity over time, but these two goals ended up coming into conflict, insofar as each seemingly utilized the same vascular networks for cutaneous breathing and blood distribution. This fuzziness

Hellenistic era, sometime after the discoveries of Erasistratus. I think the latter is more likely.

90. Gal. *Art. Sang.* 1.2 Furley = 4.704–705K points out that if the arteries were filled with pneuma, they should not spurt blood when pierced—as they appear to—but air. Erasistratus's response is that the air immediately exits the punctured artery, creating a vacuum, which draws in blood instantaneously. Galen suggests that if Erasistratus is correct, even the smallest pinprick of an artery should void the entire system of its pneuma and cause immediate death; cf. Gal. *UP* 6.17 = 3.492K; cf. *Art. Sang.* 2.3 Furley = 4.708–709K. Galen provides extensive criticism of Erasistratus's proposal that the arteries contain only air in two separate works: *Art. Sang.* and *De venes*; cf. Anon. Lond. 26.31–28.45 Manetti for pre-Galenic criticisms.

91. For a discussion of the *triploikia*, see Leith 2015.

extended to Plato's use of vacuum pressure and the movement of substances toward what is kindred to explain how tissues were replaced, since this account did not dwell on any anatomical features but instead modeled the body more loosely on the cosmos. Erasistratus, similarly, insists that any lost blood or pneuma must be replaced by an equivalent amount of new material entering the vessels according to the principle "the filling toward what is being emptied," but, by contrast, Erasistratus separates the anatomical systems that support each.[92] In short, he turns the body into a type of pressurized apparatus.

To illustrate the degree to which pneumatic devices are enmeshed with his corporeal object, we can look to the fact that when speaking about how blood moves through the body, Erasistratus suggests that the stomach contracts to push nutriment into the veins,[93] while also holding that that the veins can likewise contract to push nourishment forward through their "peristalsis and propulsive action."[94] Nevertheless, elsewhere, he insists that when nutriment exits or is used up by the flesh, this process draws an equivalent amount of blood through the sides of the vessels and into the flesh according to the principle of "the filling toward what is being emptied" [ἡ πρὸς τὸ κενούμενον ἀκολουθία].[95] In this part of his account, Erasistratus treats the flesh itself as a type of rigid vessel (much like bronze tubes) rather than as an elastic material, since (were it elastic) flesh could simply collapse in size to accommodate the lost material, rather than drawing blood into it from behind. Elsewhere, he describes the muscles expanding when pneuma enters them, just as he described the arteries expanding to accommodate the blood and pneuma pushed into them by the heart. This is another behavior not readily observable

92. Erasistratus, in his general principles, seems to spend much of his time on nourishment, or *anadosis*, since Galen supplies the greatest number of details on this subject, while in other issues (bile, kidneys), Galen often puts hypothetical arguments into Erasistratus's mouth.

93. Gal. *Nat. Fac.* 2.1 = 2.76K.

94. Gal. *Nat. Fac.* 2.7 = 2.106K.

95. According to Galen, Erasistratus explained such distribution as follows: "In the furthest, simple vessels, insofar as they are thin and narrow, presentation occurs from the neighboring vessels into the empty spaces left by matter that has been removed, and nourishment is attracted and deposited along the vessels" [Τοῖς δ᾽ ἐσχάτοις τε καὶ ἁπλοῖς, λεπτοῖς τε καὶ στενοῖς οὖσιν, ἐκ τῶν παρακειμένων ἀγγείων ἡ πρόσθεσις συμβαίνει εἰς τὰ κενώματα τῶν ἀπενεχθέντων κατὰ πλάγια τῶν ἀγγείων ἑλκομένης τῆς τροφῆς καὶ καταχωριζομένης] (*Nat. Fac.* 2.7 = 2.105K; cf. 2.8 = 2.119K). See also 1.16 = 2.63–64K, where Galen claims that Erasistratus explicitly attributed the distribution of nourishment to akolouthia. Galen suggests that akolouthia is used when the stomach contraction fails to explain how nourishment moves through the body.

in any straightforward way and one that conflicts with the notion that lost material must be replaced in the flesh through the attraction of more blood because of some static volume.[96] Similarly, Galen himself implies that Erasistratus also invoked *akolouthia* to explain how blood was drawn into the veins when some entered the flesh, despite the fact that the veins collapse when blood leaves them, which would accommodate any lost material.[97] These observations illustrate that Erasistratus is not getting his observations about body from dissection and then using these within his broader physiological scheme. Rather, the behaviors within the body are bound up with the pneumatic-like explanations that he employs to explain fluid motions. These explanations are themselves grounded in certain technological heuristics.

Erasistratus provides even more dramatic examples of physiological behaviors that manifest only within the biotechnical amalgam. For example, he also seems to accept that that when we hold our breath, the heart stops,[98] much like plugging the end of a pump makes compressing the handle functionally impossible. Later, his followers will insist that any motion that can be felt is merely the heart "oscillating" [κραδαίνεσθαι] in place, not actually pumping.[99] This account shows the degree to which technologies and the theoretical systems they both support and embody are not simply employed to explain established corporeal behaviors but actively construct the phenomena supposedly displayed by the body. In some ways, Celsus's description of Erasistratus (and Herophilus) dissecting subjects "while breath remained in them" [*spiritu remanente ea*] is not just a metaphor but a declaration of the clean distinction that had been drawn between internal and external spheres, aided by and bound up with this new pneumatic technological shift.[100]

Erasistratus was not alone in his application of pressurized devices to conceptualize and explain corporeal behaviors, often to extreme effect. For instance, two texts roughly contemporary with Erasistratus from the late fourth and early third centuries BCE both use the clepsydra to argue that we do not sweat when we hold our breath. Theophrastus's *On Sweat* and a corresponding entry from the pseudo-Aristotelian *Problemata* insist that, as the clepsydra demonstrates, the body cannot expel one substance without taking in an equal amount of another. Therefore, according to these authors, holding one's breath suspends the possibility that any flu-

96. Gal. *Us. Resp.* 2.2 = 4.707K.
97. Gal. *Nat. Fac.* 2.1 = 2.76K; cf. 2.6 = 2.95–96K.
98. See Gal. *Us. Resp.* 2.1–2.3 = 4.473–475K.
99. Gal. *Us. Resp.* 2.5 = 4.47K.
100. Celsus *Med.* 1 *pref.* 23–26 = T63a von Staden.

ids can exit from the pores (which are connected to the veins, which are themselves connected to the lungs).[101] Erasistratus, too, seems to have held some version of cutaneous breathing, although only his arteries seem to be connected to the external pores. The *Anonymous Londinensis* papyrus reports that he held the air-filled arteries to run "into the pores throughout the whole body and then through the pores in the flesh to the outside" [εἰς τὰ καθ' ὅλον τὸ σῶμα ἀραιώματα, εἶτα διεκθεῖ διὰ τῶν ἐν τῇ σαρκὶ φυσικῶν ἀραιωμάτων εἰς τὸ ἐκτός].[102]

It is striking that neither he nor his subsequent followers use these pores to imagine how air might enter through the skin to facilitate the continued pumping of the heart, even if we purposefully cease breathing. I suspect that since neither the pump nor the bellows contains pore-like openings, neither Erasistratus nor his successors thought to include pores as part of his pneuma-transmission system. Whereas clepsydrae have perforations on their bottom, so thinking with this device invites the integration of cutaneous pores, a corporeal system so thoroughly bound up with pumps and bellows does not. These two models, one active (the heart/pump) and one passive (the pores/clepsydra) illustrate how the same principle of *horror vacui* can predicate separate bodily behaviors when embodied in different totemic technologies. The cutaneous pores fail to make any significant impact on Erasistratus's pneumatic physiology because he is thinking with pumps, bellows, and pneumatic devices that do not contain them.

It is also worth noting that pneumatic devices also helped Erasistratus create a new framework to understand the etiology of disease. In his physiology, he rejects both humoral theory and innate heat as relevant concepts and instead creates pathologies around the inappropriate infiltration of one substance into the system of another. For example, the anonymous author of *On Acute and Chronic Diseases* reports that Erasistratus attributed paralysis to the liquids of the veins entering the cavities of the motor nerves, hindering their capacities.[103] Celsus reports that he claims that when blood ends up in the arteries it "arouses inflammation" [*inflammatio excitat*] and causes fever.[104] In fact, Galen reports that for Erasistratus, inflammation cannot occur without blood getting into the arteries.[105] The pseudo-Galenic *Introduction* even claims that Erasistra-

101. Theophr. *On Sweat* 25; Arist. *Pr.* 2.1, 866b9–14.

102. Anon. Lond. 23.21–23 Manetti.

103. Anon. *Morb. Ac. et Chr.* 21.1 Garofalo.

104. Celsus *Med.* 1 *pref.* 15–16; cf. *Med.* 1 *pref.* 60; *Med.* 3.10.3.

105. Gal. *UP* 6.17 = 3.493K; cf. *Ven. Sect. Er.* 3 = 9.154K; Gal. [*Hist. phil.*] 39 = 19.342–343K; *Loc. Aff.* 5.3 = 8.311K; cf. Harris 1973: 204–5.

tus ascribes *all* disease to the infiltration of blood into the pneuma-filled veins.[106] Although this claim is clearly an overstatement, the shift from "blockage" to "infiltration" suggests that Erasistratus is conceptualizing the operations of the body as parallel to those of the fine-tuned pneumatic automata around him. For example, Heron describes mechanisms that pour out liquids of different temperatures,[107] vessels that distribute wine and water either separately or mixed in different proportions,[108] a vessel that can channel different liquids through a single exit pipe in alternation,[109] and a vessel in which air and water alternatively ascend and descend,[110] among many others.[111] In all these devices, the infiltration of water into a vessel designed for air, or of wine into a vessel designed to carry water, would cause the mechanism to malfunction (see fig. 16). Each device relies on the precise balance and separation of air and water, and operates according to a type of vacuum pressure, however that may be conceived. Whereas previous Hippocratic authors spoke about the relative consistency of the humors and how it affected their buildup in various sites, and whereas Galen will later insist that the blood is thinner in some parts, thicker in others, and has different colors and smells in different people, Erasistratus mentions none of these factors. Instead, Erasistratus treats the inner liquid as not particularly important, a uniform substance subject to ἡ πρὸς τὸ κενούμενον ἀκολουθία, much like water in a pneumatic device.

In sum, while Erasistratus relies on "the filling toward what is being emptied" [ἡ πρὸς τὸ κενούμενον ἀκολουθία] to understand the body, when applying his principle to understand the mechanisms of disease,

106. Gal. [*Int.*] 13 = 14.728K. Erasistratus contends that the arteries and veins are actually connected at their ends (what we would see as the capillary venules and arterioles) by *synastomoses*, small connections through which infiltration can occur. Gal. *UP* 6.17 = 3.492–494K complains that these would serve no physiological function other than to cause disease; cf. Harris 1973: 209. See also Sor. *Gyn.* 3.1.64–68 Burg. = 3.4 Ilberg, which describes transfusion [μετέρασις] as a cause of disease, seemingly alluding to Erasistratus and the Erasistrateans.

107. Heron *Pneum.* 7.

108. Heron *Pneum.* 8, 24, 59, 65.

109. Heron *Pneum.* 22.

110. Heron *Pneum.* 53.

111. Heron also provides descriptions of two vessels, one of which distributes wine, the other water (*Pneum.* 13); drinking horns that distribute both wine and water (*Pneum.* 18; cf. 52, 64); a pipe that distributes wine in proportion to the amount of water removed (*Pneum.* 25); a vessel that distributes wine in proportion to the amount of water added (*Pneum.* 26); a vessel that distributes different kinds of wines according to the weight placed in a cup (*Pneum.* 32); and a vessel that stops distributing wine whenever water is poured in (*Pneum.* 39).

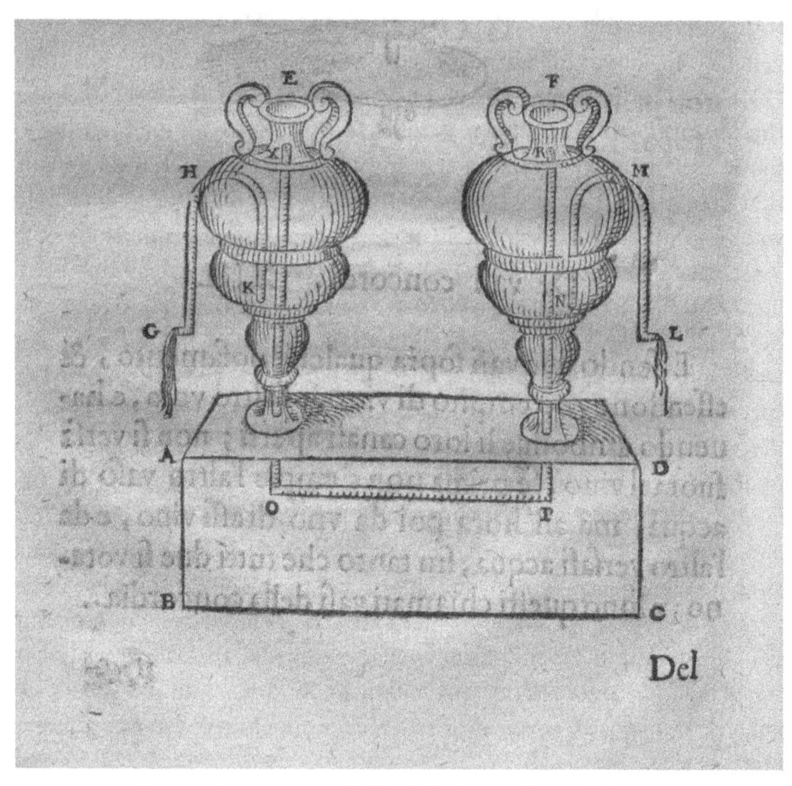

FIGURE 16. Heron, *Pneum.* 23, in *Spiritali di Herone Alessandrino*, trans. Alessandro Giorgi Hero (Urbino: appresso Bartholomeo, e Simoni Ragusij fratelli, 1592). The text accompanying the image reproduced in fig. 16 reads:

> This figure displays a dual vessel system, operating by means of vacuum pressure, that maintains separate compartments for water and air. This device describes the balanced interactions between fluids and water, and requires that wine be sealed in a vessel into which a series of siphons and outflow nozzles have been arranged. Should that seal be broken and unwanted air enter the wrong vessel, the vacuum pressure by which the mechanism functions would be lost. In other words, the infiltration of a substance into the wrong vessel causes malfunction. Many other pneumatic devices described by Hero show the same types of behavior. In fact, there seems to be a particular fascination with designing mechanisms that distribute two or more separate substances through the interconnected, but segregated actions of water and air.

he interprets its significance through the pneumatic technologies being developed around him. After all, Plato, too, denied the possibility of the void but did not ascribe all diseases to the infiltration of pneuma into the wrong vessels. Without an explicit reference to new pneumatic technologies and pipes, we cannot be certain that Erasistratus implicitly relies on these devices as material heuristics, but the fact remains that he applies his abstract principle in a way what takes for granted that his blood vessels function much like the vacuum-pressure driven automata around him.

4.5 CONCLUSION

For historians interested in the interactions of mechanics and medicine in antiquity, Alexandria in the third century BCE has provided a site of great interest. Theorists and practitioners from across the Mediterranean, the Black Sea, and the Middle East arrived in this new city, attracted by a new intellectual landscape and, in many cases, by royal patronage. This environment facilitated new types of mathematics, new aesthetic modes, and new machines. Much scholarly attention has focused on some dramatic moments in this time period where certain technological innovations influenced medical theories, such as Ctesibius's force pump supplying a model for Erasistratus's heart. Yet the vectors of influence were broader and deeper, and themselves relied on a shift in theories of corporeality. In the late fourth century, more medical theorists, such as Praxagoras, Herophilus, and Erasistratus, began to articulate the "activities," "tasks," and "function" of individual parts, as they privileged organs in their investigations. Not only did this articulation shift definitions of disease away from imbalance and toward malfunction, it also made anatomical investigations a crucial epistemic practice that was used to articulate and identify operational structures in the body. In this regard, it changed medicine as a techne.

The rise of the organism, supported and articulated by these investigative practices, set the stage for technologies to gain more influence over theories of physiology built around its tool-like parts. Theorists began to see more active mechanisms at work in the body, with tool analogies supplying key models. Some of the tools that did so were older, such as the bellows. This device had entered medical discourses as a material embodiment of air propulsion and exchange, and its totemic status made it a solution in search of problems. As such, medical theorists used its mechanics to generate explanations about the transfer and propulsion of air and blood through the body in ways previous authors had not deemed in need of account. Other technologies, such as pneumatic automata and other marvels, were new, and their utilization of pressurized, separated chambers for water, air, and sometimes wine created a template for Erasistratus's dramatic pressurization of the body, which separated blood and air into discrete vascular systems wherein the infiltration of one substance into the other produced disease. In all these instances, the chapter also emphasized that the body and its behaviors do not sit, waiting to be observed, but emerge out of the explanatory apparatuses used to elucidate them. When Herophilus refers to the alternating expansions and compressions of the lungs and thorax, or the Erasistrateans insist that the heart stops when we hold our breath, they are describing behaviors that exist in the biotechnical amalgam of their technologically supported physiology. The body displayed different features as the technological environment shifted.

To be sure, new philosophical ideas about the nature of the void emerged during this period, including the concept of the microvoid,[112] but it is hard to discern whether these new ideas led to an interest in pneumatic mechanisms or resulted from them—or, as is more likely, whether there was some reciprocal interplay between the two. In any case, abstract principles were bound to and embodied within material technologies. It was not a disembodied, abstract concept of *horror vacui* that affected the way these principles were used to understand human physiology, but the specific devices developed by Hellenistic inventors.

112. See note 66 on p. 178.

CHAPTER FIVE

The Organism and Its Alternatives

5.0 INTRODUCTION

As the organism entered medical discourses in the third century BCE, practices of knowledge production changed. More authors became interested in articulating the so-called natural activities of the body, which included corporeal processes like breathing, digestion, blood production, and pulsation. Anatomy and physiology spread as privileged modes of knowing, and Alexandria became the intellectual hub for this approach to medicine. This chapter starts by witnessing the spread of thinking in terms of corporeal organs in the decades following Herophilus and Erasistratus, as the power of this investigative methodology became clear to multiple thinkers. Despite the productivity of a tool-like corporeal teleology, however, not all medical theorists accepted this view of the body or the practices that supported it. Significant pushback came throughout the Hellenistic period, with alternative medical sects rejecting both anatomy as an investigative practice and the organism as a model of corporeality.

The most forceful critiques came in the third century from the so-called Empiricists, who insisted that the internal natural activities of the body simply could not be known. Moreover, they argued, such knowledge was not even useful for the effective administration of therapies. Instead, physicians should compare a patient's current symptoms with those of past cases and then use treatments that had proven successful in similar situations. Resemblance was all that was needed to establish successful therapy, and no speculative knowledge of functions or causes was necessary. Physiology and the theory of corporeality that it supported was useless. In the late second century BCE, another medical theorist, Asclepiades of Bithynia, likewise spurned the organism and its attendant epistemologies. In distinction to the Empiricists, however, Asclepiades did claim knowledge of the corporeal interior, championing a theory that resembled Epicurean and Lucretian notions of atoms and the void, insofar as it attributed all illness to the blockage or flow of corpuscles in the pores of the body. Asclepiades discussed breathing and other corporeal processes, but these processes no longer supported the internal activities of a living system. They simply became behaviors to be explained mechanically, without regard to function.

Starting in the first century BCE, another medical sect, called the "Methodists" after their particular "method" [ἡ μέθοδος], presented yet another view of medical epistemology and another associated view of corporeality. They insisted that knowledge of the body could not rely on hidden causes and natural activities but could rest solidly upon what they called the "manifest commonalities." These were corporeal states of constriction and laxity that they claimed were directly apprehensible, thus obviating the need for any physiological accounts. Tool-like somatic functions were once again irrelevant, and perhaps even detrimental, to the effective practice of medicine.

All these schools rejected the organized body as the proper epistemic object of medicine and conceptualized new types of corporeality to ground their medical expertise. Yet despite this hostility toward tool-like functions, the successes of the organism proved hard to avoid. Its associated terminology infused both medical and public consciousnesses and still surfaced in the arguments and formulations of these opposing sects. In this way, consolidating the body as a tool-like object was a bell that could not be unrung. Both physicians and doctors became familiar enough with organs and corporeal tasks that these concepts could now exert conceptual power even within medical theories that explicitly rejected the type of knowledge that these concepts made possible.

Alongside these intellectual currents, a major political transformation had occurred. In the third through first centuries BCE, Rome rose to power and extended its dominion across the Mediterranean. With this expansion of their rule came an associated expansion of infrastructure—notably, the aqueducts and bathhouses that became iconic emblems of empire. Although their influence is not always apparent, Asclepiades seems to have derived several of his assumptions about diseases from Roman watercourses. Moreover, he became quite popular, and although Roman authors sometimes disparagingly attribute this popularity to his liberal prescription of wine and relaxation, we might also ask whether his theories about flows and blockage were comprehensible and persuasive to a Roman public surrounded by monumental feats of engineering that themselves operated on flow and disruption. This technological environment created a cultural heuristic wherein material realities potentially made the public more receptive to Asclepiades's theories, which leveraged and aligned with the pervasive infrastructure of the empire. That is, technologies influence not only medical theorists but also the patients whom they treat.

The last section of this chapter, which deals with the Methodists, returns to women's bodies to illustrate how Soranus used the manifest commonalities to understand female corporeality, rather than using therapeutic interventions as the Hippocratics had. This approach decoupled technical interventions from basic assumptions about menstruation and, by extension, female flesh. In this regard, the chapter ends by revealing what happened when tools were given less importance in determining the nature of women's bodies. If the previous chapters examined the rise of the organism and how contemporary technologies of the fifth through third centuries BCE influenced the contours of imagined corporeal mechanisms, this chapter looks at what happens when these technologies changed over the next two hundred years, as some theorists rejected functional teleology—even as the object that it had created, the organism, held stubbornly on.

5.1 POST-ERASISTRATEAN HELLENISTIC ORGANA

Herophilus and Erasistratus helped solidify a set of investigative practices with anatomical dissection at their core, and these methodologies continued to flourish in Alexandria in the following decades. Eudemus of Alexandria (ca. 285–235 BCE), whom Galen considers to be one of the great early anatomists,[1] continued dissecting and wrote multiple treatises about anatomy, including discussion of bones, arteries and veins, hands and feet, embryonic vascular anatomy, and the nervous system.[2] Yet Herophilus and Erasistratus had not only conducted dissections, they had also outlined the functions of the parts and located those functions within a physiology interested in internal mechanisms. Fluid needed to be moved by the actions of organs, not assumed as a given capacity of the humors themselves. In the years that followed, this configuration of corporeality spread as multiple authors adopted similar ideas, with tools featuring prominently as heuristics for function-serving organs. For ex-

1. Gal. *Hipp. Aph.* 6.1 Savino = 18A.7K; *Sem.* 2.6.13 = p. 200 DeLacy = 4.646K; *In Hipp. Nat. Hom.* 2.6 Wewaldt = 15.134–135K.

2. For bones, see Ruf. Eph. *Onom.* 73 Gersh; for arteries and veins, hands and feet, see Gal. *UP* 3.8 = 1.148 Helmreich = 3.203–204K; and for embryology, see Sor. *Gyn.* 1.19.36–39 Burg. = 1.57 Ilb.; for nerves, see Gal. *Loc. Aff.* 3.14 = 8.212K. See also Manetti 2008: 308.

ample, *On the Heart*, a treatise found in the Hippocratic corpus but written in the period after Erasistratus and Herophilus,[3] discusses the parts of the body and uses function as the essential way to understand these structures. It describes how the pericardium cushions the heart as it beats,[4] how the esophagus is a funnel that receives what we consume,[5] how respiration into the lungs cools the heart,[6] and how the "auricles" or "little ears" of the heart capture air like bellows, "just as bronze smiths in the crucibles," a physiological arrangement that the author sees as "the act of a good handworker" [τὸ ποίημα χειρώνακτος ἀγαθοῦ].[7] The author also fully adopts the vocabulary of the organism, referring to the auricles as the organs [ὄργανα] that nature uses to complete this task.[8] This treatise suggests that increased specificity and attention given to anatomy rose in tandem with the conception of certain body parts as tools serving certain functions, sometimes explicitly named as such. Even when a theorist such as Philotimus, a student of Praxagoras and Herophilus, declared both the brain and the heart useless,[9] his attention to the question of corporeal utility reveals his engagement with this new theoretical mode.

The swift rise of the organized body and its accompanying knowledge practices led some authors to retroject ideas native to this type of corporeality back onto earlier theorists. For example, the pseudo-Hippocratic

3. It references both the pulse and the heart valves, which dates it to the post-Herophilean period. See Garofalo 1988: 38 and Abel 1958, as well as von Staden 1989: 174, esp. n112, which supplies alternative suggestions. For dating, see also Lonie 1973a and 1973b.

4. Hipp. *Cord.* 1 = 9.80L.

5. Hipp. *Cord.* 2 = 9.80–82L.

6. Hipp. *Cord.* 3, 5 = 9.82L, 9.84L.

7. Hipp. *Cord.* 8 = 9.84–86L; cf. *Cord.* 10 = 9.86–88L, which describes the inner membranes/valves of the heart as tightly "constructed" [ἐμηχανήθησαν].

8. Hipp. *Cord.* 8 = 9.84–86L. Although its reliability is somewhat suspect, *Vindician* (?), *Fragmentum Bruxellense* 44 (Wellmann 1901: 234) suggests that Diocles identified the auricles of the heart and supplied them the capacity to hear. For related comments about the heart, see Diocles fr. 33, 72, 74, 78, 80, 102, 108 van der Eijk. Herophilus also mentions the auricles of the heart but considers them internal (T120 von Staden). *On the Heart* expressly denies this capacity of the auricles to hear, which suggests that it comes after his declarations. We should also note that *On the Heart* also asserts that liquid moves through the lungs, which Aristotle has denied. This set of facts only helps illustrate that we cannot and should not treat the history of anatomy as the accumulation of increasingly "correct" interpretations and thus date treatises by their relative correspondence to our own anatomical models.

9. Gal. *UP* 8.3 = 1.452–453 Helmreich = 3.624–625K.

Letters 10–23 describe a fictional interaction in which Hippocrates is summoned to Abdera to treat Democritus, who his fellow citizens think has gone insane.[10] Hippocrates finds him surrounded by dissected animals.[11] Not only does this account likely assume that someone interested in investigating "the truth of human nature"[12] would naturally do so through anatomical investigations, but *Letter* 23 also adopts the idea that the organs are manufactured by nature to serve the body, another view that fits more ideally within a post-Aristotelian, teleological worldview rather than that of an atomist, such as Democritus, committed to a nonteleological view of the cosmos.[13] Later medical historians, such as Celsus, likewise claim that Hippocrates pursued the "nature of things" and assume that he must have done so through anatomy, just as Erasistratus did.[14] Galen, too, will later insist more or less the same thing.[15] Dissection became so naturalized as an investigative technique that its specific connection to functional teleology was lost.

In the same way, the notion that the body parts completed "tasks" [ἔργα] became so common that it was easy for multiple authors, even those who were not practicing physicians, to assume this idea was universal, much as we can now ask about how the body "works" without considering the metaphorical nature of this statement. Certainly, the basic framework of a functional body was capacious and flexible enough to accommodate most humoral assumptions and networks. For example, earlier Hippocratic authors tended to describe the viscera as the locus of a particular humor, the location of a particular ailment, or sequential sites through which liquids flowed. Authors now began to envision the viscera as the sites where humors were manufactured, refined, and processed—a

10. Hipp. *Ep.* 10–17 W. Smith describe the approach and scene, while *Ep.* 18–21 W. Smith recount their conversation.

11. Hipp. *Ep.* 17 W. Smith = 9.350L.

12. Hipp. *Ep.* 17 W. Smith = 9.378L; cf. *Ep.* 23 W. Smith = 9.394L, which also describes Democritus writing "about human nature."

13. Smith (1990: 29) dates the Democritus letters to 1st c. BCE–1st c. CE, whereas Craik (2006: 167) argues that these notions and practices are Democritean (so that she can recreate the Democritean contribution to medical discourses). Nevertheless, the teleological viewpoint described in *Ep.* 17–23 seems out of step with Democritus's anatomist worldview, and so, to my mind, it seems more likely that the author is making a set of anachronistic assumptions.

14. Celsus *Med.* 1 *pref.* 47. Smith (1990: 27) notes that Celsus's assertion suggests that Hippocratics engaged in systematic anatomical investigations, although this claim seems unlikely.

15. He wrote a treatise called *On the Anatomy of Hippocrates*, for example.

difference often missed by subsequent commentators who easily read these post-Aristotelian notions back into early Greek medical texts. After the Hellenistic anatomists articulated the body as the epistemological extension of Aristotle's organism, it became difficult to assess why someone would *not* open up the body to discover how it worked. Medical causality and anatomical physiology became closely wedded together.[16]

5.2 THE EMPIRICIST RESISTANCE

These ideas and practices did not proceed unchallenged. Resistance came in several forms and along different vectors. Philinus of Cos (fl. 260 BCE), a dissident student of Herophilus, founded a new medical sect, the Empiricists, who wanted to base medicine not in speculative assertions about causes or malfunctions but in the direct observations of what was manifest [τὰ φαινόμενα], which for them were the symptom sets that attended illness.[17] Unlike the Hippocratic author of *Nature of the Human*, whose purgative drugs *made* the four humors manifest, Empiricists did not articulate the body through such technological interventions. They instead insisted that medical therapeutics should be based solely on the comparison of a present illness to past cases, prescribing interventions that helped in similar circumstances.[18] This approach made it useless to speculate about the elemental components that affected the body, whether humors, breaths, or transfusions of blood into the arteries.[19] It also made it superfluous, even harmful, to engage in dissection or vivisection, and they explicitly rejected anatomical knowledge as relevant to the practice of medicine.[20] Yet, as the previous chapters have argued, investigative practices sustain epistemic objects. As such, rejecting dissection also meant rejecting the theory of corporeality that it supported. Indeed, Philinus and other Empiricist followers, including Serapion of Alexandria (fl. 225 BCE)

16. The same is the case for Simplicius, who can only fathom Empedocles's claim that random body parts assembled to form hybrid creatures by envisioning a scenario where the organs from different animals complete vital tasks for these mixed creatures but imperfectly; see *In Phys.* 371.33–372.9 = Emped. DK31 B61 (D152 LM). He does not even consider that Empedocles might not see the body as an organism with mutually supportive corporeal systems. Other authors project anatomical expectations back onto earlier theories.

17. See von Staden 1975.

18. For Empiricist divisions of medicine and their so-called "tripod," see von Staden 1989: 101.

19. Celsus *Med.* 1 *pref.* 14–15.

20. See T446 von Staden.

and Heraclides of Tarentum (fl. 3rd/2nd c. BCE), asserted that all knowledge of "hidden causes" was unnecessary and ultimately unknowable, as was knowledge of "natural actions" [*naturales actiones*], which Celsus characterizes as precisely the internal processes around which corporeal function had been built, including breathing, digestion, pulsation, sleeping, and waking.[21] Dismissing corporeal activities was tantamount to casting aside the notion that the body worked as a system to sustain life.

Even as the Empiricists rejected this new investigative approach on epistemological grounds (largely indebted to Pyrrhonian skepticism), they accepted that the body might indeed be constituted as anatomical and dietary investigations construed it. For them, however, this knowledge was simply irrelevant to medicine and medical practice. It could entertain natural philosophers but did not pertain to the success of a physician. The success of the organism made it hard to dislodge as a concept, so the Empiricists simply transferred the understanding of this entity to another discipline. Instead of leveraging causal knowledge, they drew a direct line between the manifestations of illness and the treatments thereby implied, without the mediation of physiology, physis, or function. Identifying similar symptom profiles and comparing what worked for earlier patients was the only necessary act. We might say that they refused to articulate a corporeal interior and tried to treat the body as only a surface that presented symptoms.[22] Within such a medical practice, the body became the object of technical knowledge rather than a functioning organism or fluid system in balance.

5.3 THE INFRASTRUCTURE OF ROMAN POWER

Although many of the debates about the body within the Hellenistic medical schools surrounded epistemology, another medical theorist rejected the organized body for entirely different reasons. Asclepiades of Bithynia (fl. ca. 120–90 BCE) was one of the first physicians to leave the Greek East and establish a successful practice in Rome.[23] From the third century onward, even as Ptolemaic Egypt, Seleucid Babylon, and Attalid Pergamon created or invigorated Greek cultural capitals in the eastern Mediterra-

21. Celsus *Med.* 1 *pref.* 19.

22. For example, the Empiricists admit that they can feel the force of the pulse, but not the "expansion of the artery," since the latter is "nonapparent" (Gal. *Dig. Puls.* 8.776K). See Holmes 2010a: 96–97 for similar reflections.

23. For an account of the arrival of Greek medicine in Rome, see Nutton 2004: 157–70.

nean, Rome was rising as a site of power in the west. Over the following two centuries, it came to dominate the Italic peninsula, before it swallowed up Sicily, southern Spain, parts of North Africa, and Greece into its empire. By the end of the first century BCE, it had absorbed or conquered all the former Hellenistic kingdoms. This expansion created considerable disruptions, as economies and trade networks were dismantled or reconfigured, and huge numbers of people were enslaved and moved around the Mediterranean. Athens, Pergamon, Ephesus, and Alexandria were now Rome's.

Even as Greek-speaking peoples were being conquered, the currency of Greek literature and learning retained its value, and Romans absorbed and adapted canonical texts to build a Latin literary tradition of their own. The dynamics of medicine were similar, although medicine maintained tighter associations with Greekness than did epic or lyric poetry.[24] There were certainly long-held healing practices on the Italic peninsula, but no Roman literary medicine parallel to the Hippocratic tradition seems to have developed in pre-Hellenistic times, and responsibility for health care typically fell to the paterfamilias of a household, who attended to the medical needs of the humans and animals under his power.[25] Recipes and remedies seem to have been shared knowledge rather than the technical purview of trained experts. This state of affairs shifted with the introduction of Greek healing, which occurred, if Pliny can be believed, when Archagathas of Sparta arrived in 219 BCE. Archagathas was reportedly received with much fanfare but left shortly thereafter when his methods of healing met with considerable disapproval.[26] Over the next century, the importation of enslaved Greeks brought more medical expertise, and before the first century BCE, 75 percent of doctors attested in Roman inscriptions were enslaved persons. Even two centuries later, the vast majority of physicians were still from the Greek East.[27]

Even as literary medicine was perceived as a primarily Greek expertise

24. For instance, Plin. *HN* 29.17 states that very few Romans choose medicine as a profession, and only those who use Greek terminology are treated as authorities. For discussions of the relationship of Greek and Roman medicine, see Nutton 1993; 2004: 157–70; Maire 2014.

25. For an overview of early Roman medicine, see Nutton 2004: 157–70; Scarborough 1969.

26. Plin. *HN* 29.12–13 reports that Archagathus was given a workshop and the rights of citizenship but quickly earned a reputation as "the wound man" and then "the executioner" for his therapeutic methods, which involved cutting (incisions for bleeding, most likely) and burning (scarification).

27. Nutton 2004: 165, 171.

for centuries, Rome left its own marks on health and health care. Scarborough and Nutton have examined whether doctors traveling with the Roman army spread medical ideas throughout the expanding provinces,[28] but the most dramatic impact came in the form of infrastructure. Even from the early Republican era, water technology formed a large part of Roman cultural identity, a material manifestation of civic progress and pride. Frontinus (ca. 35–103 CE) reports that Appius Claudis Crassus led the Appian aqueduct into Rome in 312 BCE, while Manius Curius Dentatus constructed the (Old) Anio aqueduct in 273 BCE. Servius Sulpicius Galba and Lucius Aurelius Cotta repaired those and added the Marcia aqueduct around 146 BCE, while around 127 BCE, the Tepula was constructed.[29] Even the practice of naming these aqueducts after those who had them constructed shows the degree to which civic infrastructure was valorized.[30]

In addition to aqueducts, public baths were a major part of Roman practices of health. Public buildings devoted to bathing had been introduced in Athens by the fifth century BCE, but these *balaneia* were relatively modest in comparison to later counterparts, involving individual tubs in which bathers only submerged their lower extremities.[31] Bathhouses developed in complexity over the next centuries, appearing across the Mediterranean as part of the cultural expansions of the Hellenistic era. Some started to incorporate submersion tubs and subterranean furnaces that first warmed small pools of water and, later, entire rooms. Still, these remained relatively small buildings, however, often privately managed, and they were not then yet the touchstones of civic architecture that they later became. By the second century BCE, Roman baths had developed their own distinct style, incorporating a heated communal pool, a hypocaust heating system that ran under the floors, and sequentially arranged rooms of increasing temperatures (named the *frigidarium*, *tepidarium*, and *caldarium*).[32] Moreover, bathhouses played an increasingly important role as social spaces, where visitors exercised, relaxed, and at-

28. Scarborough 1968; Nutton 2004: 171–86.

29. Frontin. *Aq*. 1.4–8.

30. We might also think of such later examples as Statius's *Silvae* 4.3, which was written to commemorate a road, the Via Domitiana.

31. Ginouvès (1962) provides a thorough examination of Greek baths, but for the most recent studies, see the various contributions in the edited volume of Lucore and Trümper (2013); cf. Trümper 2014. Maréchal (2020: 10–33) provides a concise historical overview of the development of baths and bathing infrastructure in the Mediterranean, as well as relevant bibliography.

32. See also Fagan 2001 and Yegül 1992 and 2010 for surveys of Roman baths and bathing culture.

tended to their basic health regimens. The appearance of these bathing complexes in Greek medical treatments mirrored their civic status, as physicians began to incorporate them more thoroughly into regimental and therapeutic practices.[33] Indeed, it seems that by the first century, Roman citizens visited these complexes daily.[34] Their importance as markers of *Romanitas* only increased, as smaller balaneia gave way to imperial *thermae*. These were enormous advertisements for empire, grand-scale statements about what it meant to be Roman, designed to accommodate thousands, if not tens of thousands, of visitors every day. Aqueducts and baths were also major cultural exports in newly conquered cities, representing the implementation of Romanness itself. As Frontinus says (perhaps self-servingly, as the Manager of the Aqueducts at Rome), "Compare our numerous, necessary structures carrying so many different waters with the obviously idle pyramids or with the other useless—but famous—works of the Greeks."[35] The huge arcades regulating the flow of water through systems controlled by *calixes*, governed by laws, and overseen by bureaucrats would have provided conspicuous material symbols of Roman power.[36] Even before these developments had been fully enacted, Rome already possessed an imposing water-delivery system and impressive baths by the late second century BCE. The technological landscape of health care had changed.

5.4 ASCLEPIADES OF BITHYNIA

Asclepiades of Bithynia left the Sea of Marmara and traveled to Rome near the end of the second century BCE and seems to have acquired a considerable patient base upon arrival.[37] Pliny considered him to be a charlatan, a rhetorician who began to practice medicine only to make a profit and who attracted patients not by the soundness of his ideas but by prescribing wine, cold water, and leisurely activities such as being rocked in a swing

33. Flemming 2013.

34. Cic. *Att.* 15.13.5; cf. Maréchal 2020: 30.

35. Frontin. *Aq.* 1.16; cf. Dion. Hal. *Ant. Rom.* 3.67.5.

36. Due to the relative paucity of relevant archaeological evidence within the city walls of Rome, information regarding its domestic supply system is interpolated from evidence found at Pompeii. For a list of the aqueducts at Rome and their construction dates, see Wilson 2008; cf. Evans (1994), who provides an in-depth account of all the aqueducts, their construction dates, routes, etc.

37. For an investigation of Asclepiades's life, see Cocchi 1758; Rawson 1982; Polito 1999.

or taking baths.[38] Celsus displays far less pique toward his predecessor but still worries that Asclepiades spent too little time with individual patients.[39] It is Galen, however, who acts as his chief detractor, frequently noting the falsity of his ideas and the ridiculousness of his claims, since his basic theories did not rely on humors or corporeal functions. Instead, Asclepiades asserted that the body was primarily composed of "unarticulated corpuscles" [ἄναρμοι ὄγκοι] and "pores" [πόροι].[40] Accordingly, Asclepiades's objection to anatomical investigations was not grounded in the argument that they distorted the natural activities that they sought to reveal. He had simply reverted to a model of corporeality that did not benefit in the same way from this type of knowledge practice.

Much scholarly debate has tried to establish what Asclepiades means by "unarticulated corpuscles" and "pores," with most recent scholars accepting that they are either equivalent or largely analogous to Epicurean "atoms" and "void."[41] In fact, Galen—our chief source for Asclepiades—states this equivalence explicitly:

> Everything is composed from atom and void according to the account of Epicurus and Democritus, or from certain masses and pores [ἔκ τινων ὄγκων καὶ πόρων], according to the physician Asclepiades (for he exchanged the names alone and said "masses" instead of "atoms," "pores" instead of "void" [ἀντὶ μὲν τῶν ἀτόμων τοὺς ὄγκους, ἀντὶ δὲ τοῦ κενοῦ τοὺς πόρους λέγων], while wanting the essence of what exists to be the same as those things).[42]

38. Plin. *HN* 26.7–9.

39. Celsus *Med.* 3.4.1.

40. The unarticulated corpuscles are variously referred to as ἄναρμοι ὄγκοι, ὄγκοι, στοιχεῖα, *corpuscula*, and *moles solidae*, while the pores are called ἀραιώματα, κενώματα, *foramina*, and *viae*.

41. Harig 1983: 44–45; Casadei 1997: 77–78, 89; Leith 2009, 2012; cf. Wellmann (1908: 695), who holds that Asclepiades's theories can be traced via Erasistratus to Aegimius of Elis; for these references and arguments, see Leith 2012: 165 n1. By contrast, Gottschalk (1980: 37–57) posited that Asclepiades more or less adopted the theory of Heraclides of Pontus, who also proposed "unarticulated masses" [ἄναρμοι ὄγκοι] as the essential constituents of matter; cf. Lonie 1964 and 1965b. The major problem with this argument is that we know even less about Heraclides than Asclepiades.

42. Gal. [*Ther. Pis.*] 11 = 14.250K, trans. Leith. Galen levels his extensive critique of Asclepiades across multiple works, including *Nat. Fac.* and *UP*, but several of Galen's treatises dealing with this predecessor are lost, including *On the Opinions of Asclepiades*; cf. Gal. *Lib. Prop.* 8 = p. 17 Singer =19.55K.

Like Epicurean atoms, the ἄναρμοι ὄγκοι are invisible and elemental, possess no qualities (which they only manifest in combination),[43] and differ only in size, shape, and perhaps magnitude.[44] They are also constantly in motion.[45] One key difference distinguishes Asclepiades's ὄγκοι from any type of atoms: his corpuscles are frangible and divisible.[46] When ὄγκοι meet each other, they either combine to form larger structures or break into smaller pieces.[47] As Caelius Aurelianus, a second-century CE Methodist physician sympathetic to Asclepiades, writes: "[These corpuscles] by their own rushing are struck by mutual blows and broken up into infinite fragments of parts" [*quae suo incursu offensa mutuis ictibus in infinita partium fragmenta solvantur*].[48] Some commentators have argued that these smaller fragments constitute a more fundamental type of matter, but Leith has shown that these *fragmenta* are simply smaller ὄγκοι,[49] different in quantity/magnitude, but not in quality/kind. As such, Asclepiades's physical theory resembles Epicurean atomism, except insofar as his corpuscles are not actually indivisible [ἄτομα].[50]

As for Asclepiades's pores [πόροι]—which sources also refer to as ἀραιώματα,[51] κενώματα,[52] *foramina*, and *viae*[53]—they do appear to repre-

43. Cf. Gal. *Elem.* 5.1.2 = 1.416–417K.

44. Cf. Cael. Aur. *Acut. Pass.* 1.105–106.

45. Cf. Cael. Aur. *Acut. Pass.* 1.105: *aeternum moventia*; Sext. Emp. *Math.* 3.5: δι᾽αἰῶνος ἀνηρεμήτων.

46. Cf. Gal. *Const. Art. Med.* 1.249K, which refers to the "unarticulated [mass] being frangible" [τὸ ἄναρμον θραυστὸν ὄν]; cf. Gal. [*Int.*] 9 = 14.698K; Sext. Emp. *Pyr.* 3.33: θραυστά.

47. Vallance (1990: 42–43) argues that this claim means the ὄγκοι are infinitely divisible.

48. Cael. Aur. *Acut. Pass.* 1.105.

49. Cf. Pigeaud 1980: 194–98; Casadei 1997: 91–101. Leith (2009: 289–90) argues that Asclepiades only ever mentions the ὄγκοι, which, when they meet each other, either break into smaller pieces or combine into larger objects.

50. To be sure, Leith (2012: 187) admits that even if Asclepiades's physical theory entailed interstitial gaps in all matter, he was only truly interested in using these theories to explain the disease and pathologies of the body, not to examine the physical or ontological implications. Leith suggests nevertheless that Asclepiades could have dealt with the physical implications of his theory in his work *On Elements*.

51. Gal. *MM* 2.4 = 10.101K; 13.2 = 10.876K; cf. *Comp. Med. Gen.* 6.16 = 13.936K, although Leith (2012: 181 n44) argues that this latter passage refers to the first-century CE pharmacologist Asclepiades Pharmacion.

52. Gal. [*Int.*] 9 = 14.698K.

53. For references, see Vallance 1990: 7; Leith 2012: 181.

sent a version of Epicurean void, as Galen suggests.[54] Indeed, several of Asclepiades's arguments about the pores seem to parallel those of the Atomists.[55] Because of this correspondence, Leith suggests that the πόροι should not be thought of as channels at all and should only be understood as "gaps" or "interstices" in matter,[56] such that they are "exactly analogous" to the Epicurean void.[57] Yet I would suggest that even if the πόροι have the strict philosophical meaning of "interstices" and operate within a physics similar to atoms and void, calling them "passageways" activates certain physiological arguments that calling them "void" [τὸ κενόν] cannot. Perhaps most importantly, claiming that the body is comprised entirely of πόροι allows for a type of scale conflation whereby the macrovessels of the body become a guarantor of the heuristic that governs the micropassageways. The divisions between parts, the examination of organs, and the tracking of pathways become far less crucial for Asclepiades, who articulates a body composed of an entirely integrated and interconnected system of microscopic, invisible openings that pervade the whole body, not dissimilar to the structure of a sponge.[58] Thus, whereas Erasistratus privileged "infiltration" in his theory of fever and inflamma-

54. Cf. Gal. *SMT* 1.14 = 11.405K; *UP* 6.13 = 1.345–356 Helmreich = 3.474K; *Hipp. Epid. VI* 4.11 = 215 Wenkebach = 17B.162K.Calcid. *In Tim.* 214 also relies on a *Placita* tradition that cast Asclepiades as a void theorist; cf. Mansfeld 1990: 3112–17; Switalski 1902: 53; Polito 2007. Similarly, both Gal. [*Int.*] 9 = 14.698K and Heron *Def.* 138.8 refer to the ὄγκοι and πόροι as having the status of elements, just like atoms and void. For all these passages, I owe credit to Leith (2012: 166–67, 171–73). By contrast, Vallance (1990: 57) has argued against this interpretation, positing that Galen was merely ascribing a doctrine to Asclepiades so as to launch polemic attacks, all while keeping his other intellectual antagonists, the Methodists, in mind, but Vallances's arguments have largely been dismissed; see, for example, von Staden 1992c and Asmis 1993.

55. For instance, Anon. Lond. 39.10–15 Manetti presents Asclepiades's doctrine of growth as supported by the fact that one body does not pass through another, which parallels earlier arguments made by the Atomists that the Epicureans pick up in turn; cf. Arist. *Ph.* 213b18–20; Them. *In Phys.* 124.4–9; Simpl. *In Phys.* 651.2–8; cf. also Leith 2012: 173–77. There are textual issues regarding the passage in Anon. Lond. That prevent outright acceptance of his argument. In addition, Leith (2009: 300–305) argues that Asclepiades's soul atoms may have been "smooth and fine," just like the Atomists' soul atoms—a fact that further supports his claims; cf. Calcid. *In Tim.* 215.229–230 Waszink.

56. Leith 2009: 181–82.

57. Leith 2012: 164.

58. There is some precedent in this comparison, since Gal. *Nat. Fac.* 1.13 = 2.32K compares Asclepiades's model of the bladder to a sponge.

tion, Asclepiades reverts back to a blockage model, ascribing all disease to either the obstruction and blockage of these pores [ἔμφραξις, ἔνστασις, *statio*, *obstrusio*, *coacervatio*][59] or excessive flow.[60] Among the diseases he ascribes to types of flow, he includes phrenitis,[61] pleuritis,[62] pneumonia,[63] sore throat,[64] and cholera.[65] Diseases caused by various types of blockage include heart attacks,[66] headaches,[67] and diarrhea.[68] While it may seem counterintuitive to attribute diarrhea to blockage, Asclepiades reportedly argued that the crowding of the corpuscles in one part led to their overflow in another. Moreover, either flow or blockage can evidently cause fever,[69] since Galen states that Asclepiades reduced all fevers ultimately to "stoppage" or "impaction" [ἐμφράξεις].[70]

If these are the diseased states, how do the corpuscles move within the body in non-pathological ways? Asclepiades argues that they adhere to the principle of "the movement toward what is sparse" [πρὸς τὸ λεπτομερὲς φορά], which describes the movement of corpuscles from areas of greater relative density to areas of lower relative density.[71] Whereas Erasistratus made his prevailing physical principle that of "the filling of what is being emptied," which refers to a type of vacuum pressure, Asclepiades

59. Celsus *Med.* 1 *pref.* 15–16; Sext. Emp. *Math.* 3.5; Gal. [*Int.*] 13 = 14.728K; Cael. Aur. *Acut. Pass.* 1.106; Vallance 1990: 7 for this terminology.

60. Despite the assertion at Gal. [*Int.*] 13 =14.728K that Asclepiades ascribed all illness to such a circumstance, some entries in Cael. Aur. suggest otherwise.

61. Gal. *Med. Exp.* 28 Walzer and Frede state that for Asclepiades, phrenitis is caused by fever and blockage in the cerebral membrane; cf. Cael. Aur. *Acut. Pass.* 1.6.

62. Cael. Aur. *Acut. Pass.* 2.89.

63. Cael. Aur. *Acut. Pass.* 2.142.

64. Cael. Aur. *Acut. Pass.* 3.5.

65. Cael. Aur. *Acut. Pass.* 3.188.

66. Cael. Aur. *Acut. Pass.* 2.163.

67. Cass. Iatr. *Pr.* 77.

68. Cael. Aur. *Acut. Pass.* 3.220.

69. Cael. Aur. *Acut. Pass.* 1.8.

70. Gal. *Trem. Palp.* 7.615K; cf. *Di. Dec.* 9.798K. Along with a physiological theory, Asclepiades's model of flow might have epistemological ramifications as well. Sext. Emp. *Math.* 8.6–7 links it to the flux theory of reality first posited by Heraclitus, stating that Asclepiades claimed that the same river cannot be pointed out twice "because of the speed of the stream" [δύο δείξεις διὰ τὴν ὀξύτητα τῆς ῥοῆς]. Similarly, Gal. *Sect.* 1.75–76K connects this notion back to the idea that the corpuscles move perpetually, and thus the body stays in constant flow.

71. Vallance (1990: 123–47) has also tried to stress that Asclepiades's πρὸς τὸ λεπτομερὲς φορά both derived from and responded to Erasistratus's ἡ πρὸς τὸ κενούμενον ἀκολουθία.

characterized the movement of corpuscles in a way more akin to diffusion. That is, ὄγκοι move as though liquid, flowing both into the body and through its inner pores to areas that have lower concentrations and are less crowded. Since he considers all flesh porous, the corpuscles can move either through the macrovessels of the body or, more often, through the tissues themselves. That said, a sponge does not get clogged or blocked up, since there are so many passages to reroute liquid, so his pathology seems, at some level, to rely on the conceptual work that the visible vessels do for him. Nevertheless, he does not place a great deal of importance on determining the particular pathways for the corpuscles or delineating one type of transmission from the other.

This configuration of corporeality has baffled and befuddled both ancient and modern commentators. Galen, for example, sometimes addresses Asclepiades's claims with arguments and lengthy experimental rebuttals, while at other moments he simply dismisses Asclepiades's assertions as laughable and unworthy of anyone who has even a rudimentary knowledge of anatomy.[72] Modern scholars have likewise been confused about Asclepiades's theories since they seem so out of step with anatomical knowledge even in his own era. Some recent attempts, including those by David Leith, have tried to rehabilitate and rescue Asclepiades, especially from the hostile sources that preserve his doctrines, often through dismissive criticism of them. Much of the bafflement and disbelief, however, stems from Asclepiades's reconfiguration of the body not around functional parts but around the substances inside it. Within his model, the body becomes almost like a waterfall, such that it becomes meaningless to ask what function the rocks serve over which the water flows. Internal parts are material facts, not functional objects.[73] His explanatory apparatus, structured as it was on atomist-like physics and corporeal flows, privileged neither organs nor their investigative corollary, anatomical investigations. That said, Asclepiades did apparently try to demonstrate that the food entering the stomach does not turn into "something good" by cutting open an animal and tasting its stomach contents.[74] Since these contents did not in fact taste good, he could conclude that digestion did not take place. This example illustrates that even if Asclepiades conducted

72. For the latter, see Gal. *Nat. Fac.* 1.13 = 2.32–35K; cf. Gal. *Us. Resp.* 2.11 = 4.483–484K.

73. Gal. *Nat. Fac.* 1.13 = 2.32–35K declares that Asclepiades could not have conducted even basic anatomical investigations and suggests that his arguments imply that multiple viscera serve no function whatsoever.

74. Gal. *Nat. Fac.* 3.7 = 2.166K.

a type of dissection, he did not do so to conduct anatomical investigations but to reject the very notion of natural activities, as he attempted to reject the notion that the stomach was an organ that digested food.

We might compare Lucretius's discussion of the body in *De Rerum Natura* book 4, written only a half century later, which displays many similar features and can likewise confound modern expectations built around tool-like functions. Like Asclepiades, who was influenced by Epicurean atomism, Lucretius (ca. 99–55 BCE) was an Epicurean who consolidated the body around its internal pores, passageways, and gaps [*viae*, *foramina*, *rara*], through which atoms flow. He describes some localized operations, notably in the sensory parts, where vision occurs in the eye through the entrance of *eidola*, hearing in the ear through the entrance of sound atoms, and flavor in the tongue through the interaction of atoms with certain pore shapes.[75] Nevertheless, there seems to be nothing unique about the structures of these parts,[76] and at other times, such as when dreams and vision enter the "interstices of the body" [*corporis rara*],[77] or when the whole body feels hunger, Lucretius implies a more holistic sensory mechanics. Food does enter the stomach and, contrary to Aslcepiades's claims, is "digested" [*concoctum*]; this food is then given to the limbs, but details are scarce.[78] When Lucretius speaks about sleep or illness, both deviant states, he emphasizes the disruption of the body as a whole. An overflow of bile or the power of a disease can cause fever and throws the "whole body" [*totum corpus*] into disarray by rearranging the corporeal passageways.[79] Similarly, we press out voice "with our body" [*cum corpore nostro*] and the lungs are not mentioned.[80] In short, like Asclepiades, Lucretius does not privilege functional organs in his corporeal mechanics, even if individual parts appear in his explanations.

This approach should be understood as part of Lucretius's rejection of teleology in nature, which Asclepiades shares, insofar as he insists that animals utilize body parts because they already exist, not because they are created for certain uses. Artificial things, such as shields, mattresses, and cups, were invented for use cases [*utilitas*], but the sensory parts and limbs are different types of objects.[81] For moderns accustomed to the

75. Lucr. *DRN* 4.617–657.
76. See esp. Lucr. *DRN* 4.642–657.
77. Lucr. *DRN* 4.730.
78. Lucr. *DRN* 4.627–632.
79. Lucr. *DRN* 4.663–672.
80. Lucr. *DRN* 4.549–552.
81. Lucr. *DRN* 4.843–856.

arguments of evolution, it is easy to assume that this rejection of intelligent design simply implies that Lucretius accounted for the complex functions of the body through the mechanical interactions of atoms. That is, we assume that he believed in something akin to our organism, even as he rejected divine design as responsible for creating it. It is crucial to recognize, however, that Lucretius's rejection of natural teleology and divine intention also implies the rejection of structural teleology. That is, by insisting that parts were not made by nature with a particular function in mind, he also stops using function to understand corporeal behaviors.

This rejection appears perhaps most visibly when he explains voluntary animal movements, which he argues are caused by images striking our mind [*animus*], which then, after an act of intellection, moves the atoms of the spirit [*anima*]. These spirit atoms then pour through the entire body and push it in certain directions. At the same time, the body grows less internally dense [*rarescit*] and thus air "in abundance comes and penetrates the opened passageways [*patefacta foramina*] and is thus spread into the smallest parts of the body."[82] These two things, spirit and air atoms, move the body "like a ship by sails and the wind" [*ac navis velis ventroque*].[83] In other words, gone are sinews, muscles, or any individuated operational parts. What is left is a passive vehicle where individuated active structures play little to no role. This paradigm extends to his understanding of life itself, which results from the presence of soul and spirit atoms, making the body a type of container, rather than a functioning machine.[84]

Asclepiades's theories can be seen in a similar light. He likewise rejected teleology both at the level of design and, by extension, at the level of structure. Accordingly, his conflation of the visible and invisible vessels creates little problem for him, since he is not particularly interested in tracing the internal boundaries of the parts in order to cordon off the operations that they complete. This approach does not make him merely

82. Lucr. *DRN* 4.877–897.

83. Lucr. *DRN* 4.897. What is interesting is that Lucretius then claims that there is no need to be surprised that such small atoms can move the large body, since small rudders turn entire ships and machines can lift huge weights with slight motions on treadmills—all without applying any of the mechanical principles to his own model; see *DRN* 4.898–906.

84. We might compare this understanding to even earlier assertions of Democritus, who argued that in the uterus, the external parts of the embryo develop first, which likewise suggests the body is first and foremost a container; see Arist. *Gen. an.* 2.5, 740a14–17; cf. 740a34–b1.

ignorant of internal anatomy so much as it renders him uninterested in both anatomical investigation and the epistemic object that it creates.

Despite a nonteleological framework and a body that did not make organs the crucial inner divisions, Asclepiades still included several aspects in his accounts that seem more native to the organism, or at least to the philosophical tradition from which it emerged. For example, he seems to have spoken about both the pulse and respiration.[85] This account illustrates how important these corporeal behaviors had become, such that even someone without an interest in explaining their function still felt the need to account for them. In fact, when Asclepiades explains the latter with reference to two technological analogies, he does so to explain only the dynamics of inner processes, not what purpose breathing serves. As Aëtius reports:

> Asclepiades understands the lung to be like a funnel [χώνης δίκην], but he posits the cause of respiration to be the low density in the thorax [τὴν ἐν τῷ θώρακι λεπτομέρειαν], toward which the outside air flows and moves because it is of higher density [παχυμερῆ ὄντα]. And it is driven back out when the thorax is no longer able to receive or contain any more; but since some small quantity of low density always remains in the thorax (for not everything is expelled), toward this the remaining bit travels back up out toward the outside heaviness again. Indeed, these things are similar to what happens in cupping glasses. And he says that voluntary breathing happens when the thinnest passageways in the lung are gathered together [συναγομένων τῶν ἐν τῷ πνεύμονι λεπτοτάτων πόρων] and the throat contracts; for these things comply with our will.[86]

The funnel, generally used as a unidirectional device, is here treated as facilitating bidirectional flow, moving back and forth according to some dynamic relative density. The lungs thus first become little more than a passive gateway mediating the exterior and interior spaces and facilitating "the movement toward what is sparse." The basic expansion and contraction of the lungs is nowhere to be found, and his second technological analogy, this time to a cupping vessel, likewise configures them as static in the case of nonvoluntary breathing. Lastly, in both cases, the function that these technologies generally serve—to help channel liquid and to attract or extract bodily fluids, respectively—do not here imply a clear purpose.

85. See entry for Herophilus T157 von Staden.

86. Aët. 4.22.2.

According to Galen, however, Asclepiades elsewhere claimed that breathing brings soul atoms.[87] Nevertheless, this purpose should not be quite understood as a *physiological* function, since these soul atoms do not operate within an interdependent corporeal system. In other words, Asclepiades's account does not truly explain how the body works, since it is only the mere presence of these soul atoms that explains life. The body becomes a mere bag of the right type of matter, not an operational system.

We should also note that Asclepiades also mentions voluntary inhalation, which he attributes to a separate mechanism: the active contraction of the passageways within the lung and the throat. He thus uses tools to explicate the physical mechanisms involved in the regular, automatic activities of the body but does not invoke any such technologies to describe a voluntary action. Indeed, Asclepiades is often treated, with Erasistratus, as a "mechanical" thinker, because he explained all (or almost all) corporeal dynamics with a basic physical principle applicable to both living and nonliving entities. The voluntary contraction of the throat and the passageways of the lungs complicates this assertion, and since Atomism and mechanics were never treated as intrinsically related in antiquity, we may reevaluate whether this terminology truly applies to him. Although he does construct a body whose parts are passively subject to physical forces, doing so moves the parts further away from their conceptual role as corporeal tools with active mechanisms.

5.5 ASCLEPIADES AND AQUEDUCTS

Because none of Asclepiades's treatises are extant, it is hard to assess how important technologies were for explaining internal fluid dynamics more generally beyond this one example, especially since respiration already provides a relatively complicated picture with no straightforward relationship between the funnel and cupping vessel and the act of respiration. That said, interpreting his theories within the context of the second century CE, where water infrastructure was becoming increasingly sophisticated, monumental, and integrated into daily lives, can provide a way to assess how important technologies still were to his medical theories. In fact, several features of contemporary water technologies mirror his basic pathological particulars. For example, Cassius Iatrosophista, a second- or third-century CE follower of Asclepiades and author of a medical treatise entitled *Physical Problems*, asks why it is that when someone stubs

87. Gal. *Us. Resp.* 2.11 = 4.483–484K.

their foot, bruising only occurs nearby, but areas further away can also be affected, such as when distant glands swell after the injury. He provides a potential answer:

> In his *On Wounds*, Asclepiades says that material is first carried to the parts that have been struck. The material is carried there and approaches the affected parts in proportion to their ability to accept it. When they are full, and can take in no more, the matter carried there flows out, and since it has not been accepted by the parts to which it was born, it is then carried on. If it reaches hollow places, it stays in them, as happens in the case of water, which, as long as it is borne along on a level surface, uses an even motion, but on coming across hollow places it remains in them. The same thing happens to the material that is borne toward the parts that have been struck. As much material as can be accommodated moves into them, but the rest goes into hollow spaces, and especially into the glands with their fine pores, making them swell. This is a convincing argument.[88]

The passage does not declare what type of channel this water is flowing in before it spills over into the "hollow spaces," but the hollows within a sponge clearly do not fit the model. Instead, Asclepiades here seems to have exploited, or at least conflated, the notions of a "void" and of a "passageway" in the human body, since if the body was simply a network of interstices, bruising and soreness should affect surrounding areas in direct proportion to proximity. His use of πόροι allows him to involve watercourses as part of his conception of the corporeal interior. Indeed, what should we make of the comparison to water flowing across a "level" surface? Given the amount of technical sophistication that was required to level aqueducts and the ostentatious straight lines that aqueducts produce across the horizon, we might be more inclined to conclude that Asclepiades is imagining some sort of man-made, open-air watercourse.[89] Indeed, this type of masonry conduit formed the largest portion of these aque-

88. Cass. Iatr. *Pr.* 41, trans. Vallance with emendations. See Vallance 1990: 87; cf. Cass. Iatr. *Pr.* 1, which discusses why a round wound does not heal as fast with recourse to the observation that a river is stronger at its center, and thus the flow of material in the center of the wound must keep that spot from healing quickly.

89. Vitr. *De arch.* 8.5 mentions the *dioptra* and the *chorobates* as two instruments used to level aqueducts.

ducts,[90] which drew their water largely from springs (although the Tepula drew from a lake, and rivers were also a possible source).[91] These watercourses were most often covered, but only just below ground level. Even so, they were generally large enough for a human to enter so as to conduct maintenance. They were not designed to run full, and water flowed only about a third of the way up the wall. As Hodge states: "It helps to consider the aqueduct almost as an artificial river rather than as a water main."[92]

Although aqueducts running into Rome did not need to incorporate inverted siphons thanks to the gradient of the surrounding land,[93] many lines elsewhere in the empire did include them.[94] Once the water reached the city, it was fed into a settling tank [*piscina*] that filtered out large debris mechanically.[95] These settling tanks were generally connected to a *castellum divisorium*, which diverted water to the three main delivery points: private homes, public fountains, and the baths, the last of which was the largest consumer of water.[96] In private homes, water was channeled to the *impluvium* in the atrium, where it was connected to a lead box fitted with taps that supplied domestic fountains, kitchens, and other such areas.[97] In the baths, water ran into boilers and cisterns through a series of pipes, which allowed waters of different temperatures to be mixed to suit each pool. After use in the various rooms, the water would have been used

90. Although Vitr. *De arch.* 8.6.1 cites only three types of channels (masonry conduits, lead pipes, and terra-cotta pipes), water systems incorporated other materials as well, including stone-cut channels, earthen trenches, clay-lined gullies, and wooden pipes; cf. Hodge (1992) 2002: 106. The surface channel, however, made up some 80–90 percent of the total length of Roman aqueducts; cf. Hodge (1992) 2002: 93.

91. Surface water from lakes and rivers was used far less often, although both the Anio Vetus and Anio Novus drew from the Anio river. Frontin. *Aq.* 1.4 asserts that Rome drew its water from the Tiber for its first four centuries, although springs and wells supplemented this source. Springs generally supplied better quality water.

92. Hodge (1992) 2002: 2.

93. Hodge (1992) 2002: 17.

94. For information about inverted siphons during the Roman period, see Ortloff 2009. This technique may have been imported from their contact with Pergamon; cf. Lewis 2000a: 647.

95. Kleijn 2001: 31.

96. Vitr. *De arch.* 8.6.1 describes a *castellum* that seems largely idealistic given the archaeological evidence. Frontin. *Aq.* 78.3 states that there were 247 castella in his time.

97. Jansen 2000: 115; cf. Hodge (1992) 2002: 115; Humphrey 2006: 47.

to irrigate the toilet troughs. In general, any water that was not consumed eventually helped the sewage and drainage system.[98] Aside from the few uses in the home that could be turned off with taps, the whole water system of Rome largely operated without valves on a principle of constant outflow, so while individual pieces of equipment could be disengaged, the water itself generally needed to be turned off at the source, lest the incoming stream simply overflow its channel and burst pipes. Although it cannot be determined with any degree of certainty, to my mind, it is very telling that blockage within Asclepiades's body leads to spillover into other parts, rather than a simple stoppage of flow, as would happen in a closed system. His body thus mimics the constant flow of Roman water infrastructure in this basic regard.

To be sure, water runs in rivers as well as aqueducts, so no certain connection can be made between Asclepiades's application of fluid dynamics and the Roman water supply system. Nevertheless, other passages seem to indicate that Asclepiades incorporates characteristic features of water engineering into his accounts. For instance, Caelius Aurelianus preserves a passage that mentions again how blockages occur:

> However, [Asclepiades] says that a blockage of these [corpuscles] is caused by the size, shape, multitude, or speed, or he says that different affections arise by the bending of the passageways and by the *blockade of little scales in different places and passageway* [*conclusione atque squamularum *exsputo* varias inquit fieri passiones locorum aut viarum differentia*].[99]

Editors have had considerable trouble with this passage, especially with "little scales" [*squamulae*] and have offered multiple emendations as a result. Drabkin emended the second sentence to read "*conclusione corpusculorum effecto*,"[100] while Vallance suggests that "little scales" [*squamulae*] refers to fragmenta, and thus indicates corpuscles that have broken off the sides of the passageways.[101] I would like to suggest that no emendation of

98. For instance, the most famous drain, the Cloaca Maxima, probably built by the Etruscans in the sixth century BCE, had other storm drains and sewers connected to it, which it fed into the Tiber; cf. Humphrey 2006: 49; Camardo et al. 2011; Pérez et al. 2011.

99. Cael. Aur. *Acut. Pass.* 1.107, emphasis mine.

100. Drabkin 1950; cf. Vallance 1990: 115 for references to other editions.

101. Vallance 1990: 114–15.

squamularum is necessary and that the term likely comes from a common and well-described problem within Roman waterways: sinter deposits.

One of the major concerns in designing aqueduct channels is determining their gradient. A steep slope causes a swifter current, which causes considerable wear on the aqueduct itself. A gentle slope produces a slower current and less wear but leads to a rapid accrual of calcium carbonate deposits, or "sinter." This buildup constricts the flow of water and can completely block pipes.[102] In fact, sinter accumulated with such speed that workmen needed to chip it away before the channels became clogged and overflowed. Frontinus describes large crews of workmen who were tasked with this type of repair:[103]

> These [necessary repairs] arise from two reasons: either lime, which sometimes hardens into a crust, thickens, and the path of the water is constrained, or the walls crumple, whence leaks necessarily damage the sides of the stream and the substructures.[104]

Despite its massive and extraordinary scale, the whole system was under threat of blockage and disruption. The pipes needed to be cleaned, maintained, and replaced lest malfunctions cause considerable damage to property and public health.[105] It was a massive, complicated, and conspicuous apparatus that required constant attention and upkeep. Asclepiades, as it seems, transferred this notable "pathology" of Roman waterways onto the watercourses of the body, which mimic the breakdowns, block-

102. The minimum gradient needed for gravity-fall conduits was 1:200 according to Vitruvius and 1:4800 according to Pliny. Physical evidence shows that the slopes are as little as 1:1200 and as great as 1:95; cf. Humphrey 2006: 44. Hodge ([1992] 2002: 216) puts the average gradient of flow somewhere between 3.0 and 1.5 m per km fall.

103. Frontin. *Aq.* 2.116–121 describes the crews of workmen and the constant upkeep to which they attended. He mentions two groups, those left by Agrippa to Augustus, numbering 240, and those of Caesar, originally organized by Claudius and numbering 460. Among these groups there were overseers, pavers, plasterers, etc.; cf. *Aq.* 2.96. For a fuller account of sinter and its accumulation, see Hodge (1992) 2002: 98–105, 228 and Wilson 2008: 299–300.

104. Frontin. *Aq.* 2.122.

105. Cf. Frontin. *Aq.* 1.7, which suggests that the Appian and Anio had already fallen into disrepair by the mid-second century BCE. Sewer systems and public toilets suffered this same threat. Despite the fact that water from the baths was diverted through attached public toilets in periodic deluges, much of the excrement would have required manual removal; cf. Camardo et al. 2011.

ages, and bursts of aqueduct infrastructure. *Squamulae* would thus refer to flakes or "scales" of sinter that break off from the sides of vessels.[106]

Another feature of Asclepiades's disease etiology reflects Roman water infrastructure. It occurs when he explains dropsy:

> And not all diseases arise from the blockage of corpuscles, but certain ones do, like phrenitis, lethargy, pleurisy, and severe fevers; fluid diseases are caused by the disturbance of liquid and pneuma. In this way, Asclepiades thinks that bulimia arises because of the size of the pores in the stomach and in the abdomen; he says that fainting and the flux of the body, as well as uncontrollable laxity, arise because of the looseness of the pores; in the same way, *dropsy arises by boring a small conduit into the flesh* [*perforatione carnis in parvam formulam viarum*] which is able to turn the accustomed nutrition of the body into water.[107]

This language of "boring" [*perforatione*] a small conduit [*parva formula*][108] into the flesh reflects a difficulty that Frontinus later describes in *De aquaeductu*—namely, that people illicitly bored holes into water pipes to draw water for their own personal use without paying taxes for the connection. He describes this fraud as a major problem that he intends to solve:[109]

> But many landowners whose fields the aqueducts flow around *bore into the conduit* of the stream [*formas rivorum perforant*], whence it happens that private citizens stop the public watercourses just to use it for their gardens.[110]

It does not take any particular doctrinal commitment to interpret dropsy—the visible accumulation of fluid—as an unsanctioned boring into a fluid-carrying vessel. Nevertheless, such an interpretation shows precisely the type of scale conflation that can occur between Asclepiades's micro- and macropassageways in the body, as he moves between the unseen pores within the body and their larger, visible counterparts. The effects that aqueducts had on the history of physiology are hard to pin down, especially

106. Pliny mentions "iron scale" [*squama ferri*] "from a sharp edge or point" (*HN* 34.46.154).

107. Cael. Aur. *Acut. Pass.* 1.107–108, emphasis mine.

108. Cf. Frontin. *Aq.* 1.37 for this use of *formula*.

109. Volk (2010) examines this type of theft and its poetic use in Manilius.

110. Frontin. *Aq.* 1.75; cf. Frontin. *Aq.* 2.87; 2.115; 2.128.

since none of the principles involved in aqueducts are new but simply expand technology already implemented in the classical and Hellenistic worlds. Nevertheless, the increasing conspicuousness of water infrastructure and its particular social and technical challenges seems to have provided a useful template for Asclepiades to describe the pathologies of a body built from flowing corpuscles and the passageways that they occupy.[111]

The difficulties in securely identifying moments in which technologies altered Asclepiades's theories might invite us to ask another question instead: Why was Asclepiades's particular physiology so *popular* in first-century BCE Rome? If his medical theories simply deployed Epicurean ideas to explain human physiology, why were his arguments so well received? To paraphrase Cassius Iatrosophista, why did patients think that his pores, corpuscles, blockages, and flows amounted to a "convincing argument"? Asclepiades was so popular that Pliny felt he had to fabricate an explanation for it, and he was influential enough for Galen to attack almost three hundred years later. Cicero cites his rhetorical skill as superior to that of other physicians,[112] while Pliny claims that Asclepiades was trained as a rhetorician before he turned to pursue a fraudulent career in medicine and that he relied on his power of persuasion to attract patients.[113] The other commonly cited reason for Asclepiades's popularity is that he pampered his patients (in contrast to other Greek physicians, who were seen as knife-happy butchers). Yet Celsus suggests that Asclepiades was not always so gentle after all, but "when he had exhausted the patient for three days with total abstinence, he prescribed food on the fourth" [*Asclepiades ubi aegrum triduo per omnia fatigarat, quartum diem cibum destinabat*].[114] Although more lenient treatments such as moderate wine and gentle rocking were prescribed thereafter, Asclepiades's remedies do not seem wholly comparable to spa treatments. His popularity therefore does not seem to be fully explained by the fact that he prescribes solely pleasurable remedies. Moreover, even if his rhetoric won him patients in the short term, it does not explain how he established a medical succession, with disciples across the Roman Empire for several hundred years after his death.[115] We cannot attribute this legacy solely to his personal mel-

111. In this way, his return to and expansion upon a corporeality of flux mimics the way aqueducts expanded on previous water supply technologies.

112. Cic. *De or*. 1.62.

113. Plin. *HN* 26.12.

114. Celsus *Med*. 3.4.6.

115. Papyrological evidence attests to a school of doctors proclaiming themselves to be Asclepiadeans up until at least the third or fourth century CE; see

lifluousness. Instead, we can ask whether his explanations of blockages and flow modeled the human body on the everyday imperial infrastructure, which already itself formed a core part of their health regimens. This model of fluid dynamics thus appeared extremely persuasive and comprehensible to a Roman populace already attending to their bodies in public baths fed by large-scale aqueducts. In this way, Asclepiades's arguments—whatever their specific medical or philosophical merit—seem to tap into a shared cultural heuristic already associated with health and illness. Imagining that scientific ideas get developed in some philosophical vacuum, wherein the influence of one's predecessors is the only palpable force, would be to miss how medical ideas need to be not only logical but also accessible and convincing.

5.6 METHODISM AND ORGANIC ACTIVITIES

In the first century BCE, another Greek medical sect emerged in the Roman world. These Methodists, so named because they championed a particular "method" [μέθοδος] of medicine, was reportedly founded by Themison of Laodicea (ca. 120–40 BCE) but was "completed" by Thessalus of Tralles (fl. ca. 20–70 CE) in the century following. The sect became quite popular, and over the following century rose to become one of the most prevalent medical schools in the Roman Empire, with notable practitioners including Mnaseas, Menemachus, Olympicus, Soranus, Julian, and Caelius Aurelianus.[116] The Methodists bring with them a complex, complicated history, with internal disagreements that can be hard to parse, especially since texts about their core ideas have been lost and their ideas are reported to us by sources often deeply hostile to their project.[117] In general, though, it is safe to say that the Methodists adapted a basic pathological system that loosely resembled the blockage and flow espoused by Asclepiades but dispensed with his corpuscles and pores, claiming that all such "hidden causes"—whether atoms, elements, or humors—were fit for philosophical speculation but ultimately useless for the physician. Although

Leith 2016 for references. See also Cael. Aur. *Acut. Pass.* 3.113 and Plin. *HN* 29.6, who name other Asclepiadeans.

116. See Gal. [*Int.*] 14 = 4.684K = fr. 283 Tecusan for a full doxographical list; cf. Gal. *MM* 1.2 = 10.7–8K = fr. 162 Tecusan. For dates of these figures, and other Methodists, see Tecusan 2004: 14; cf. Berrey 2017b.

117. See Tecusan (2004: 1–43), who provides a concise overview of Methodism and emphasizes these difficulties. For other accounts, see Webster 2015; Nutton 2004: 187–201; Gourevitch 1991; Pigeaud 1991: 9–50; Rubinstein 1985; Frede (1982) 1987; Edelstein 1967: 173–91.

this approach may at first seem to align them with the Empiricists, the Methodists argued against these predecessors, too, insisting that basing treatments solely on the resemblance between present and past cases was mere guesswork and could not serve as the foundation of a medical techne. Instead, they maintained that even without causes, proper treatment could be known with logical certainty. Illnesses could be identified and categorized, insofar as all corporeal affections were instantiations of one of three "manifest commonalities" [φαινόμεναι κοινότητες]: "constriction" [στέγνωσις/*strictus*], "flow" [ῥύσις/*laxus*], or some mixture of the two.[118] These affections could be apprehended directly [καταλαμβάνειν] by the trained physician without the need for intervening causal knowledge or investigative practices.[119] Accordingly, they put forward a distinctly nonetiological medicine that they claimed treated disease with both efficacy and certainty.

Their nonetiological method altered the vocabulary of illness, as they left aside the nomenclature of "diseases" [νόσοι]—likely because it both reified the disease and transformed it into a quasi-causal agent—and shifted toward the terminology of "affections" [πάθη], or, more accurately, "affections contrary to nature" [τὰ παρὰ φύσιν πάθη].[120] Using this designation—likely influenced by Stoic epistemology[121]—the Methodists could remain neutral on the question of both disease ontology and causes. These affections were many and took over disease names with long currency in the medical tradition, such as phrenitis, lethargy, dropsy, and more, but Methodists now defined these ailments by sign sets that identified each affection and allowed them to apply the appropriate therapy. That is, although it acknowledged only three commonalities, Methodism presented a robust and nuanced medical practice that relied on far less information.

The implications and contours of this approach to medicine are complex, and the notion of rejecting causes both angered ancient antagonists and confuses modern commentators. To clarify some issues, it helps to recognize that when the Methodists rejected "hidden causes," they seemingly reject several different types of causality together. First, they

118. E.g., see Gal. *Sect.* 6 = 1.80K.

119. On the complicated sign sets used to see these "manifest" states, as well as the ontological complications that followed from a nonetiological medical system, see Webster 2015.

120. Galen charges Thessalus with using νόσοι and πάθη interchangeably; see *MM* 1.7 = 10.52K = fr. 162 Tecusan.

121. See Frede (1982) 1987: 261–78 for the relationship of Methodism to Stoicism.

downplay or ignore the so-called precipitating causes [τὰ προκαταρκτικὰ αἴτια] that both Hippocratic and Empiricist physicians had spent great efforts cataloging, such as bouts of drinking, excessive eating, or overexertion, considering these sorts of causes too numerous to evaluate and unnecessary for treatment decisions.[122] Second, they also dismissed the relevance of hypothetical interior "causes" to medicine—causes such as Asclepiades's pores as well as humors, breaths, and other traditional pathological substances. Third, they also scorned anatomical investigations and the type of inner physiological picture that such investigations had articulated, likewise holding these inner activities to be hidden and unknowable. For instance, Soranus of Ephesus (98–138 CE), a Methodist author of the first century CE,[123] wrote a treatise on *Gynecology* in which he rejects "physiology" [τὸν φυσικὸν] as "useless" [ἄχρηστον] for his purposes, insofar as it is "an ornament for fancy learning" [φερέκοσμον πρὸς χρηστομάθειαν] and unnecessary for his account about women's health.[124] He then repeats this claim, making clear that he sees dissection and anatomy as a core part of this medical approach.[125] It has confused commentators and translators that Soranus then goes on to display precise anatomical knowledge of the uterus and female reproductive anatomy in the sections that follow, especially as he frames learning anatomy as largely performative, so that "we shall not provoke the suspicion that we denigrate something accepted as useful because of ignorance."[126] Yet the knowledge Soranus displays seems to be more than a simple set piece, and he is not the only Methodist to demonstrate anatomical acumen. Other texts describe surgeries that include detailed knowledge of vessels, muscles, and nerves.[127] Generally, this juxtaposition strikes commenta-

122. See Gal. *AA* 3.1 Singer = 2.343–344K = fr. 112 Tecusan; Gal. *Sect.* 7 = 1.84–87K = fr. 203 Tecusan; Gal. [*Ther. Pis.*] 14 = 14.277–280K = fr. 208 Tecusan; Celsus *Med.* 1 *pref.* 54 = fr. 100 Tecusan; Sor. *Gyn.* 3.1.66–67 Burg. = 3.4 Ilb.; Cael. Aur. *Acut. Pass.* 3.19.190, 2.13.87; and Cael. Aur. *Tard. Pass.* 2.14.19. See also Galen's *On Antecedent Causes*. Although they devalued this information, the Methodists would recognize *procatarctic* causes if they changed therapy; see, e.g., Cael. Aur. *Acut. Pass.* 3.1.4–6, 3.9.99; *Tard. Pass.* 2.10.125.

123. For bibliography on Soranus, see Tecusan 2004: 2 n1.

124. Sor. *Gyn.* 1.1.31–33 Burg. = 1.2 Ilb.

125. Sor. *Gyn.* 1.3.37–45 Burg. = 1.5 Ilb. uses almost the same language as the previous passage, stating that dissection is useless [ἄχρηστός] and undertaken for the sake of fancy learning [χρηστομαθείας ἕνεκα]. See also Celsus *Med.* 1 *pref.* 64 = fr. 100 Tecusan; Sor. [*Quaest. med.*] 51 = fr. 295 Tecusan. For a discussion of this fact, see Lloyd 1999: 189.

126. Sor. *Gyn.* 1.3.44–45 Burg. = 1.5 Ilb.

127. Tecusan 2004: 31.

tors as either hypocrisy or contradiction. Others ascribe it to the resolutely practical approach of the Methodists, for whom anatomy is not useful—except insofar as it is. Yet this interpretation, too, creates a tension in Soranus's more totalizing negative assessment and the Methodists' general, nonetiological medical framework.

These combined complications have made the Methodists' position difficult for many to comprehend, let alone accept. Understanding their position involves situating it within Hellenistic epistemology, especially Stoic debates about *kataleptic* impressions, but it also requires recognizing that in the Hellenistic period, several things had been bundled together, most notably anatomical investigations and the structural teleology that underpinned a tool-like organism full of mechanisms and functions. By taking up an epistemological position that distrusted the investigative techniques used to reveal inner activities, the Methodists also devalued the type of corporeality that these investigations supported. That is, to reject a certain type of epistemic practice was also to relinquish the epistemic object that it sustained. Yet one of the major difficulties for understanding the Methodists' notion of corporeality is assessing whether they are staking an ontological position or merely an epistemological one—or whether such positions can ever truly be separated. Are they claiming that the body very well may be organized in the ways that philosophers and philosophically aligned physicians claim, but we can neither gain sufficiently certain knowledge of it, nor is this information useful? Or are they saying that the body is actually arranged according to some other logic, perhaps closer to, although not identical with, Asclepiades's model of flow and obstruction?

Some aspects point to the latter, since neither Soranus nor the other Methodists explain the physiological processes in the body. Similarly, they stopped discussing "organs" and returned to the vocabulary of "affected parts" and "places." Moreover, as part of this nonetiological orientation, they likewise treated "affections" as states of the whole body, rather than the breakdown of certain inner processes. Disease could manifest in certain corporeal localities, but they did not translate into physiological malfunctions, only spaces of constriction or flow. The Methodists also display no commitment to teleology at a cosmic level and in no way suggest that the body is designed by either nature or a demiurgic god to function in certain ways. Yet certain patterns of thought that were native to thinking about the body as a teleologically arranged organism also surface in their accounts, which may imply a tacit acceptance of corporeal organization. For example, Thessalus and other Methodists, according to Galen, "say that health is the stability and strength of the activities in accordance with nature" [τὴν μὲν ὑγείαν εὐστάθειαν τῶν κατὰ φύσιν ἐνεργειῶν εἶναί φασι καὶ ἰσχὺν], even though they did not define illness as a disruption of these activities but as either "a certain state of the body" [διάθεσίν τινα σώματος] or "the body disposed somehow in a certain state" [πως διακείμενον σῶμα].[128] Soranus likewise uses the vocabulary of corporeal "activities" [ἐνέργεια]—including that of the womb—which is an assertion more native to functional models of corporeality.[129] Elsewhere he describes why drunkenness can affect fertility, saying, "The body in its natural condition is productive of its proper enactments [τῶν ἰδίων ἐνεργημάτων], but it is not in a natural condition when it has been set in drunkenness and indigestion; and so just as no other natural task [φυσικὸν ἔργον] can be managed in such a state, neither can conception."[130] Even as the Methodists reject an etiological account of disease that relies on knowing the functions and physiological activities of the body, they still tacitly accept inner mechanisms, activities, and tasks. In some sense, they present a body that contains functions without physiology.

This tension, I would argue, arises from the strength that the organism had gained as an operational model of the body in the Hellenistic period. The Methodists shifted the vocabulary slightly to accommodate the tool-like body within their epistemology, but the notion that the body parts performed vital tasks was not something that they dislodged. Indeed, even the focus on affected parts rather than substances reflects the success of the organism in directing medical attention toward corporeal structures.[131] Although they did not seek to articulate the body as a functional object, nor to delineate the corporeal mechanics required to sustain life, neither did they reject this theory of corporeality outright.

128. Gal. *MM* 1.7 = 10.51K = fr. 162 Tecusan. Similarly, Galen reports that Olympicus of Miletus, another Methodist physician of the first century CE, defined a state of pathological affection as "an enduring change from the natural to unnatural," insofar as health constitutes a condition defined as "the stability and strength of the natural enactments [εὐστάθειαν τῶν κατὰ φύσιν ἐνεργημάτων καὶ ἰσχύν]" (*MM* 1.7 = 10.54K). See also [*Def. Med.*] 133 = 19.386K, which includes a related definition of disease as an "impediment of natural capacities" [τῶν φυσικῶν δυνάμεων παραποδισμός]. For the first-century date of this treatise, see Kollesch 1973: 60–66. Salas (2020b) outlines the history of such definitions.

129. For the function/activity of the womb, see Sor. *Gyn.* 1.9.56–58 Burg. = 1.31 Ilb. For other uses of this term, see *Gyn.* 1.14.32–34 Burg. = 1.43 Ilb.; 2.10.107 Burg. = 2.27 Ilb; 2.11.15 Burg. = 2.28 Ilb.

130. Sor. *Gyn.* 1.12.84–87 Burg. = 1.38 Ilb. Soranus likewise uses the term "respiratory organs" [τὰ ἀναπνευστικὰ ὀργάνα] (see *Gyn.* 2.13.76 Burg. = 2.39 Ilb.), but only once.

131. Cf. Tecusan 2004: 11.

The role that technologies play within such a medical system to articulate and imagine the body is relatively limited. Since the Methodists did not expound upon inner physiology, they did not use tool analogies to explain either inner mechanisms or functions.[132] Moreover, they did not use material instruments or therapeutic substances as investigative interventions to reveal what was hidden and instead argued that the body largely discloses itself to those properly trained, at least as it relates to medicine. In this regard, their version of corporeality is sustained by the success of their therapeutics. To this end, they initiated a relatively standardized practice of care that centered on a three-day cycle (counted inclusively) called the *diatritus*. This cycle started with an initial day of rest and fasting, which targeted the whole body, before food was given on alternating days along with harsher and more directed treatments called "metasyncritic" remedies that were applied to individual affected parts and regions.[133] Their therapeutics incorporated traditional pharmaceuticals but were directed toward relaxing or tightening the body and were generally more moderate than Hippocratic treatments. They created a new pharmacological logic and, with it, supported a different medicalized body.

5.7 SORANUS AND FEMALE CORPOREALITY IN METHODISM

One place where a different set of material tools reflects a different corporeality is in Methodist gynecology. In the fifth and fourth centuries BCE, Hippocratic gynecological authors had argued that women's bodies were intrinsically different from those of men, insofar as female flesh was porous and more wool-like in composition, which caused it to soak up excess blood. Menstruation was required to purge this extra fluid, and the failure to do so regularly caused illness. This view of female corporeality as intrinsically overdamp brought with it a program of therapeutic interventions. Hippocratic treatises such as *Diseases of Women* were replete with emmenagogues, including irritating uterine suppositories, where the distinction between normal menstruation and artificially produced bleeding was all but erased. Female corporeality necessitated an accompanying set of medical technologies, and these therapeutic tools sustained the notion that women were essentially different from men. By contrast, Soranus's *Gynecology*, written in the early second century CE, notes that some

132. When describing the uterus, Soranus does invoke the comparison of the uterus to a cupping vessel, but in doing so seems to refer solely to its shape. This same comparison is made by Ruf. Eph. *Anat.* 64 Gersh and Gal. *Sem.* 2 De Lacy = 4.516K.

133. See Leith 2008 for a description of the *diatritus*; cf. Nutton 1990. Scarborough (1992) outlines Methodist pharmacology; cf. Tecusan 2004: 10 n13, 34.

of his predecessors—among whom he notably does not list Hippocrates—claimed that women have their own affections [ἴδια πάθη], and he cites the Empiricists, Diocles, Athenion and Miltiades the Erasistrateans, Lucius the Asclepiadean, Demetrius of Apameia, Aristotle, and Zenon the Epicurean.[134] By contrast, Soranus joins Herophilus, the Asclepiadeans, Themison, and Thessalus in asserting that women are not afflicted by a unique set of pathologies, nor are women's bodies essentially different as a whole [τῇ ὅλῃ φύσει διαφέρει], although he clarifies:

> We say that unique affections in accordance with nature [κατὰ φύσιν ἴδια πάθη] do exist for women (for example, conception, birth, milk production, if one wishes to call these tasks [τὰ ἔργα] affections), but there are no affections contrary to nature unique in kind, but only unique in presentation and in the part that they can affect.[135]

In other words, Soranus claims that women have no unique diseases, since stricture and flow still encompass all their affections, but that these commonalities manifest differently in the uniquely female components. For Aristotle, treating the body teleologically meant creating a hierarchy of corporeal goals, and since reproduction was one of those goals, it meant arranging male and female bodies hierarchically according to their different contributions to generation and different functional parts. Since the Methodists did not treat the body as a teleologically structured object, they did not define women by any inferiority, even though they accepted the inferior status of women as another type of fact. Still, Soranus seems to negotiate teleology and function in other ways, notably making a distinction between what is healthy and what is natural, claiming that childbirth and menstruation harm the female body even though they are natural acts.[136]

Since Soranus is light on philosophical musings, it is difficult to trace a direct line from this stance to other features in his gynecological practices. Yet Hippocratic gynecology prescribed a narrow normative range of two cotyles in menstrual volume, which, as chapter 1 highlighted, is

134. Sor. *Gyn.* 3.1 Burg. = 3.1–5 Ilb.

135. Sor. *Gyn.* 3.1.86–90 Burg. = 3.5 Ilb.

136. Sor. *Gyn.* 1.8.51–69 Burg. = 1.28 Ilb. This claim extends back to Herophilus's theory that childbirth and menstruation are helpful to some women, harmful to others (T203 von Staden). Soranus also occasionally adopts quasi-Aristotelian and Stoic terminology by referring to conception as a "natural act" [φυσικὸν ἔργον] and "natural enactment" [φυσικὸν δὲ ἐνέργημα] (*Gyn.* 1.13.2–4 Burg. = 1.42.1 Ilb.). For an exposition of these dynamics, see Freidin, forthcoming.

about half a pint. Not only is this amount far more than what we now consider normal, it likely reflects the volume of douches used to stimulate menstruation or purify the uterus to facilitate conception. By contrast, for Soranus, two cotyles represents the upper limit of healthy menstrual volume, rather than a norm. Moreover, he accepts a far wider range for normal menstruation and only treats its absence when it is pathogenic or problematic. That is, he acknowledges that some women are too young or too old to have a period, but he also notes that some women fail to menstruate because they exercise a lot or conduct strenuous vocal exercises. Such amenorrhea would have been seen as catastrophic by the Hippocratics, but Soranus insists that the Hippocratic understanding need not be correct, since the women in question are not even suffering an "affection." Unless a woman wants to conceive and is having trouble doing so, no treatment is necessary.[137] In short, when Soranus decoupled female corporeality from the constant technical interventions required by Hippocratic theories, menstrual behaviors stopped resembling therapeutic tools.

Soranus advises limited medical intervention designed to provoke menstruation and rejects the excessive use of emmenagogues, claiming that "the ancients made a mistake inserting so-called emmenagogues to draw down blood and prescribing drinks able to do the same things, not seeing that the drinks damage and upset the stomach, and the vaginal pessaries ulcerate the womb by scraping it."[138] He goes on to describe how these suppositories cause scar tissue to form inside the uterus, which ends up having a negative effect. In contrast with Hippocratic authors, Soranus advocates for remedies that are soothing, relaxing, and productive of flow, and these are not that different from any other treatment that should encourage laxity.[139] That is, for a Methodist, female corporeality does not necessitate a radically different set of medical tools. For example, when constriction affects the area around the womb, Soranus outlines a course of treatment that starts with the diatritus, which require the patient to rest in a warm room and fasting. He then recommends applying warming remedies, such as warm cloths, towels, and wool, warming pans, and so on. Soft pure wool is soaked in sweet, warm oil, and then pressed externally to the pubic region, epigastrium, and hips.[140] If the pains persist into the last day of the diatritus, he recommends venesection, a sitz bath with fenugreek, linseed, or mallow, and a vaginal suffusion of the

137. Sor. *Gyn.* 3.2.59–70 Burg. = 3.9 Ilb.
138. Sor. *Gyn.* 3.2.187–191 Burg. =3.12 Ilb.
139. Sor. *Gyn.* 3.2.83–84 Burg. = 3.9 Ilb.
140. Sor. *Gyn.* 3.2.99–124 Burg. = 3.10 Ilb.

same medicaments whipped in egg and poured into the vagina. After this, light food can be given. If this first diatritus does not succeed, he recommends continuing the cycle, alternating between fasting on one day and light food and the application of cupping, leeches, or poultices to draw blood downward on the next.[141] In short, the same logic applies to both male and female corporeality. There are some specific remedies that only make sense for application to female parts, but gynecological tools neither articulate nor sustain a different type of body.

The relationship that technology bears to female corporeality within this gynecological framework can be discussed largely in terms of its absence. That is, Hippocratic authors constructed female corporeality with an intimate integration between menstruation and their therapeutic interventions, such that women's bodies in their natural state were pathologized, with regular—and regulated—menstruation required to prevent illness. Suppositories and other emmenagogues for the Hippocratics can thus be seen as prosthetic extensions of a woman's physis, with the distinction between natural and artificially induced menstruation blurry. Soranus's acceptance of a far wider normative range and his insistence that failure to menstruate was not itself an illness in need of intervention shows how he decoupled women's bodies from a therapeutic interface. To be sure, he, too, described "relaxing" suppositories and other means of inducing menstruation, but in accepting a wider range for what counted as normal and making painful or unwanted effects the marker of an affection in need of treatment, he depathologized women's bodies. The Methodist system articulated corporeality, not therapeutic interventions, and both female bodies and the remedies specific to their parts and operations were simply reflections of their broader system.

5.8 CONCLUSION

This chapter began with the spread of the organized body within medical discourses, as more authors after the successes of Erasistratus and Herophilus started referring to inner corporeal organa and took for granted that anatomical investigations were a natural way to gain knowledge about the body. Yet structural, tool-like teleology and the techniques that supported it met with epistemological objections, and the rise of Empirical and Methodist medical schools in the late Hellenistic period led to a devaluation of the organism as the epistemic object of medical inquiry. This devaluation no doubt both contributed to and is evinced by the shift

141. Sor. *Gyn.* 3.2.125–186 Burg.= 3.11 Ilb.

away from using anatomical investigations as a way of knowing. Some medical practitioners still appear to have conducted anatomical instruction in Alexandria through the Hellenistic period and early Roman imperial periods, but the practice declined in importance. As we saw, without anatomy, the use of tools as a heuristic with which to conceptualize the body parts also decreased, as mechanisms, not function, became the primary reason to invoke them as analogies.

Yet this chapter also illustrated how technologies still shaped the corporeal conceptions of those who rejected investigating the natural activities of a functional body. Features native to the organism appeared in the accounts of both the Empiricists and the Methodists, despite their explicit rejection of causal knowledge based on physiology. With Asclepiades we saw another view of corporeality altogether, one concerned with pores, corpuscles, blockages, and flows, but interested in explaining the movement of these substances within the body. The chapter argues that several notable features of Roman aqueducts appear mirrored in Asclepiades's pathologies, as he toggles between macro- and micropathways of the human form. Moreover, baths and water infrastructure gained prominence and acquired social meaning as bathing became an essential part of Roman practices of corporeal care. This development surely had an impact on contemporary assumptions about the body as a system subject to blockages and spills that could be treated with heat, cold, and other remedies designed to relax or constrict its inner fluid movements. The technological environments of Roman health care left their imprint on physicians and patients alike. In short, because of the multiple vectors along which influence runs, technologies continued to influence medical theories and basic assumptions about the body in multiple, overlapping ways. The tools of the organism still exerted their power, even as alternative technological systems now supported non-organized corporealities.

CHAPTER SIX

Galen and the Technologies of the Vitalist Organism

6.0 INTRODUCTION

After Herophilus, Erasistratus, and Eudemus of Alexandria conducted their research in the third century BCE, little writing on anatomy seems to have been produced for the next two hundred and fifty years, as Empiricism, Asclepiadeanism, and Methodism gained greater popularity in medical communities. As a corollary, the theory of an organized body likewise lost popularity, and these medical sects dismissed the tool-like functionality implied by physiological accounts. In the first and second centuries CE, the situation started to change. Anatomy enjoyed a resurgence both in medical schools and in the wider public, as civic competitions showcased anatomical acumen alongside other medical skills.[1] With this rise, surgery likewise gained in status, and practitioners developed a wide range of cutting, ligating, and probing tools to conduct these operations. As a result, these physical implements became the preeminent icons of working physicians. With the return of anatomy came greater emphasis on tools and tool heuristics to conceptualize and discuss the parts and structures of the body. Authors such as Rufus of Ephesus began to reassert the importance of anatomical knowledge, while characterizing the inner structures as the essential tools of the physician. Other physicians, such as Marinus, Quintus, Lycus, Satyrus, Numisianus, and Pelops, practiced dissection and wrote lengthy treatises about it. Tools and the organism that they supported returned to the forefront of medical discourses.

It was this medical environment that Galen of Pergamon entered. Except for perhaps Hippocrates, Galen was the most influential medical author in antiquity, and in his prolific writings he brought together anatomy and physiology as he detailed the inner workings of the human body. Since the beginning of this investigation, a certain specter has loomed in

1. See both Bubb 2022 and Salas 2020a for accounts of this return of anatomy.

the background, occasionally making itself visible, but never fully coming into view: the question of vitalism and mechanism. Multiple philosophers have addressed how technologies influenced theories of living things. For many, this question feels most securely addressed by analyzing where ancient authors drew boundaries between natural and artificial objects. Most often, this analysis leads to the origins of motion and focuses on whether biotic matter contains special powers of self-movement that the nonliving materials of artificial objects do not. In other words, to address the question of technology and the body they ask whether living things possess some explanatorily basic "vitalist" force that cannot be reduced to so-called mechanical principles, or whether all living processes can be explained with reference to nonliving physical interactions.

For these philosophers, answering this question reveals whether living things can be, even theoretically, reproduced artificially. By extension, this approach becomes the most penetrating way to address the degree to which technologies can be said to influence ideas about biotic life-forms. Yet, as the previous chapters have emphasized, orienting an examination of biotechnical interactions on the presence or absence of vitalist forces supplies only a partial analytic frame. Tools and technologies can structure, articulate, or influence theories of the body in many ways at multiple levels. Reducing these dynamics to the impetus of motion occludes and distorts these relationships. This chapter addresses these topics most directly, insofar as it focuses on Galen of Pergamon, whose theories of "natural capacities" provide the foundational concepts for so much of the vitalist debate.

Like many of his teleologically committed predecessors, Galen uses tools and technologies as an essential structuring metaphor to understand the corporeal whole. He argues that bodies are composed of purpose-built functional parts that nature has engineered in the best possible way. To him, the body is an ideal biotic artifact. As a result, his accounts are replete with tool comparisons, and he uses organa to understand the body at both the holistic and local levels. At the same time, however, Galen is careful to insist that living tissues operate in ways that nonliving tools cannot, insofar as they possess unique "natural faculties" that nonliving substances do not have. In this regard, he makes the most forceful argument for vitalism in the ancient medical tradition. Yet this chapter will illustrate that it is precisely because Galen takes the body to be tool-like that he insists it must have its unique vitalist powers. In other words, in the dichotomy between mechanism and vitalism, tools and technologies align with vitalist accounts, not the other way around. Yet the integration of tools into his vitalist, teleological organism goes further still.

In her work *The Body Multiple*, Annemarie Mol describes how physicians, pathologists, physiotherapists, and patients all mean different things when they speak about a supposedly single disease, since they have different practices and referents when they talk about the effects, diagnosis, pathology, cause, and so on. She argues that each version of a disease is therefore *enacted* by a unique, if overlapping, set of practices.[2] A similar insight can be used to understand how technologies were integrated into Galen's theoretical account of the body. Attributing unique capacities to living tissues can often strike modern readers as antiquated and unsophisticated, akin to positing quasi-mystical forces. To Galen, however, these living capacities were logically demonstrable powers, visible through the application of certain tools and techniques. In fact, he developed a full array of purpose-built dissection tools, ligatures, dissection boards, and specialized knives to do just this. In this regard, Galen participated in a technological flourishing that reached its apogee in the Roman imperial period.

At the same time, Galen also moved well beyond most of his predecessors by privileging the vivisection experiment as an essential investigative practice. This type of direct intervention created malfunction and was therefore essential, in his opinion, for knowing the functions of the organs, as well as for determining the unique natural faculties of many bodily tissues. Following Mol, we might say that Galen's vivisection experiments enacted his natural faculties (at least those that the demonstrates), insofar as those faculties became real through the practices that performed them. In this way, the body and behaviors demonstrated by vivisection experiment cannot truly be extracted from the technologies and techniques that revealed and supported them. Understanding the relationship of tools to Galen's theory of corporeality requires recognizing that his organism remained inseparable from the complicated set of material devices used to demonstrate its powers. Producing the vitalist body was itself a technological feat.

This chapter closes by reflecting on the reciprocal relationship that Galen's increasingly specialized set of anatomical instruments thus bore to his own self-fashioning. As part of his general philosophy, Galen argues that the divine demiurge is the perfect architect of life, and he dares his reader to design any corporeal feature better than Nature already has. Moreover, to display the technological skill encoded within living bod-

2. Mol 2002: 32. As was outlined in the introduction, the broader goal of her argument is to stress the multiplicity of realities or objects that come from granting validity to each of these enactments of atherosclerosis.

ies, he uses an extensive tool kit filled with many instruments that he designed and had manufactured himself. His organism and the tools that articulate it thus maintain a reciprocal relationship, as the specialized implements used to break the body down mirror its characterization as a highly engineered object. In so doing, Galen implicitly situates himself in the role of a divine craftsman, designing tools and reverse engineering the organism that Nature had built. This approach merges medical expertise with the skill of Nature itself, albeit in a different way than Hippocratic physicians had. Once again, techne and physis form a recursive loop as they support, sustain, and structure one another.

6.1 THE RETURN OF ANATOMY

In the first century CE, Dio Chrysostom (ca. 40–115 CE) writes about a rich culture of public spectacle in the Roman Empire, with performances of various types that entertained teeming crowds. Both the Hippocratic treatises and inscriptional evidence suggest that public medical lectures were performed in both the Hellenic and Hellenistic periods.[3] Yet Dio describes something that appears new: performances of anatomy, where physicians would sit in the middle of a crowd and "go through the connections of the joints and bones, their similarities, differences, and other such things, passageways, breaths, and percolations."[4] He distinguishes between these men and "true physicians," which reveals his disdain for public medical performances, but Dio's comments illustrate that anatomy was capturing a wider audience again in the late first century CE.[5] Contemporary medical authors reflect this trend, including Rufus of Ephesus (fl. ca. 70–110 CE), who practiced in the eastern part of the empire and reportedly lived during the reign of Trajan, which makes him an older contemporary of Soranus.[6] Rufus was a prolific author, and seven extant treatises in Greek have been attributed to him, although Ibn Abī Uṣaybiʿah

3. Many Hippocratic treatises appear to have been composed as public-facing speeches; see esp. Jouanna 2012 for analysis. For inscriptions attesting to public medical lectures, see IKPerge 12 (Perge, 2nd c. BCE); IScM 1.26 (Istros, 2nd c. BCE).

4. Dio Chrys. *Or.* 33.6; cf. Gersh 2012: 175; Salas 2020a, esp. ch. 2.

5. See Edelstein 1935 and Lloyd 1983: 149–67 for a historical overview of ancient anatomy.

6. Suidae Lexicon, *s.v.* Ruphos, claims that Rufus was active during the reign of Trajan (98–117 CE), but Scarborough (2008: 720) argues that he lived a full generation prior; cf. Gal. *At. Bil.* 1 = 5.105K, who calls him one of the "recent doctors" [τῶν νεωτέρων ἰατρῶν]. See also Abou-Aly 1992: 25.

lists fifty-eight texts in his *Literary History of Medicine*, which suggests Rufus produced far greater output.[7] Rufus incorporates humors, breaths, and waters into his pathological accounts, and scholars have aligned him with multiple different schools, including Hippocratic humoralism and Pneumatism, although most consider him an eclectic thinker not entirely aligned with an individual sect.[8] His text *Medical Inquiry* promotes the value of asking the patient questions to help discern the cause of their illness, an etiological approach that had been devalued by Asclepiades and the Methodists. As part of this turn back to causes, Rufus also looked to anatomy. Multiple anatomical works are attributed to him, including a treatise called *The Names of the Parts of the Human*, in which he instructs his reader how to demonstrate and name the external parts on an enslaved model and then the internal parts on a dissected monkey to an audience of students, or perhaps to a crowd in the manner Dio describes. Rufus, too, seemingly participated in this broader trend.

Other details about the status of dissection can be gleaned from Rufus's treatise. For instance, a substantial part of *The Names of the Parts of the Human* is spent in disambiguation, wading through various terms for anatomical features in Doric, Attic, and other dialects, and Rufus cites Aristotle, Herophilus, Hippocrates, Cnidias, Cleitarchus, Praxagoras, Philistion, and Homer as part of his investigations. In the case of cranial sutures, Rufus even complains that their names have been given by "Egyptians who speak Greek poorly."[9] On the one hand, this account suggests that Rufus might have studied in Alexandria, where a living anatomical tradition likely survived in some form. On the other, G. E. R. Lloyd's impression of the "terminological anarchy" into which Rufus waded suggests the low status that anatomy had reached as a technical medical discipline by his era.[10] Indeed, Rufus's work remains at a relatively macro level, insofar as proper names are only given to structures that are apparent or important,[11] but even if it sometimes seems like he is engaging in

7. Ibn Abī Uṣaybiʿah, *Uyūn al-anbā* 4.1.10.2 Savage-Smith, Swain, and van Gelder.

8. See Abou-Aly 1992: 57 n2 for bibliography on these various scholarly assertions. Abou-Aly himself thinks that Rufus is a Hippocratic even in his assertions about pneuma.

9. Ruf. Eph. *Onom.* 133 Daremberg and Ruelle.

10. Lloyd 1983: 163. Gersh (2012) places Rufus's project next to that of the medical lexicographers Xenocritus of Cos, Bacchius of Tagara, and Erotian, all of whom wrote Hippocratic lexica. Her broader argument seeks to locate Rufus within the literary culture of the Second Sophistic.

11. Ruf. Eph. *Onom.* 180.

performance or education, he does make discoveries about the fallopian tubes, or at least clarifies Herophilus's position.[12]

The naming practices that Rufus outlines draw on comparisons with food, geological features, and some human-made devices (esophagus as trumpet, scrotum as spear case, uterus as cupping vessel, pylorus as gates),[13] but in general, he does not leverage anything except resemblance in finding appropriate names and does not call parts *organa*. As such, it is hard to discern the type of interplay between mechanics and medicine at work in Alexandria during the time of Erasistratus and Herophilus. Nevertheless, when Rufus begins his treatise, he explicitly foregrounds the conceptual link between anatomy and tools, stating that knowing the body parts is the same as the cithara player knowing the cords, the grammarian knowing the letters, and the metalworker, leather-cutter, and carpenter knowing their tools:

> What do you learn first in cithara playing? To pluck and name each of the cords. What do you first learn in reading and writing? To recognize and name each of the letters. And so, the bronze worker, the leather-cutter, and the builder begin to learn the other arts in the same way, starting with the names, first the name of iron and pail and any of the other things belonging to the art.[14]

In other words, for Rufus, learning a techne starts with learning its basic instruments, and these basic tools are not cooking, drugs, or surgical implements but the body parts themselves. A second, closely related anatomical treatise attributed to Rufus called *The Anatomy of the Parts of the Body* goes even further in connecting anatomically articulated parts to technological implements.[15] In the introduction, the author aligns himself with the teleological approach, which sees the body parts as crafted to complete certain tasks:

> For a human seems like a small cosmos to the philosophers, closely imitating the heavenly order, having intricate workmanship [ποικίλην δημιουργίαν] in their completion and in the preparation of their parts,

12. Ruf. Eph. *Onom.* 186.

13. Ruf. Eph. *Anat.* 64 does say that the uterus is similar in shape to a cupping vessel, just as Hipp. *VM* does (see p. 54); cf. Gersh 2012: 156–58 for a discussion of these naming practices.

14. Ruf. Eph. *Onom.* 1–5.

15. Lloyd (1983: 151) takes *The Names of the Parts of the Human* (*Onom.*) as authentic, but suspects Ruf. Eph. *Anat.* is a reworking derived from it.

> and in the completion of their tasks. Therefore, just as with the other medical topics, one must learn the understanding according to anatomy. And so, making the principles of our art as instructions like a framework, we shall establish the construction and name that nature provides to the parts.[16]

Although understanding the human as a microcosmic reflection of the greater cosmos can be traced through different lineages, including to Democritus's assertion that "man is a small cosmos,"[17] Rufus's formulation seems to be Platonic in outlook, insofar as it aligns this cosmic mirroring with the intricate workmanship that went into crafting corporeal components suited for their own functions. In other words, anatomical interest and structural, tool-like teleology once again appear together.

If Rufus's anatomical practices focused on relatively large somatic structures, the next few decades saw even greater interest in discovering and detailing smaller features, especially in Alexandria. In fact, Marinus (fl. 100–130 CE), who taught at Alexandria, wrote a practical manual of dissection across twenty books, which dealt with the nerves, the muscles, the skull, the brain, the voice, and so on.[18] His investigations were as fine-grained as detailing the musculature of the cheek. Galen credits him with no less than reviving anatomy as a medical discipline.[19] Indeed, Marinus left more than books. He also established a lineage of pupils who continued to promote this investigative approach. These pupils included Quintus, who in turn taught Lycus, one of Galen's chief foils and author of an enormous work on anatomy of around fifteen thousand lines.[20] Quintus also taught Satyrus and Numisianus, who taught Pelops,[21] and Galen, writing in the late second century CE, claims all these last three

16. Ruf. Eph. *Anat.* 1 Gersh.

17. DK68 B34 (D225 LM) = David. *Proleg.* 38.18.

18. Gal. *Loc. Aff.* 8.212K; cf. Singer 241 n32. Galen seems to have learned much from Marinus (see Singer 230) and speaks about his works multiple times, including in *Lib. Prop.* 3 = p. 11–13 Singer = 19.25–30K and at *AA* 9.2 Singer = 2.716K in regard to the meninges of the brain.

19. Gal. *AA* 2.1 Singer = 2.280–283K.

20. To Galen's eye this length did not translate into comprehension or accuracy (*AA* 4.10 Singer = 2.469–470K). He claims that Lycus is ignorant of the function of multiple facial muscles (*AA* 4.6 Singer = 2.449K; cf. *AA* 4.10 Singer = 2.466–467K) and makes many derogatory comments throughout his corpus; see also Galen's extended critiques in *Ad. Lyc.*

21. Galen claims to have been educated by all these men, save Lycus, whose ideas he knew only through writings (see *AA* 8.2 Singer = 2.660K). Rocca (2008: 243–44) outlines this lineage.

as teachers. Galen himself notes the growing interest in anatomical investigations (especially of the viscera), stating with at least some degree of hyperbole that the enthusiasm for this type of anatomical knowledge "increases daily."[22] By the early second century CE, then, there was a growing number of anatomists, many of whom had worked or trained in Alexandria, engaged in investigation of the structures of the body and their function.[23] With the resurgence of anatomy came the return of tools and instruments to the forefront of articulating the human body and its component parts. Nowhere is this return more visible than in the works of Galen of Pergamon.

6.2 GALEN OF PERGAMON AND *ON THE FUNCTION OF THE PARTS*

Galen of Pergamon (129–ca. 215 CE) was born in the eastern part of the Roman Empire in a wealthy city engaged in an extensive building program. Pergamon had become a Roman province upon the death of Attalus III, who had bequeathed the city and its territories to the Romans in 133 BCE. After more than two centuries of eventual decline, the city was undergoing a revival when Galen was born, adding aqueducts, baths, shrines, a theater, an amphitheater, and a new forum to a cityscape already featuring massive monuments, an extensive library, and a Temple of Asclepius.[24] Galen's father, Nicon, was an architect who participated in these construction projects.[25] Rather than push his son straight into his

22. *AA* 4.1 Singer = 2.417K. Although Galen himself places considerable importance on the viscera and their role in corporeal function, he laments that others are not paying due attention to the hands, legs, bones, muscles, vasculature, and nerves, which are more crucial to understand for those conducting the type of topical surgeries more common in the second-century Roman world; see *AA* 2.3 Singer = 2.288–291K; cf. *AA* 3.1 Singer = 2.346K. At the same time, Galen also recommends that physicians interested in learning about anatomy travel to Alexandria, where they can easily and effectively study osteology in particular (*AA* 1.2 Singer = 2.220–221K). He also insists at *Foet. Form.* 4 = 4.677–678K that anatomical knowledge had greatly improved in comparison to previous generations, such that much anatomical knowledge (like the brain being the seat of consciousness and sensation) was all but universal in his day.

23. For the resurgence of anatomy in the first half of the second century CE, see Grmek and Gourevitch 1994; Nutton 2020: 21–22.

24. Nutton 2020: 6–9.

25. Inscriptional evidence points to two separate architects named Nicon (Julius Nicodemus/Nicon and Aelius Nicon), one of whom was no doubt Galen's father, although it is not known which (Nutton 2020: 9–10). Galen mentions many

own discipline, Nicon had Galen educated in classical rhetoric, mathematics, and philosophy before having a dream that made him switch his son's training to that of medicine when Galen was sixteen.[26] His father's professional interest in mechanics, engineering, and building technologies seems to have left its mark on Galen, who brings up architecture and its associated technologies within his works multiple times as a positive model for medicine.[27] Nevertheless, the impact that technologies had on Galen's ideas about the body as an epistemic object cannot be easily summarized, not least because of the vast size of his corpus. Across more than 120 known treatises, amounting to more than ten thousand pages, Galen variously speaks about the body as an object to be understood, interrogated, dismantled, observed, felt and listened to, maintained with healthy practices, treated with drugs, and nurtured as a vehicle for the soul. Accordingly, despite a high degree of systematic consistency, the way that technologies relate to the epistemic object under consideration varies from text to text. Nevertheless, this chapter will focus on explicating how Galen uses technologies as a structural model to understand corporeality and will illustrate how the tools he uses to demonstrate the behaviors and properties of such a body both rely upon and form a recursive relationship with the object that they help reveal, demonstrate, and assemble.

Because Galen presents himself as the legitimate successor to Hippocratic medicine, adopts crucial features of Plato and Aristotle, and treats the consolidation of the body as an organism as an indisputable fact, it often becomes hard to see the history of medicine outside of the vectors that he uses to describe it. Nevertheless, a close examination of his treatises can reveal the degree to which even compared to the predecessors with whom he aligns himself, he refashioned his epistemic object, interrogating and articulating it in new ways through the use of a new set of investigative technologies. To this end, we can look at the ways that tools underpin his explanatory apparatus.

Galen is an avowed humoral theorist, adopting a doctrine derived from the Hippocratic treatise *Nature of the Human*, with blood, phlegm, yellow

---

biographical details across his works, perhaps most notably in *Lib. Prop.*, *Prop. Plac.*, and the introduction to *AA*. For details of Galen's biography, see Schlange-Schöningen 2003; Hankinson 2008b; Boudon-Millot 2012; Mattern 2013; Nutton 2020.

26. On his education, see *Aff. Dig.* 8 = 5.41–42K; *Ord. Lib. Prop.* 3–4 = pp. 26–28 Singer = 19.57–60K.

27. He uses sundials, eclipse prediction, and water clocks, which he believes belong to architecture, as paradigmatic devices to support his own epistemology; see *Lib. Prop.* 11 = p. 18 Singer = 19.40K; cf. *Aff. Dig.* 3 = 5.66–88K; see p. 258.

bile, and black bile as the four major bodily fluids. At the same time, he also expanded this theory to explicitly assert that these humors act as vehicles for the four qualities (hot, cold, wet, and dry), which needed to be kept in balance. Even as he foregrounds his allegiance to Hippocrates, though, Galen describes a body that falls more in line with Aristotle's corporeal configuration: a body composed of tool-like organs, all functioning to sustain life. In this regard, he fully integrated humoral theory into this organism, leveraging the conceptual framework of tool-like parts to explain balance and imbalance within a functional body. He insisted that the body itself produced the humors, adapting the Aristotelian notion of residues [περιττώματα] to characterize the humors as either the healthy result of digestion (blood), or as the residual by-products of imbalance and malfunction (phlegm, yellow bile, and black bile).

As has been well noted by commentators, Galen promoted a cosmic teleological framework also adopted from Plato, Aristotle, and the Stoics, by holding that "nature" or a "divine demiurge" has designed the body and its parts.[28] Galen's teleology, however, goes even further than his predecessors insofar as he takes the dictum "nature does nothing in vain" to an extreme, demonstrating in painstaking anatomical detail how each and every part within the human body has been perfectly designed by Nature to complete its particular function.[29] Indeed, for Galen, accepting that

28. For discussion of Galen's teleology and arguments for a divine creator, see Hankinson 1989; Flemming 2009 (taken from Nutton 2004: 229 n48). See also Flemming 2000; Kovačić 2001; Frede 2003; Jouanna 2003; Gill 2007: 101; see Flemming 2009.

29. For example, see Gal. *UP* 1.5 = 1.6.18–22 Helmreich = 3.9K. For a discussion of the "Panglossian" nature of Galen's teleology, see Hankinson 1989; cf. Schiefsky 2007: 23–24. For Galen, since biotic design can be perfected, human bodies alone are ideally designed by nature, with all other animals being steps away from functional perfection. Aristotle, by contrast, accepts a comparative anatomical approach, such that each animal's body is suited for its particular lifestyle and habits. Against even a fully Panglossian interpretation of human anatomy, it is worth noting that Galen holds that the jejunum serves no special function but is only necessitated as a result of the formulation of the stomach and intestines, etc.; in this way, it is not really a part but an accident [σύμπτωμα] (*UP* 5.3 = 1.255 Helmreich = 3.348K; cf. *UP* 5.3 = 1.257 Helmreich = 3.351K). Moreover, despite attacking Aristotle for a similar position, Galen suggests that certain material realities bind Nature to make trade-offs with some drawbacks in design (*UP* 5.4 = 1.260 Helmreich = 3.355K). He asserts that the divine demiurge must work with the limits of matter, and he juxtaposes this construct with the God of Moses (*UP* 9.14 = 2.158–159 Helmreich = 3.906–907K; cf. Flemming 2009).

Nature is purposive requires accepting that the human body simply could not have been constructed in a better manner, and he devotes a lengthy treatise, *On the Function of the Parts*, to demonstrating this idea.

Started after his arrival in Rome in 162 CE, *On the Function of the Parts* was not completed until the beginning of the next decade.[30] As was the case in previous such accounts, *On the Function of the Parts* connects its teleological outlook to techne and technologies both structurally and essentially. Galen uses the vocabulary of organa to describe functional corporeal components, often at a far more individuated level than Aristotle,[31] and although one might suspect that the metaphorical resonances of organa might have worn off after five hundred years, Galen explicitly makes tools paramount in his account of corporeal function and perfection. In fact, he begins both *On the Function of the Parts* and his related anatomical handbook, *Anatomical Procedures*, with in-depth accounts of the hand, precisely because for him the hand is the essential human organ, or, in his words, a "tool for tools."[32] For Galen, we have hands to engage in all the technai, including forging and wielding weapons, constructing defensive walls, making ships, building musical instruments, and manufacturing tools.[33] Nature designed the human body and made it upright precisely so that we might use technologies, an act that he sees as the paradig-

30. Flemming 2009: 61–62.

31. Like Aristotle, he defines organs as structures that complete an entire action, "just as the eye is of vision, the tongue of speech, and the legs of walking" (Gal. *MM* 1.6 = 10.47K). Galen, like Aristotle, also calls the body as a whole the "tool" [ὄργανον] of the soul (*UP* 1.2 = 1.1 Helmreich = 3.2K). When identifying an organ as unique, however, we need to examine its structure through dissection to see some component, feature, or quality that resembles nothing else in the body. We can identify this distinct feature as the cause of the organ's unique action. All of its other features are those common to all organs (*UP* 4.12 = 1.217 Helmreich = 3.296K). Despite the clear and explicit Aristotelian precedent, Galen actually extends the use of the term "organ/tool" down into much smaller structures than Aristotle ever did. For instance, while Aristotle refers to the entire hand as an organ, Galen refers to each of the fingers as a tool (*UP* 1.7 = 1.11 Helmreich = 3.14–15K), as he does the bones "along the carpus and metacarpus" (*UP* 2.8 = 1.91 Helmreich = 3.124K). Elsewhere, he notes that the fingers and thumb are allies, engaged in the same complete task of grasping (cf. *UP* 1.5 = 1.7 Helmreich = 3.10K).

32. Gal. *UP* 1.4 = 1.6 Helmreich = 3.8K; cf. *UP* 3.1 = 1.123 Helmreich = 3.168–169K, which returns to the hand as the tool for tools. Galen had previously written a version of this text, *Anatomical Procedures in Two Books*, that followed a different order. Galen starts his later work, *Anatomical Procedures*, with the hands and explicitly defers to the project of *UP* as precedent (see *AA* 1.3 Singer = 2.234K).

33. Gal. *UP* 1.2 = 1.1–3 Helmreich = 3.2–5K.

matic expression of human intelligence.[34] This account differs from that of Plato, insofar as Plato asserted that humans stood upright to gaze at the stars and facilitate contemplation (rather than to free our hands up to swing hammers).[35] Conversely, Aristotle argued that humans are the only animals equipped to use hands because they are the only animals with intellect,[36] although elsewhere he also suggested that humans stand upright so that the weight of the body will not weigh down upon the intellect.[37] For Aristotle, technai are not the ultimate expression of human rationality, and therefore tool use is not the ultimate *telos* of the body. For Galen, by contrast, employing instruments is the primary corporeal goal. The body is made for tools.

Tool use underpins Galen's teleological theories in another way, as well, providing him with the use cases through which to test his arguments about human corporeal perfection. Why, after all, it is preferable for humans to have two legs and two arms, rather than four legs and two arms, as centaurs do? Galen's answer mixes dietary concerns (would the top part eat cooked barley, the lower grass and hay?), physiological quandaries (the centaur has two chests; does it therefore have two hearts?), and teleological concerns about the perfection of the human form. He insists that a centaur's body is not an improvement on a human's, since if it were, it would imply that Nature could have designed human bodies better. Centaur bodies are inferior, according to his claims, since they prohibit engaging in human crafts. Centaurs cannot build houses or construct ships, since they cannot climb ladders or climb the rigging on boats. They cannot farm, since they cannot climb trees to pick fruit. Moreover, centaurs cannot cobble, smith, weave, mend clothes, or write books. Why not? Because their bodies are not suited to use the tools developed to complete these tasks, since some implements require foot pedals, while others require practitioners to sit down.[38] He then outlines the anatomical minutiae that facilitate humans' upright stance—namely, the unique oblique orientation of our acetabulum (hip socket) and our curved femur. These features allow us to support our weight, to sit down, and to do all the tasks "we do when seated, such as writing while spreading a book

34. Gal. *UP* 1.3–4 = 1.4–6 Helmreich = 3.5–9K.

35. Gal. *UP* 3.3 = 1.131–134 Helmreich = 3.179–184K ridicules Plato's position.

36. Arist. *Part. an.* 4.10, 686a25–b28.

37. Van der Eijk (2014: 91–93) collapses all three arguments together.

38. Gal. *UP* 3.1 = 1.123–128 Helmreich = 3.168–175K; cf. *UP* 3.3 = 1.131–134 Helmreich = 3.179–184K.

spread out on our lap."[39] In other words, he argues that Nature has designed our body to accommodate certain human tools, rather than accept that craftspeople have designed tools to accommodate the human body. The modern equivalent would be for an anatomist to assert that Nature has designed the scaphotrapeziotrapezoidal joint of the thumb and basal carpometacarpal bone so that humans can text on their phones.

Just as the phone is a technology wedded to the current moment, the implements and technai that Galen describes belong to a second-century CE Roman context. For instance, he mentions weaving as something centaurs could not do, and this limitation arises because Roman weaving techniques of the second century CE incorporated the vertical double-beam loom (invented in the previous century), which required the weaver to sit (which centaurs could not do) rather than stand (which they could).[40] In other places, he describes how Nature designed the esophagus and its associated muscles so as to facilitate vomiting, because it saw that physicians would recommend purging this way once a month.[41] Nature designed the body with both the specific tools and therapeutics of Galen's age in mind. As basic as Galen's technical activities may sound, they still reflect the specific tools and practices of his era.

Other moments within *On the Function of the Parts* present instances of what might be called prosthetic integration, where technologies used to extend or supplement our body get reciprocally integrated into Nature's corporeal design. When discussing the soles of our feet, Galen argues that soft, elongated feet with toes are perfect, since they allow humans to stand upright. Yet, as anyone who walked outside barefoot knows, the soft soles of human feet feel like a distinct disadvantage, since they hinder running over even moderately rough terrain. Why, then, are human soles not hard like hooves? Galen argues that soft soles require humans to make sandals. Although this state of affairs may seem a disadvantage, he insists that it is actually better than having tougher feet, since "whenever an old sandal is broken, it is easy for us to strap on a new one in its place; but if feet had some sort of sandal that grew on them, such as horses have with hooves and cows have with their cloven hooves, we would immediately go

39. Gal. *UP* 3.9 = 1.57 Helmreich = 3.214K.

40. See Prevosti 2013; cf. Ciszuk and Hammarlund 2008: 124–25. This design contrasts with that of earlier, Greek, two-beam vertical warp-weighted looms, which involved standing up to weave (which centaurs could potentially do). Prevosti notes that Seneca (*Ep.* 90.20) mentions both types of looms and considers the vertical double-beam loom more refined [*subtilus*].

41. Gal. *Nat. Fac.* 3.8 = 2.173K.

lame if we injured it."[42] In other words, the tools that humans use to augment their corporeal capacities are included within the divinely designed nature of the body.

What is more, for Galen, this relationship forms a reciprocal heuristic, insofar as sandals both operate as a prosthesis for the foot and act as a guide for understanding pedal anatomy. When a hypothetical interlocutor asks why the skin on the palms of our hands and soles of our feet is not more loosely bound to the tendons and ligaments, Galen replies, "so you will agree that a sandal applied to the foot clearly must be bound on all sides, if it's going to fulfill its function well, but not that one growing on the foot must be bound and held much more firmly, tightly joined to the parts under which it lies?"[43] Galen thus uses sandals—prosthetics supplemental to and integrated within Nature's design—to explain why the intricate physiology of the sole's ligaments is ideal, which then serves as a type of "natural sandal."[44] The fact that Roman sandals of Galen's age need to be bound to the feet with tendon-like straps and are made out of animal skins stacked into a sole helps reinforce the fact that shoes are imitations of human anatomy made from actual body parts, as well as augmentations of our feet. This account binds technological solutions to corporeal limitations and establishes a body that both requires tools physically and integrates tools conceptually.

Such integration of a contemporary technological milieu into the fabric of the body extends elsewhere in *On the Function of the Parts*, notably through several extended comparisons between corporeal features and mechanical devices. For example, in book 10, Galen describes the chiastic structure of the optic nerves as a "divine tool" used to level the eyes, and his geometrical exposition of this anatomical feature seems to resemble a *groma* leveling device.[45] In book 7, he lavishes praise on the demiurge for an ingenious "device" [μηχανή] that allows the tracheal muscles to be

42. Gal. *UP* 3.4 = 1.136 Helmreich = 3.186K; cf. *UP* 1.3 = 1.4–5 Helmreich = 3.5–7K, where Galen states that humans can augment their corporeal capacities with the technologies of cloaks, corslets, swords, hunting nets, walls, towers, writing, and laws, all of which Nature has considered when designing humans.

43. Gal. *UP* 3.10 = 1.173 Helmreich = 3.236K.

44. Donna Haraway's cyborgs spring to mind (see Haraway 1991: 149–82), insofar as Galen's corporeal configuration integrates attachments and implements into the physis of the body. Whereas Haraway's cyborgs reveal how enmeshed our somatic lives are with technological systems and other organisms (see especially her conception of *sumpoeisis* in Haraway 2016), Galen uses these moments to consolidate the body and its parts as an individuated autonomous form.

45. Gal. *UP* 10.12–13 = 2.92–105 Helmreich = 3.812–831K.

reached by nerves extending from both above and below their endpoints (which thus allows the muscles to contract both up and down on their respective sides).[46] Galen compares the overall effect to that of a *glossocomion* or "box splint" used to reset broken bones and dislocations by creating two loops pulled in opposite directions by the turn of a single crank (fig. 17). He makes a great deal of this comparison, stating that both architect-engineers [μηχανικός] and device-using doctors [ὀργανικοί] frequently employ this mechanical implement to reverse force by means of pulleys,[47] and he even treats his own description of his discovery as equivalent to initiating his readers into a mystery cult.[48] He then proceeds to provide a thorough description of the box splint, including tips for how to construct one most successfully. Despite the careful description of the pulleys and shared axle, the isomorphism between this device and the laryngeal nerves is limited to the reversed path of the cords/nerves and the oppositional force thereby created. That said, this resemblance to the box splint proves relatively crucial, since the notion that the nerves need to approach from the direction toward which the muscles contract is indebted to the assumption that the nerves pull like straps, rather than simply supply pneuma and the motor impulse, as he elsewhere holds.[49] Nevertheless, Galen does not invoke such a deliberately marked mechanical device

46. Gal. *UP* 7.14–15 = 1.415–424 Helmreich = 3.571–584K. The "device" that Galen lavishes so much praise upon is the way that the vagus nerve sends off a branch to the upper tracheal muscles from above (which thus contract upward), and then extends beyond the six lower muscles of the trachea, only for a branch (recurrent laryngeal nerves) to curve around the aorta on one side and an oblique arterial branch on the other (right subclavian artery) to extend back up to the laryngeal muscles from below (which thus contract downward). This form allows the nerves to find a stable and safe course around these arterial vessels, using them as "pulleys" or "turning posts" in their journey.

47. Gal. *UP* 7.14 = 1.415–416 Helmreich = 3.572K. By *organikoi* he seems to be referring to surgeons and physicians specializing in broken bones, dislocations, and other trauma-based injuries, which often rely on mechanical implements to be reset.

48. Gal. *UP* 7.14 = 1.418–419 Helmreich = 3.576K. He elsewhere describes learning and explaining the functional features of the body as a "true hymn to the Demiurge" (*UP* 3.10 = 1.174 Helmreich = 3.237K).

49. The nerves, although never explicitly operating as leather reins, tend to be thought of along these mechanical lines—i.e., although it is the pneuma passing through them that causes the contraction, not the pull of the nerves in a particular direction, identifying the position of the nerves' connection point as crucial to the direction of contraction makes them operate in accordance with the mechanical logic of ropes.

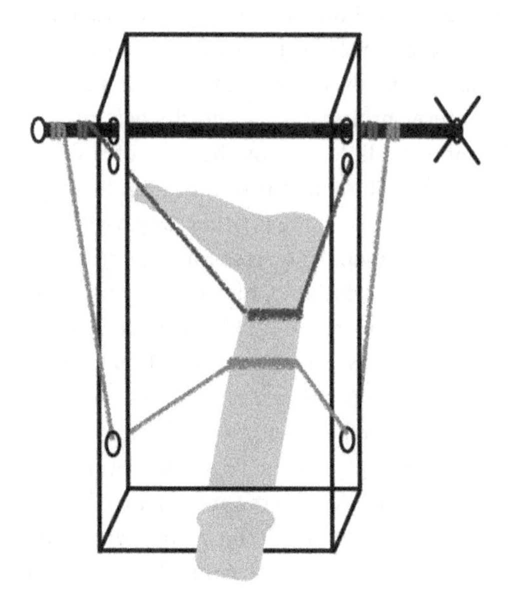

FIGURE 17. Galen's *glossocomion*, as described in *On the Function of the Parts*, book 7.

for its immense explanatory value, but as part of his broader program to cast the body as a highly engineered object whose internal mechanisms are comprehensible by comparison to human technologies.

In *On the Function of the Parts*, tool analogies abound elsewhere as well. He compares the joints to hinges.[50] He likens the tendons in the hands to puppet strings,[51] while also saying that they pass through tunnels of tendons on the fingers, called pulleys, which work like reins on a yoke passing through metal rings.[52] He compares the spine to a keel,[53] the sesamoid bone in the metacarpus of the foot to "certain supports or foundations,"[54] the bladder to a cistern,[55] the production of voice to a reed,[56] and so on. Many of these comparisons are short didactic (or figural) comparisons that serve to make his point vivid or more comprehensible without any thorough commitment to extensive isomorphism. In many instances, he explicitly limits these statements and their import. Nevertheless, Na-

50. Gal. *UP* 1.15 = 1.30 Helmreich. = 3.42K.
51. Gal. *UP* 1.17 = 1.34–35 Helmreich = 3.48K.
52. Gal. *UP* 1.17 = 1.1.41 Helmreich = 3.56–57K.
53. Gal. *UP* 3.2 = 1.131 Helmreich = 3.179K.
54. Gal. *UP* 3.8 = 1.148 Helmreich = 3.202K.
55. Gal. *UP* 5.5 = 1.265 Helmreich 3.362K.
56. Gal. *UP* 7.13 = 1.411 Helmreich = 5.566K.

ture is characterized as an architect-like demiurge who employs technai to craft a body designed specifically to use tools, and the parts that facilitate tool use or operate according to this schema are understood by applying human engineering as a conceptual rubric. Technology is the template against which Galen understands the body and evaluates it for its excellence. He even encourages his readers to try to engineer corporeal features themselves, only so that they must then admit that the demiurge has perfectly designed the body when they themselves cannot devise better structures or internal implements. Even when the functional structures within the body do not bear resemblance to any recognizably technological devices (grooves, spatial arrangements, protective coatings, edges on joint sockets, reduplicated muscles, etc.), Galen still frames these features as devices perfectly engineered to balance the need for protection, range of motion, strength, and so on, which demonstrates the precise technical skill of Nature.[57] In other words, the body can only be understood against the pattern of human technology. It is, for him, a biotic artifact.

6.3 TECHNOLOGIES AND THE NATURAL FACULTIES

It is crucial to recognize that despite using tools as a dominant structural heuristic to understand what type of object a body is and what purpose it serves, Galen establishes a fundamental distinction between nonliving tools and living organisms, such that bodies can never be reduced to an arrangement of human-produced technologies.[58] Whereas abiotic tools operate by passive material forces, Galen argues that living tissues possess active powers or "capacities" [δυνάμεις] that allow them to complete their various biotic functions.[59] Galen insists that all organs must, by definition, have "functions" [χρεῖαι] the fulfillment of which are their "activities" [ἐνέργειαι].[60] The cause of these activities he simply labels a "faculty" [δύναμις]."[61] The nature of these faculties is complex, especially since Ga-

57. See, e.g., Gal. *UP* 8.14 = 1.496 Helmreich =3.683K.

58. Berryman 2002b, 2009.

59. As he states, "The living differs from the nonliving by its participation in a capacity, not by its containment of a substance" (Gal. *Art. Sang.* 8.2 Furley = 4.732K).

60. Galen says that he will use these terms more or less interchangeably at times; see *MM* 1.5 = 10.43K. For a brief overview of Galen's notions of function and activity, see Debru 2008: 265.

61. As he states, "I define an activity as productive change and a capacity as the cause of this" [ἐνέργειαν δὲ τὴν δραστικὴν ὀνομάζω κίνησιν καὶ τὴν ταύτης αἰτίαν δύναμιν] (Gal. *Nat. Fac.* 1.3 = 2.6–7K).

len uses this vocabulary to describe the general capacity of a living thing to grow, the intrinsic powers of individual organs, and the specific potencies of medicinal substances (as multiple predecessors did).[62] It is clear that Galen's faculties and the activities that they complete are clearly indebted to the Aristotelian concept of "potentiality" [δυνάμις] and its correlate "actuality/activity" [ἐνέργεια]. At the same time, Galen also treats these faculties as specific forces adherent in the physical matter of the body, as well as a type of power that flows like a fluid.[63] As Hankinson has suggested, "[Galen's] metaphysics of powers is coarse-grained" and "he is happy to conceptualize powers promiscuously, as more or less generic or specific as the occasion demands, without apparently needing to settle on any particular level as being privileged in some way."[64] They drive at least three tiers of corporeal operations: the construction of the organism (just discussed), the growth and nourishment of the parts, and the individuated activities of the organs. At each level, these distinctly non-tool-like capacities are enmeshed with and sustained by tools and technologies, whether tools are structuring principles, limit cases, or, more importantly, the material means by which these powers are demonstrated to exist.

In the case of ontogenesis, the faculty governing the growth of a living thing from an embryo replicates the actions of a technician. In *On the Formation of the Foetus*. Galen outlines how insemination results in the orderly growth of an organism, and he insists that the sperm contains some faculty which causes the growth of the animal and its component parts in a way that human technologies could never replicate. Yet Galen gives several names to this force, calling it "nature,"[65] the "demiurgic faculty" [ἡ δημιουργοῦσα δύναμις], and "forming faculty" [ἡ διαπλαστική/ διαπλάττουσα δύναμις], as well as considering it potentially a type of soul.[66] While remaining agnostic about the philosophical specifics of such a self-guiding force, Galen describes this power as governing the

62. Hippocratic authors had already begun to use the vocabulary of *dunamis* to describe a plant's potency, and this nomenclature became even more formalized in the pharmacology of Dioscorides; see note 134 on p. 57.

63. Hankinson 2014.

64. Hankinson 2014: 969. This fluidity of conceptualization becomes especially apparent when acknowledging that Galen uses the same term to speak about the "powers" (i.e., medicinal effects) of plants and drugs, as well as the "powers" of the soul.

65. Gal. *Foet. Form.* 6 = 4.687–688K.

66. Galen ultimately claims that since he has not *demonstrated* this faculty anatomically, and only by inference, he ultimately remains agnostic as to its precise nature (*Foet. Form.* 6 = 4.700–702K). At *Temp.* 2.6 = 79.20–80.6 Helmreich = 1.635K, Galen refers to it as the δύναμις διαπλαστική.

growth and articulation of an organism by attracting (and integrating) different types of matter (such as blood and pneuma) at different times in different amounts.[67] He likens it to a laborer possessing the plan of the divine craftsman and then working continuously to manufacture the correct shape and position of all the organs and parts, as well as their unique qualities, mixtures, and faculties.[68] Nevertheless, he is careful to distinguish this process of ontogenesis from nonliving material acts, by insisting that, contra Aristotle, an embryo cannot possibly grow in the same way that a theatrical automaton performs its elaborate motions, whereby a single original impulse leads to a cascade of preprogrammed effects. The embryo seems to respond to the dynamic conditions of existence, even while displaying a type of adaptive, almost animalistic, rationality (which Galen is loath to grant a basic biotic force, let alone a nonthinking automaton). Moreover, the number of functional parts and engineered aspects that compose the body must be in the tens of thousands and thus the body is *too* intricately engineered to be constructed by a preprogrammed process initiated by a single impulse.[69] Nature wields techne, as does some fundamental nature or forming capacity inherent in sperm, but the organism does not operate by the same forces as lifeless tools.[70]

On the Natural Faculties, completed in the years following Galen's return to Rome in 169 CE, provides the longest sustained treatment of the capacities, especially as they relate to individual parts, demonstrating why such powers must necessarily exist and how they operate. Galen begins by trying to prove that a set of general powers possessed by all living tissues must exist in order to account for growth and sustained nourishment. He describes four fundamental capacities, by means of which living organs bring the right type of matter to themselves, make this nourishment part of their own tissue, keep it there as long as needed, and then expel what is harmful: he calls these the powers of attraction, assimilation, retention, and expulsion.[71] Human craft cannot replicate this fundamental biotic power of growth, possessed even by plants, since this power, according to Galen, requires qualitative alteration of nourishment, not simply spatial rearrangement.[72] Tissues thus change and assimilate what they attract to

67. Gal. *Nat. Fac.* 2.3 = 2.84–85K.

68. Gal. *Foet. Form.* 5 = 4.682–683K.

69. Gal. *Foet. Form.* 6 = 4.688K.

70. References to Nature using craft are frequent; see, e.g., Gal. *Temp.* 2.5 = 69.14–22 Helmreich = 1.619K.

71. Cf. Nutton 2020: 85.

72. In fact, Nature *expands* upon the capacities of human sculptors by being able to construct the interior of bodies by changing one substance into another (Gal. *Nat. Fac.* 2.3 = 2.84K).

themselves, they do not simply weave new material together with the old. This process supplies Nature with the power to extend bodies in three dimensions without their tissues getting thinner. Children can inflate a bladder to expand it in all directions, but Nature alone can cause the qualitative alteration required for growth.[73]

In addition to these four general faculties, Galen also insists that each of the organs possesses its own unique capacity or power to actively perform its particular corporeal function. This power derives from its tissue's particular physical mixture of the four basic qualities.[74] In fact, we can identify both the uniqueness of an organ and its unique power by recognizing some perceptible quality that it alone displays, and Galen declares that this unique material substance reveals the means by which the organ completes its own proper task.[75] Some organs operate by being able to attract specific substances at a distance, such as the kidneys, which have the intrinsic capacity to attract and extract yellow bile/urine from the blood, even across the body. Galen compares this process to the action of magnets attracting iron, bags of grain absorbing moisture, and certain purgative drugs collecting certain fluids within the body (as well as, quite strangely, dogs being particularly attracted to human feces).[76] Other organs can attract their respective substances by dilation, such that

73. Gal. *Nat. Fac.* 1.7 = 2.17K.

74. Gal. *Nat. Fac.* 1.6 = 2.14K; 2.8 = 2.118K; cf. 2.9 = 2.126K. See also *Temp.* 2.6 = 79.29 Helmreich = 1.636K, where Galen calls these mixtures the tools [ὄργανα] with which the shaping capacity exercises its power; cf. van der Eijk 2014: 123.

75. Galen names a series of parts that have unique tissues (excluding the arteries, veins, and nerves within them), such that "the remaining tissue is simple and elemental for each organ, as regards perception" [τὸ ὑπόλοιπον σῶμα τὸ καθ' ἕκαστον ὄργανον ἁπλοῦν ἐστι καὶ στοιχειῶδες ὡς πρὸς αἴσθησιν] (*Nat. Fac.* 1.6 = 2.13K). Gal. *Foet. Form.* repeats this claim. In cases where organs manifest two different tissues, each tissue has its own elemental substance, and each coat is associated with a different faculty. Each of these tissues is also associated with a unique "alterative faculty" [ἀλλοιωτικὰς δυνάμεις], since it needs to be able to turn nourishment into matter of the same type.

76. Gal. *Nat. Fac.* 3.15 = 2.210K states that there are two types of attraction, the type that occurs through "the filling toward what is being emptied" [τῇ πρὸς τὸ κενούμενον ἀκολουθίᾳ], which occurs through the expansion of a vessel, and the type that occurs because of the shared qualities [οἰκειότητι ποιότητος] of substances. The former can act over long distances, such as through a long pipette, while the latter only operates at short distances, as in the case of magnets. He also notes that vacuum pressure attracts light material sooner than heavy (*UP* 7.9 = 1.397 Helmreich = 3.546–547K). For dogs being particularly keen on human excrement, see *Nat. Fac.* 3.9 = 2.177–178K.

proximate fluid or air of any type will be drawn in (rather than a substance of a particular quality). For example, the arteries have a capacity to dilate and contract, which draws in vaporous blood and expels smokelike fumes, while the thorax dilates and draws air into the lungs. Galen compares this type of action to the bellows, operating according to the principle of "the filling toward what is being emptied."[77] Still others operate through attraction such that the objects are physically drawn closer to their delivery point by contraction of a given organ, as in the case of the esophagus.[78] Indeed, across his anatomical treatises, Galen spends considerable time discussing the fibers [ἶνες] that compose each nutritive organ (esophagus, stomach, intestines, etc.), claiming that the longitudinal fibers contract to produce attraction (i.e., pull objects closer), oblique fibers contract to retain substances (i.e., hold them tight), while the transverse or perpendicular fibers contract to expel substances (i.e., squeeze them out).[79] Other faculties possessed by the organs are less about moving matter and more about creating certain substances, such as the liver's blood-making capacity, which resides in its bloodlike material substance, or the lung's capacity to initiate the refinement of air into vital pneuma, and so on. Galen's faculties are powers adherent in the material composition of the organs and should thus, by his rubric, be understood as physical forces. His powers are explanatorily basic, necessary, and invoked at multiple levels of the organism.

Scholars have long reacted to Galen's general framework by again em-

77. For a full account of Galen's theory of respiration, see Debru 1996; Salas 2020a, ch. 6.

78. Despite the fact that at *Nat. Fac.* 3.15 = 2.210K Galen only mentions two types of attraction, at *Nat. Fac.* 1.16 = 2.62K Galen takes issue with Erasistratus's claim that the cavity exerts no active attraction (i.e., is passive) and argues that it expands and contracts to pull food downward, and so *does* have an attractive faculty. "Attraction" here must refer to the mechanical capacity to move food, not attraction at a distance from sympathy. Galen may be thinking about Aristotle's assertions at *Part. an.* 3.3, 664a20–25, where he claims that an esophagus is not necessary for food intake, since not all animals have one and it does not prepare the nourishment.

79. Galen provides a succinct description of these three fibers relative to the stomach at *UP* 5.11 = 1.281–283 Helmreich = 3.384–387K. He also supplies a quasi-geometrical argument to limit the number of possibilities to three at *UP* 5.14 = 1.290 Helmreich = 3.396–397K. Although elastic substances outside of the body can clearly replicate this basic motion with external impetus to start the motion, the power to *spontaneously* contract resides only in living tissue. Accordingly, Galen ascribes some active innate capacity to control *when* these fibers contract to the organs.

ploying the division supplied by seventeenth- and eighteenth-century mechanical philosophy, insofar as they characterize him as a "vitalist" who is arguing against the "mechanist" tendencies of his main two antagonists in *On the Natural Faculties*, Asclepiades and Erasistratus. This stance finds some support in Galen's own characterization, insofar as he takes issue with Asclepiades's reliance on the principle of "the movement towards what is sparse" to explain corporeal dynamics, and Erasistratus's use of "the filling towards what is being emptied." Two related arguments get made. One is the assertion that because Galen did not reduce the body to material forces alone, his reliance on technological heuristics is only "piecemeal."[80] The second is that Erasistratus and Asclepiades, in contrast to Galen, relied solely on "physical" explanations, and thus ought to be aligned with mechanical principles and technologies. These characterizations misconstrue the contours of the debate behind *On the Natural Faculties* in a few ways.

First, as has been illustrated, it is Galen, not his opponents, who claims tools as native and essential to his theories of the body. He has expended tremendous effort to affiliate himself and his conception of the body with mechanics and engineers, using the language of "material weights" [αἱ τῶν ὑλῶν ῥοπαί], not mechanics, to describe the theoretical alternative.[81] Second, and more importantly, the debate is different with each of his predecessors, although Galen lumps them together in many regards. His dispute with Asclepiades (and Epicurus) primarily concerns whether the body parts function *at all*. Although Asclepiades does account for cer-

80. See Berryman 2002b: 250.

81. Gal. *UP* 9.4 = 2.15 Helmreich = 3.703K. Indeed, it is also often said that Asclepiades and Erasistratus used only "physical" explanations, yet this assertion too can misconstrue the terminology of the debate, since Galen insists that his natural faculties are physical in nature and abide by all physical laws. In fact, his argument is that Epicurean atomism—and the variant that Asclepiades promotes—precludes the very possibility of "natures" [φύσεις], since atoms simply become attached to one another by spatial proximity and material entanglement, and thus groups of them can never possess an intrinsic physis. Without any such underlying essence, the possibility for organs to have intrinsic activities is also lost, as is the possibility to possess any active faculties. Moreover, although Galen would like to establish his disagreement with his predecessors as one of fundamental difference, insofar as he asserts that they reduce the operations of the organs to passive reactions to inanimate physical forces, Erasistratus still suggests that the heart has an innate capacity to pump, and Asclepiades seems to have accepted that the arteries have an innate capacity to expand. This commonality leaves open the real possibility that both of these "mechanistic" thinkers incorporated some forces that might now qualify as "vitalist" into their accounts.

tain internal processes (such as growth), he describes multiple organs as useless and without any function.[82] Therefore, although Galen does not frame it as such, the real debate with Asclepiades is not simply whether material or vitalist forces are responsible for explaining how the organs function, change, and work, but whether the parts are organs in the first place. Why Asclepiades appears so obviously ludicrous to Galen—and to modern readers—is that he consolidates the body not around its organs and functions but around flows and pores. In short, Galen and Asclepiades have not only different types of explanations but different types of explanatory objects.

Erasistratus, by contrast, did incorporate function as a rubric to consolidate the body.[83] Yet Galen considers him to be insufficiently committed to the organism as a functional, divinely designed object, insofar as there are some parts, such as the gallbladder, that Erasistratus holds to be useless. Galen also takes issue with the fact that Erasistratus explicates corporeal processes solely according to spatial rather than qualitative alterations.[84] In general, Galen does not deny that vacuum pressure

82. Asclepiades explains certain internal processes, including blood production and "anadosis"/nutritive absorption (see Gal. *Nat. Fac.* 1.13 = 2.39K), and he explains yellow bile as generated in the bile ducts (Gal. *Nat. Fac.* 1.13 = 2.40K), although he argues that many other substances elicited through specific purges are actually *produced* by the medicaments themselves, not pulled from the body (Gal. *Nat. Fac.* 1.13 = 2.35K). For Asclepiades's rejection of function as a way to consolidate the body, see ch. 5 on pp. 202–18.

83. Erasistratus's *General Principles*, according to Galen, discussed the "natural activities," what they are, and how and when they occur, but Galen claims that he did not go far enough, since Erasistratus was not explicit enough about how the kidneys accomplished urinary secretion (Gal. *Nat. Fac.* 1.16 = 2.63K). This critique, to my mind, shows how much overlap existed between their basic corporeal framework.

84. The bulk of Galen's argument in *On the Natural Faculties* is launched against Erasistratus's and Asclepiades's explanations of nourishment, since the former argues that vacuum pressure can explain how each of the parts grows and sustains itself through "uptake" [ἀνάδοσις] and the latter claims it is diffusion (cf. Gal. *Nat. Fac.* 1.16 = 2.63–64K). Neither posits an "assimilative faculty," only suggesting that new material gets woven into the fabric of the body. As Galen states, "[Erasistratus] thinks that animals grow like a woven net, rope, sack cloth, or basket, for each of which addition occurs when more material similar to the original composition is woven around the edges." (*Nat. Fac.* 2.3 = 2.87K). This, for Galen, cannot explain growth or self-sustaining nutrition, and is equivalent to asserting both that the organs are passive and that Nature has no technical skill. That is, he makes the acceptance of qualitative alteration as the mechanism of

or diffusion are operative forces in the body, since he is more than willing to acknowledge that these (or related) forces do play a role in corporeal functions. Instead, he objects to making these forces totalizing such that they make organs *passive* rather than *active*.[85] To put this in other terms, Galen argues that if Nature is purposeful, the parts/organs must serve functions [χρεῖαι], rather than be useless. Performing a function is a type of activity [ἐνέργεια] that, for him, is an essential corollary to the claim that they have corresponding capacities [δυνάμεις]. For Galen, the vitalist "capacities" are thus intrinsically linked to conceiving of the body teleologically, inseparable from attributing technical skill and teleological purpose to Nature's designs.[86] Function implies natural activities, and natural activities, for him, logically imply natural capacities. Committing to tools as a structural heuristic for the organism requires accepting that organs provide their own means of alteration and change in a remarkably un-tool-like way. For some theorists, such as Canguilhem, the essential question is what forces operate the tools. Stepping back can reveal that not every corporeal theory has tools in the first place.

Although the above ideas have been presented somewhat dogmatically, Galen's organism should not be taken as emerging out of a set of philosophical commitments or existing as a network of beliefs about teleology. Rather, his corporeal theories were constructed with and sustained by a particular set of epistemological and investigative practices that involved a number of material tools. The last three chapters have argued that the body as an explanatory object does not exist outside the particular apparatus that assembles and articulates it. This argument, again, is not to assert that physical bodies with certain behaviors do not exist, but that theorists have to make choices about what types of bodies, parts, and substances count as informative (animals/humans, male/female/intersex, old/young, idealized/disabled, racial type, etc.); how many individual bodies to examine; what behaviors to include; in which ways they are observed or manip-

---

growth a necessary corollary of organic teleology. It is also worth noting that in his *General Principles*, Erasistratus seems to have spent much of his time discussing *anadosis*, since Galen includes the greatest number of proposed details about this subject, while in other issues (bile, kidneys), Galen often puts hypothetical arguments into Erasistratus's mouth (see, e.g., *Nat. Fac.* 1.16 = 2.63–64K).

85. See Adamson 2014: 202–4 for a similar assertion. Galen also takes issue with Erasistratus claiming that the cavity passively receives food via the esophagus and insists that the esophagus actively pulls it downward (see *Nat. Fac.* 1.16 = 2.62K).

86. See claims to this end at Gal. *Nat. Fac.* 1.12 = 2.26–27K; 2.3 = 2.80K.

ulated; whose reports can be trusted; and so on. The phenomena being explained are thus emergent properties of the apparatus that reflects these decisions. For Galen, one of the most essential investigative techniques, absolutely essential to both his explanatory apparatus and his conception of the organism, was anatomical investigation and vivisection experiment. These techniques require an entire assembly of specialized tools.

6.4 VIVISECTION AND THE VITALIST BODY

Galen reports that when he arrived in Rome (162–165 CE), he began to conduct anatomical demonstrations publicly, much in the manner described by Dio. His first such display came at the invitation of Boethus, who invited Galen to demonstrate "how speech and breath are produced and by what organs."[87] The years in Rome that followed were immensely productive for Galen, and during this time he wrote a number of treatises about these practices and his associated ideas.[88] These texts included *On the Anatomy of Hippocrates and Erasistratus* (lost), *On the Causes of Respiration*, *On the Voice* (lost), an early version of *Anatomical Procedures* (lost only a few years later, and then rewritten), and, lastly, *On Vivisection* (lost).[89] Galen claims that after a few years, he gave up these demonstrations in the period 163–169 CE when he was back in Pergamon, although he returned briefly to them at the prompting of others when he arrived back in Rome in

87. Gal. *Praec.* 2 = p. 80.25–27 Nutton = 14.612K. This vivisection experiment, or extensions thereof, appear multiple times in Galen's works, including *AA* 8.4 Singer = 2.669K and, presumably, the lost treatise *On the Voice*, dedicated to Boethus. He also mentions it at *Praec.* 5 = p. 94 Nutton = 14.627–628K; and *Opt. Med. Cogn.* 107, 3–11 Iskandar; cf. Grmek 1997: 154–56. In fact, Boethus was the also the eventual dedicatee of *On the Opinions of Hippocrates and Plato*, *On the Function of the Parts*, *On the Causes of Breathing*, *On the Voice*, *On the Anatomy of Hippocrates*, *On the Anatomy of Erasistratus*, *On Vivisection* (lost), and *On Dissection of Dead Bodies* (lost); see Gal. *Lib. Prop.* 1 = pp. 5–6 Singer = 19.13–16K and *AA* 1.1 Singer = 2.215–218K).

88. On his public demonstrations, see Gal. *Lib. Prop.* 2 = p. 9 Singer = 19.20–22K. At *Lib. Prop.* 1 = pp. 5–6 Singer = 19.13–16K, Galen mentions several public disputes with the anatomist Martialius, which lead to *On the Anatomy of Hippocrates* and *On the Anatomy of Erasistratus*. For these references, see Hankinson 2008b: 11.

89. Gal. *AA* 1.1 Singer = 2.215–216K; see also *On My Own Books*. Galen wrote his first work on anatomical dissections for Boethus, who took it with him to Palestine, where he died. This first version followed the order of Marinus's text, whereas he rearranged the second edition, *AA*; see note 32 on p. 329. Galen mentions that he wrote a text *On Bones* as well as *De muscolorum dissectione* not long before (*AA* 1.3 Singer = 2.227K).

169 CE.[90] The anatomical and experimental practices established during these years also underpinned his magisterial works *On the Opinions of Hippocrates and Plato*, *On the Function of the Parts*, *Anatomical Procedures*, and *On the Natural Faculties*, as well as many additional anatomical-focused treatises. Anatomy was crucial for understanding the body, its intricate functionality, and, by extension, the nature of the divine.

Galen all but exclusively conducted his research on animals, except when chance provides him with the opportunity to glimpse a human cadaver (as when he saw the abandoned body of a murdered brigand or bodies wounded in battle).[91] Galen happily uses animals (with their inferior bodies) as proxies for humans, but he places limits on what the dissection will reveal. Even beyond questions of heteromorphism between human and animal parts, he notes that any dead subjects, whether animal or not, will fail to provide an accurate window into corporeal activities.

In his early work *On Medical Experience*, he claims that anatomical investigations can reveal the shapes of organs but declares that they cannot disclose natural functions or natural kinds.[92] This stance seems quite similar to the Empiricist arguments discussed in the last chapter that cutting the body damages it, and a dead body cannot tell you that much about the living. Yet after this period of public dissections in Rome, Galen's attitude seems to have shifted, or at least been refined. In *Anatomical Procedures*, Galen claims that anatomical investigations of dead subjects are indeed limited to reveal the "manifest" [τὰ φαινόμενα] things in the body, including the position, number, unique substance, size, and construction of the parts. Nevertheless, it is only vivisection that discloses the activities in a straightforward way, or, at other times, "the premises for discovering them" [εἰς τὴν ταύτῃ εὕρεσιν λήμματα]."[93] To understand the function and activities of organs, so crucial to Galen's conception of the body, he insisted that vivisection experiments were necessary to manipulate, damage, and impede the forces that he attributed to the organism.[94] In other words, the production of malfunction was essential to reverse engineer the function and the faculties of the organs. Indeed, without such

90. See Lloyd 2008: 37.

91. Cf. Celsus *Med.* 1 *pref.* 40–44. Gal. *MM* 2.4 = 10.100K criticizes the Empiricists for relying on anatomy as it pertains to experience, but not knowing the normal activity of the body parts; cf. *AA* 1.2 Singer = 2.221K; 2.3 Singer = 2.288–289K.

92. Gal. *Med. Exp.* 140–141 Walzer; see Hankinson 1994: 1843 for this observation.

93. Gal. *AA* 9.1 Singer = 2.707–708K.

94. See Gal. *Foet. Form.* 5 = 4.678–679K.

vivisection interventions, he often holds that the capacities were merely posited, not demonstrated.[95]

Galen describes several vivisection experiments across his anatomical treatises, some of which are mentioned in multiple places. *Anatomical Procedures*, written in the 170s CE, contains instructions for numerous such displays, with the methods, tools, and arrangements required for each demonstration. In order to discern whether any air escapes through the walls of the lungs and into the chest cavity, he cuts a hole through the chest of an animal and attaches an inflatable bladder.[96] He manually squeezes the exposed heart of the animals that he vivisects to demonstrate that an animal will perish when it is hindered. Elsewhere, he cuts off the pericardium of an animal and applies both heat and cold to observe the effects of thermic change on the vital heat.[97] In *On the Opinions of Hippocrates and Plato*, he describes a series of vivisection experiments designed to locate the *hegemonikon* in the brain. He opens animals' skulls while they are alive and both presses on and slices into various ventricles of the brain to see the damage done to motor and sensory operations.[98]

95. He sometimes seems to treat the designation of a "power" as a placeholder for more essential causal knowledge, although it is unclear whether he thinks such knowledge can ever be attained. For example, Galen states, "As long as we remain ignorant about the essence/substance of the cause of the activity, we call it a faculty" [καὶ μέχρι γ' ἂν ἀγνοῶμεν τὴν οὐσίαν τῆς ἐνεργούσης αἰτίας, δύναμιν αὐτὴν ὀνομάζομεν] (*Nat. Fac.* 1.4 = 2.9K; cf. Hankinson 2014). As Hankinson (2008a: 168) notes, Galen rejects induction [ἐπαγωγή] as useless for scientific demonstrations. For example, when discussing the function of breathing of the body, Galen employs an analogy to a smoky oven to suggest that exhalation expels certain smokelike vaporous by-products of innate heat "burning" our blood, but he insists that these analogy-based arguments are not to be seen as conclusive or having been proven by an "incontrovertible demonstration" [ἀναγκαίαν τὴν ἀπόδειξιν] (*Us. Resp.* 3.10 = 4.492K). See also his rejection of reasoning by analogy at *Thras.* 5.812–813K.

96. Gal. *AA* 8.12 Singer = 2.703–705K. He claims to observe the bladder inflating and deflating, thus demonstrating (incorrectly) that air does pass through the membranes of the lungs.

97. Gal. *Us. Puls.* 2.8 = 5.158–159K. This experiment is inspired by an instance where he saves a wounded slave by removing his infected septum (thereby dangerously exposing his heart).

98. Gal. *PHP* 7.3.14–22 De Lacy = 5.604–606K; Gal. *AA* 9.4 Singer = 2.726K. Galen also applied pressure to the ventricles during trepanation to the same effect (cf. *Loc. Aff.* 2.10 = 8.128K; *Inst. Od.* 2.886K = *CMG* Suppl. 5, 64.1–3 Kollesch; *UP* 8.10 = 1.481–482 Helmreich = 3.664K). This experiment is discussed by Rocca (2003, esp. 196–98, 249–53). Rocca (2008: 249–53) describes the various ligations and sections of the ventricles of the brain. See also *Loc. Aff.* 4.3 = 8.230–233K.

Elsewhere, he takes an animal (likely a goat or dog) and ligates its carotid arteries to see whether the pneuma flowing to the brain from the heart is crucial to the animal's survival, only to find that the animal runs around all day, seemingly unharmed.[99] He severs the spinal cord at different vertebrae to demonstrate which portions of the body become paralyzed.[100] He ligates the nerves at various points along the spine to demonstrate the source of the faculties (the brain) and the medium by which they are conveyed (pneuma). When discussing the voice, he not only severs nerves but also cuts out intercostal muscles and excises individual ribs to demonstrate how they each contribute to the production of sound.[101] In one of his most theatrical demonstrations (likely a version of the demonstration made at Boethus's request), Galen makes incisions along the spine of a pig and exposes the nerves that run to the intercostals using a specialized hook. He then loops thread around the nerves and ligates them, which paralyzes the muscles and renders the previously shrieking pig silent and voiceless.[102] The attendant crowd, too, falls silent and awestruck, only to be even more shocked when the linen ligatures are loosened and the pig explodes back into screams. The crowd, in a macabre mirror of the pig's squealing, would likely have shouted out (praise) in return.[103]

These vivisection demonstrations were far from quiet, literary affairs.

99. Gal. *Us. Resp.* 5.2–4 = 4.502–506K; cf. Gal. *Us. Puls.* 2.1 = 153–154K. When the animal (supposedly) runs around all day with these arteries ligated, Galen determines that a surplus of psychic pneuma must reside in the rete mirabile structure within the brain, which makes the pneuma received through the carotids supplementary rather than vital. Although animals, including humans, can survive brief periods with our carotid arteries ligated, Galen's observations appear incorrect. Grmek (1997) suggests that Galen was not the first to conduct such an experiment. Debru (2008: 269) states that this animal is a dog.

100. Gal. *AA* 8.3 Singer = 2.661–666K; *AA* 8.5 Singer = *AA* 2.675–678K; 8.6 Singer = *AA* 2.683–684K. See also comments on this practice at Gal. *Foet. Form.* 5.1–4 = 6.678–679K.

101. Gal. *AA* 8.7 Singer = 2.684–689K; cf. *AA* 8.3 Singer = 2.666K, where Galen describes a vivisection experiment whereby he cuts the fibers of the intercostal muscles at both ends, which refines another—potentially previously existing—version that simply cuts into the muscles or damage one side.

102. Gal. *AA* 8.4 Singer = 2.669K; *Praec.* 2 = p. 80.25–27 Nutton = 13.612K Galen repeats that he has conducted the nerve ligature intercostal experiment many times in public with pigs; he then offers a slight variation to illustrate how to immobilize the entire thoracic cavity, including the diaphragm (*AA* 8.8 Singer = 2.690K).

103. See Gleason (2009), who emphasizes the theatricality and dramatic impact of Galen's experiments. For more recent discussion of this aspect, see Salas 2020a.

Galen mentions performing them to crowds of people, having practiced his techniques in private to ensure a successful show. Some were elaborate, multiday events,[104] with many important figures in attendance (Galen mentions Eudemus the Peripatetic, Alexander of Damascus—another Peripatetic in Athens—Sergius Paulus the consul, and his friend Boethus).[105] In *Anatomical Procedures*, Galen even describes dissecting an elephant heart that he was able to procure after the animal had been imported and killed in the arena.[106] Galen seems not to have been the only one to perform such events publicly (or to have been present at such a dissection), since he describes being handed a stylus pointing at a passage about anatomy in a predecessor's book about which he was then expected to lecture spontaneously, "as is custom."[107] Hankinson embeds these vivisections within the epistemological arguments of Galen's age and expands their theological consequences regarding the divine demiurge.[108] Grmek has discussed the rhetorical and argumentative force of these displays,[109] while Rocca mentions their marketing power.[110] Gleason has emphasized the performative, spectacular nature of these vivisections, which, as mentioned above, Galen sometimes explicitly developed for maximum dramatic impact and awe.[111] Most recently, Salas has emphasized the polemics implicit in Galen's methods and the interplay between literary and performative contexts.[112] Yet as well as being carefully orchestrated performance pieces, philosophically significant experiments, and argumentative cudgels, Galen's vivisections were technolog-

104. Galen mentions public demonstrations involving the thorax that take several days (*AA* 8.5 Singer = 2.677K).

105. See Gleason 2009.

106. Gal. *AA* 7.10 Singer = 2.619–620K. See Hankinson 1988 and Salas 2020a, ch. 3.

107. Gal. *Lib. Prop.* 1 = p. 5 Singer = 19.14K; cf. *AA* 7.16 Singer = 2.642–643K, where younger practitioners challenge Galen to anatomical displays. Hankinson (2008b: 11), Rocca (2008: 245), and Gleason (2009) all argue that public scientific or philosophical displays were commonplace competitions, and they situate this practice within the culture of spectacle prevalent within the "Second Sophistic." For a discussion of this culture of spectacle, see Kollesch 1981; Brunt 1994; von Staden 1997; Lloyd 2008; Gleason 2009. On anatomical demonstrations in particular, see also Hankinson 1994.

108. Hankinson 1994.

109. Grmek 1997, esp. 170. See discussion on p. 258 below.

110. Rocca 2008, esp. 245–46.

111. Gleason 2009. See also Barras, Birchler, and Morand 1995: vii; Hankinson 2008b; Lloyd 2008; Mattern 2008: 69–97.

112. Salas 2020a.

ical interventions into animal bodies. They were investigative practices sustained by precise techniques and an accompanying set of tools.

6.5 LOGICAL AND MATERIAL TOOLS OF THE LEMMATIZED BODY

Even if he participated in a broader culture of spectacle and debate, Galen's specific explanatory and investigative practices, as well as the epistemologies that support them, are unique and complex. Galen tries to model his method of proof, which he calls demonstration [ἀπόδειξις] or the "demonstrative method" [ἡ ἀποδεικτικὴ μέθοδος],[113] on axiomatic mathematical proofs in the style of Euclid (although while also owing a debt to Aristotelian method of demonstration).[114] In *On My Own Books*, Galen mentions that he all but succumbed to Pyrrhonian skepticism until he found geometrical proofs.[115] Indeed, in *Method of Healing*, he uses geometrical proofs as the template from which to assess the correct therapeutic practice.[116] Yet these demonstrations rely not on universal maxims but on sensory information, and Galen insists that these axioms can rest on an empirical footing.[117] In fact, Galen idealized water clocks, sundials,

113. Gal. *PHP* 2.5.96 De Lacy = 5.262K. Galen mentions a treatise called *On Demonstration* in which he establishes this method and its methodology (see *MM* 1.4 = 10.37K). Although it is lost, Galen provides a summary, claiming that "in that work [*On Demonstration*] it was shown that the origins of all demonstration are those things which are plainly apparent to the senses and to the intellect, and how in every inquiry into something it is necessary to replace its name with a definition" (*MM* 1.5 = 10.39K). He explicitly claims to have modeled this approach on geometrical proofs (*Lib. Prop.* 11 = p. 18–19 Singer = 19.40–41K). Although this *apodeictic method* [ἀποδεικτικὴ μέθοδος] is uniquely his, Galen attributes it to the "ancient" physicians, such as Hippocrates, Diocles, and Praxagoras (*MM* 1.2 = 10.9K). For a discussion of *apodeixis/apodeiktike theoria*, see Tieleman 2008.

114. Much scholarly work has examined the relationship of Galen's logical system to his mathematical and philosophical predecessors. For mathematics, see Kieffer (1964) 2020 ; for Galen's debt to Aristotle, see Singer 2013; and for an overview of Galen's logic in general, see Barnes 1991, 1993, 2003; Morison 2008. For the relationship of Aristotle's syllogism to geometrical axiomatic-deductive proofs, see McKirahan 1992; Goldin 1996. See K. Miller forthcoming 2023 for these references.

115. Gal. *Lib. Prop.* 11 = p. 18 Singer = 19.40K.

116. Gal. *MM* 1.4 = 10.32–34K.

117. Gal. *MM* 1.3 = 10.29K sets up a polarity between two "criteria of confirmation" [τοῦ πιστώσασθαι κριτήρια]: reason and experience [λόγος καὶ πεῖρα]; cf. *PHP* 2.5.96 De Lacy = 5.262K; cf.*Temp.* 2.2 = 50–51 Helmreich = 1.588–590K and

and eclipse prediction as ideal demonstrative proofs, since they involve geometrical demonstrations that are then empirically verified.[118] The vivisected body should thus operate like just such an artifact as an empirical instantiation of his physiological theories.

As is often overlooked, or simply smoothed over, the "experience" that Galen establishes as part of his demonstrative method is not passive observation, but far more often interventionist experimentation designed to induce certain corporeal behaviors or manufacture malfunctions of a particular type.[119] In regard to vivisections, this experimentation often amounts to establishing a series of if/then statements that provide the "lemmata of discovery," discussed above, that structure his demonstrations.[120] For example, when trying to experimentally test whether the arteries contain blood or air, Galen sets the parameters of his experiment:

> If, when arteries are wounded, blood is observed to be voided, then either it was contained in the arteries themselves or it is transferred

---

SMT 11.459–61K. *Aff. Dig.* 559K even seeks to ground the mathematical statement "twice two is four" in empirical confirmation by verifying it with pebbles, distinguishing between strong and weak assent, based on the demonstration. Hankinson (1991: xxv) argues that Galen establishes a quasi-Stoic position, insofar as the senses can provide the "natural criteria" for knowledge for more details; he argues that Galen assimilates Carneades's Academic epistemology and the notion of the plausible [τὸ πιθανόν] with a version of the Stoic evaluative stance toward *cataleptic* impressions (he cites *PHP* 9.7.119 De Lacy = 5.777–782K; cf. *PHP* 9.1.1–13 De Lacy = 5.722–724K). Tieleman argues that Galen bridges Empiricist with Rationalist investigative methods: "in fact, he more or less presents a compromise between two positions" (2008: 54). Tieleman describes Galen's methodology as an axiomatic-deductive structure that starts from empirically accepted first premises and then tests them (as in *PHP*); the therapeutic model moves outward to causes and then back to first premises. He also argues that it bears its greatest debt to Aristotle (although Platonic *diarhesis* features).

118. Gal. *Lib. Prop.* 11 = p. 18 Singer = 10.40K; cf. *Aff. Dig.* 5.66–88K.

119. Hankinson (2008a) outlines Galen's basic epistemological stance. He notes a distinction between experience [ἐμπειρία] and test [πεῖρα] but does not make a real distinction between therapeutic evaluation and anatomical/vivisection experiments. Similarly, Tieleman (2008) argues that the methodologies underpinning therapeutic apodeixis and anatomical apodeixis are in fact the same, without commenting on the radical shift in epistemology required by active, interventionist experimentation of the type practiced by Galen. More recent contributions can be found in Hankinson and Havrda's (2020) edited volume.

120. For the vocabulary of the "lemmata of an apodeixis," see Gal. *Us. Resp.* 5.7 = 4.509K.

from elsewhere. But, when the arteries are wounded, blood is observed to be voided, and it is not transferred, as we shall demonstrate.[121]

Sometimes, such conditional statements require further lemmata to account for potential objections, but Galen moves through these until he reaches universally accepted premises, or "self-justifying primary indemonstrable propositions" [προτάσεις ἀναποδείκτους πρὸς τὸν λόγον].[122] By starting with these, Galen wants to demonstrate his conclusion exhaustively "in a manner allowing of no evasion."[123]

As has been argued throughout these chapters, the application of a new explanatory apparatus alters the contours of the epistemic object in question. In this case, the body emerging from this particular argumentative and investigative practice takes on properties necessitated by Galen's logical tools. For example, in *On the Opinions of Hippocrates and Plato*, Galen's assertion that the hegemonikon, or commanding part of the soul, is the principle of both perception and voluntary motion leads to the next "logical" claim that bodily tissues (i.e., nerves) must transmit this power from the site of perception, such that "where the origin of the nerves is, there lies the commanding part."[124] Galen establishes an if/then flow chart of the relevant vessels to discern where the hegemonikon resides in the body by tracking the flow of sensation through these pathways by ligating them at various points, one by one. As persuasive as his demonstrations become, Galen functionally lemmatizes the body, such that its vessels themselves become a linear decision tree.[125]

121. Gal. *Art. sang.* 1.2 Furley = 4.705K, trans. Furley and Wilkins.

122. Gal. *MM* 1.4 = 10.33–34K.

123. Gal. *Art. Sang.* 1.4 Furley = 4.705K; cf. *PHP* 6.5.21–22 De Lacy = 5.543K, where Galen similarly states that "from these things, you would understand very well that through anatomy what is manifest [τὸ φαινόμενον] itself forces even those proposing opposite ideas to agree unwillingly to the truth."

124. Gal. *PHP* 8.1.21 De Lacy = 5.655K.

125. In *On the Natural Faculties*, Galen stages a series of lemmatized anatomical interventions to disprove Asclepiades's assertion that the bladder absorbs urine from the general bodily cavity as vapor that passes directly through pores in its walls (much like a sponge or ball of wool). Galen hopes thereby to show that the kidneys must, by process of elimination, possess a natural capacity to attract urine from the blood, which it passes to the bladder through the ureter. Galen makes several arguments against Asclepiades's assertions, noting the thickness of the bladder's coats as preventing such absorption and its position near the bottom of the cavity as unconducive to collecting vapors, which would more naturally rise. He then outlines a series of vivisection experiments to make a demonstration [δεῖξις]. Step one: he instructs the reader to cut the peritoneum

He ligates and cuts arteries and, more importantly, nerves, to reveal that the blockage of the nerves coming from the brain, not the arteries from the heart, produces a loss of sensation and motor functions. On the one hand, this argument seems perfectly reasonable and logically sound. On the other, for this logic to obtain, certain physical attributes need to be posited to the substance occupying or flowing within these pathways. The result is that pneuma from our perspective (a hypothetical substance) is demonstrated to (a) transmit the capacity for sensation and motion; (b) flow *only* through (or along) the vessels of the body, rather than moving through voids; (c) move instantaneously; and (d) collect together again, even after it has been dispersed through an incision. The physical properties attributed to pneuma are certainly not unprecedented in previous physiologies, but it is impossible to disentangle these physical properties of a hypothetical substance (especially [d]) from the specific needs of his demonstrative methodology. The physical behaviors and properties displayed through *apodeixis* of Galen's physiological theories cannot be separated from his experimental methods.

There are other ways in which the corporeal phenomena that Galen seeks to demonstrate (e.g., the faculties) are inseparable from his technical interventions into the body. Any discussion about the "mechanism" of Galen's body should acknowledge that he himself manipulates the powers of living tissues as though he were dismantling a machine. For example, Galen's vivisection experiment demonstrates that the arteries possess a pulsatile faculty that allows them to expand and contract, and that this faculty, even if it resides in the unique fibers of the arterial membranes,

---

of a (male) animal, expose the ureters running from kidneys to the bladder, and then ligate them. Bandage the animal back up and let it run around for a while, which will show that it does not now urinate. Step two: open the wound back up, which will expose that the ureters are distended, almost to bursting. Step three: loosen the ligatures and observe the urine running into the bladder. Step four: ligate the penis before the animal has a chance to urinate, then manually squeeze the bladder. Doing so reveals that no urine runs back up through the ureters toward the kidneys but remains in the bladder. Step four: undo the ligature on the penis and let the animal urinate. Step five: ligate *one* of the ureters, leaving the other free to flow into the bladder. The ligated ureter will fill and distend, while the free one will fill the bladder. Step six: puncture the distended ureter to reveal that it is full of urine. Step seven: cut through both ureters and sew the animal back up. After sufficient time has elapsed, open up the animal to reveal that its abdominal cavity is filled with urine, as though the animal were suffering from dropsy. This sequential set of anatomical interventions operates like a decision tree of if/then or either/or clauses, transforming the interior passageways of the body into logical lemmata.

ultimately comes from (or is activated by) the heart. Galen conducts an anatomical experiment on a living animal, likely a goat. He instructs the reader to expose an artery (preferably the one in the groin) by making an incision in the skin. Then, the experimenter must ligate the upper portion and compress the bottom of the exposed portion with the fingers. At this point, he should make a longitudinal incision into the arteries, through which he must insert a short bronze tube or reed quill about as long as a finger and as close to the diameter of the artery's interior wall as possible. Here is his description in full:

> If you will expose any one of the large and obvious arteries, freeing it from the skin and then from the matter that lies over it and around it, so as to be able to put a ligature around it; then open it along its length and insert a hollow reed or small bronze tube into the artery through the opening, so as to mend the wound with it and prevent hemorrhage; then so long as you study it in this condition, you will see the whole artery pulsating. But when you put a ligature round and press the coat of the artery against the reed, you will no longer see the part beyond the ligature pulsating, although the passage of the blood and the pneuma through the hollow of the reed proceeds as before to the end of the artery. If the arteries got their pulsation in this way, then even now pulsation would be continuing in the parts beyond the ligature near their ends. Since this does not happen, it is clear that the faculty that causes the movement of the arteries is transmitted to them from the heart through their coats.[126]

If it is a tight enough fit, the blood flow will resume once the ligature and your fingers are removed, as will the pulse both above and below the tube. Next, he instructs his reader to apply ligatures near each end of the tube or press tightly on the tube. Galen states that at this point the dilation and contraction above the tube will keep pulsing, but the artery on the other side of the ligatured sections will stop moving, even though blood moves through it. He argues that this phenomenon demonstrates that the pulsatile faculty flows from the heart through the coats and is halted by the ligatures, whereas if the incoming blood expanded the arterial walls, the pulse would occur on either side of the ligatured tube. The tube lets one liquid (vaporous blood) flow, while the ligatures stop another quasi-liquid (pulsatile faculty) from flowing through the arterial coats.

126. Gal. *Art. sang.* 8.4 Furley = 4.733–734K; cf. *AA* 7.16 Singer = 2.646–647K, trans. Furley and Wilkins.

Indeed, despite Galen having elsewhere asserted that the faculties adhere in the specific mixtures that compose an organ's unique tissues and act independently, as if by a type of intrinsic reason, this experiment of his transforms the pulsatile faculty into a liquid-like force that moves through the coats of the arteries. In Mol's terminology, the experiment *enacts* this faculty and makes its behaviors manifest. Under these circumstances, the faculties are not mystical powers but physical substances that can be mechanically manipulated, damaged, and channeled by technical means. Powers flow like fluids down pathways that can be cut off or restored. As such, this particular anatomical vivisection does more than "make visible what is invisible," as Celsus put it, since it is not direct observation alone that reveals these powers and their effects, but the deliberate obstruction and interruption of function that demonstrates their existence.[127] Even more, it is worth noting that Galen's vivisection experiment demonstrates the exact opposite of what we now believe to be the case. That is, we now accept that the heart forces blood into the arteries, thereby causing their dilation, and not, as Galen here claims, that the arteries dilate, thereby actively drawing blood into themselves. Moreover, he holds that arteries can do this because of a power transmitted to them from the heart. This discrepancy between what Galen thinks he is demonstrating with certainty and our own countervailing ideas makes particularly clear that the fluid features of his pulsatile faculty cannot be witnessed outside his particular technical interventions into the body. The corporeal phenomena that he demonstrates exist as structured by the parameters of his experiments and the particular material tools he uses to enact them. Galen demonstrates the pulsatile power through the use of ligatures, tubes, scalpels, and other tools and thereby enacts this faculty and its particular properties.

6.6 THE MATERIAL TECHNOLOGIES OF VITALISM

Galen's experimental methods were a real shift from his predecessors' investigative methods. In *Anatomical Procedures*, he speaks as though his predecessors did not engage in experimental vivisection, especially not of the interventionist method that Galen himself promotes, which involves discerning function through the production and induction of malfunction.[128] To be sure, Galen was not the first to conduct any such experi-

127. Cf. Grmek 1997: 144–47.

128. "Many such facts have been discovered throughout the body, which the anatomists disregarded, shirking detailed discussion and content with plausible ideas. It is thus no wonder that they were ignorant of many things in the living

ments.[129] He mentions other anatomists who want to determine whether the active motion of the arteries is contraction or expansion by placing a section of an excised artery from dead animals into hot water to relax and restore them into either their expanded or contracted natural (resting) state.[130] He alludes to predecessors who either ligated or cut the carotid arteries to manipulate the flood of pneuma,[131] and he seems to suggest that Erasistratus conducted the same bronze-tube experiment, (albeit to opposite conclusion).[132] Nevertheless, Galen certainly formalized and

animal. For if they pass as unimportant what is demonstrable only by careful dissection, would they trouble to cut or ligate parts of the living animal, to discern the function thus impeded?" (Gal. *AA* 1.3 Singer = 2.232K), trans. Singer.

129. We might consider the earliest example of inducing a type of malfunction as the recommendation to cross one's legs so as to block the flow of sensation-producing pneuma in Hipp. *Morb. Sacr.*

130. Gal. *Us. Puls.* 6.2 = 5.169–170K. This account evinces an experimental method that attempts to obviate the epistemological problems associated with investigating living processes in a dead animal (and reveals the degree to which this question about the arteries was an ongoing debate). Although Galen will himself warm and moisten tendons of the animals that he is in the process of dissecting to keep the tissues pliable (*AA* 1.5 Singer = 2.251–252K; cf. *AA* 4.2 Singer = 2.427–428K), his experimental methods involving vivisection are of a distinctly different kind.

131. Grmek (1997) suggests that Galen was not the first to conduct the experiment that ligated the carotid arteries, based on Galen's comments that the "majority of Hippocrates's successors" mistakenly asserted that blocking or cutting these arteries would render the animal voiceless and stupefied, "because of having dissected badly" [κακῶς ἀνατεμόντων] (*PHP* 2.6.7 De Lacy = 5.264K). Yet to my mind it is unclear whether they conducted such anatomical experiments or Galen is fabricating a cause to explain their difference of opinion.

132. Gal. *AA* 7.16 Singer = 2.648K. This suggestion, too, might be a projection of what Erasistratus *would* have said, should his theory have been placed within Galen's experimental parameters. Not only does it seem strange that Erasistratus would examine the operation of pneuma-filled arteries through incision, which would introduce blood into their passageways, the Greek is somewhat unclear, insofar as Galen states, "Erasistratus described it in the opposite way, saying that the part below the quill appeared to be moving. So great is the daring of some making overzealous declarations about things that they have never observed" [διηγεῖτό γε μὴν ἐναντίως ὁ Ἐρασίστρατος ὑπὲρ αὐτοῦ, φαίνεσθαι λέγων κινούμενον τὸ κάτω τοῦ καλάμου. τοσαύτη τίς ἐστιν ἐνίων τόλμα, προπετεῖς ἀποφάσεις ποιουμένων ὑπὲρ ὧν οὐδέποτ' ἐθεάσαντο]. Not only does this account leave one with the impression that Galen does not think that Erasistratus conducted this experiment, since he never would have observed what he said he did, διηγεῖτό could be either imperfect ("Erasistratus described") or optative, in which case the sentence would read "Erasistratus *would* describe it in the opposite way"

extended vivisection experimentation far beyond anything seen or mentioned in earlier extant sources. He made these technical interventions a core part of articulating the vitalist organism. Even so, Galen's methods should not be seen in isolation from broader technological shifts within the practice of medicine, especially since the material tools of healing had undergone a considerable shift in the first and second centuries CE.[133]

Galen describes a full array of tools necessary to conduct his dissections and vivisections, and only a few such examples can illustrate the technological particulars that his dissections and vivisection experiments require. When discussing how to successfully remove a rib during dissection, he mentions, "Here [at the sternum] [each rib] changes its nature, becoming cartilage instead of bone, and can be cut easily with a sharp, strong lancet. Such there are in the instrument cases [κεφαλικαί] called *deltoi* by the ancients. Veterinary surgeons also have such lancets. You should have instruments like these handy for cutting the cartilages, as you see in my own equipment."[134] He also recommends using a "cutting block" [ἐπίκοπον] to help excise ribs, a "myrtle scalpel" to scrape off the membrane surrounding it, and a *meningophylax*, or flat spatulary probe, underneath the ribs to avoid perforating the pleura, before using a chisel to sever the rib and remove it.[135] He mentions the particular size of the bend in the hook used to pull out a nerve from beside the spine so as not to damage it.[136] He speaks about using suitable instruments to remove the liver: "some narrower, some broader."[137] Across his anatomical treatments he mentions dozens of specialized instruments, all used to disassemble animal bodies and induce malfunction. Moreover, Galen also produced multiple tools himself, largely bronze instruments, which were designed to complete particular tasks of dissection. For example, he mentions a "sharp-pointed bistoury" as a tool of his own design, made so as to excise the vertebrae of a live piglet when demonstrating the function and faculties of the intercostal muscles.[138] Famously, in *On the Avoidance of Grief*, Galen laments losing casting molds of specialized tools that

---

with the emendation of μὴν to ἄν. Salas (2020a: ch. 6 and 7) provides an overview of this ligation procedure but is more confident than I am that Erasistratus developed the experiment.

133. For catalogues of ancient medical tools, see Bliquez 2015; Künzl 1983; Milne 1907.

134. Gal. *AA* 7.7 Singer = 2.607K, trans. Singer.

135. Gal. *AA* 8.6 Singer = 2.685–687K.

136. Gal. *AA* 8.4 Singer = 2.667K.

137. Gal. *AA* 6.2 Singer = 2.575K.

138. Gal. *AA* 8.6 Singer = 2.682–683K.

he had invented in the fire at the Temple of Peace, where he had stored them.[139] This type of innovation, however, cannot be attributed solely to his genius, since medical implements had undergone considerable development in the preceding century.

Although the Hippocratic Corpus mentions multiple tools, archaeologists have recovered few material implements aside from cupping glasses, which, as Lawrence Bliquez notes, "are the sole type of [instrument] of which we have indisputable pre-Roman survivals."[140] Bliquez also points out that Hippocratic authors use more generic terms, such as "knife" [μάχαιρα], "little knife" [μαχαίριον], "iron tool" [σίδηρος] or "little iron tool" [σιδήριον], when describing cutting tools and will often employ devices originally intended for other purposes.[141] He states, "If it is true that not infrequently the Hippocratic employed items not intended for medical purposes, or even created what he needed on the spot, then there were simply fewer instruments specifically created for surgical purposes."[142] He then cites the Hippocratic text *Joints*: "You always have to use whatever is at hand."[143] This flexible attitude changed starting in the Hellenistic period. Celsus mentions that Diocles of Carystus invented his own eponymous scoop for extracting missiles embedded in flesh.[144] Other figures of the next generations invented their own "virtuoso" tools. Meges developed a specialized lithotomy knife; Heron designed a diaphragm-driven cupping glass; Heraclides of Tarentum made a special file for leveling teeth; and Marcellus produced a trident cautery.[145] By the time Celsus wrote his encyclopedic chapters on surgery in the first century BCE, medical implements had developed dramatically. References to specialized knives, scalpels, and scoops abound in his text. These technological developments might have started with the increased interest in dissection that rose in Alexandria. At the same time, they also reflect the expansion of surgical interventions that physicians performed by the first century BCE.[146]

139. Gal. *Ind.* 4–5 Pietrobelli; cf. *Lib. Prop.* 11 = p. 19 Singer = 19.41K. See also *Comp. Med. Loc.* 12.871K.

140. Bliquez 2015: 25.

141. Bliquez 2015: 28; cf. Lopez-Salva 1999: 311–31.

142. Bliquez 2015: 50.

143. Hipp. *Artic.* 7 = 4.94L; cf. Gal. *Hipp. Art.* 18A.344K.

144. Celsus *Med.* 7.5.3. Diocles's tool does not appear to have been widely adopted (Bliquez 2015: 141–43).

145. Bliquez 2015: 7, 15–16. See Gal. *Comp. Med. Loc.* 12.871–872K for the reference to Heraclides.

146. Whereas the Hippocratics shy away from the knife, Paul of Aegina, writing in the seventh century, lists over 120 different surgeries performed during

Archaeological evidence for medical tools seems to confirm this growing body of medical technologies, since finds of such implements increase dramatically from this point onward, largely thanks to the fact that physicians began to be buried with their tool kits. Although it is difficult to determine precisely what this interment practice means, at the very least it reveals the increased importance that surgical tools now played in the self-image of medicine.[147] Medical iconography suggests such a shift occurred in the centuries before Galen. For Greek physicians of the fifth century BCE, although other medical implements do occasionally appear alongside them in iconography, the cupping glass tended to be the medical emblem par excellence, appearing on the *Peytel Aryballos* and carved reliefs (figs. 2 and 3), as well as on coinage and votive offerings.[148] By the Hellenistic period, however, specialized surgical instruments started taking on a larger role.[149] This shift continues in Roman visual evidence as well, and by the second century CE, burial plaques display a tool kit of specialized knives and probes as a more common icon of a practicing physician (figs. 18, 19a, and 19b).[150] Although cupping vessels never dis-

his era, and these seemingly reflect practices that had mostly been developed by Galen's time. Celsus *Med.* 7 *pref.* 1–3 says that the practice developed substantially in Alexandria, and he names Philoxenus, Gorgias, Sostratus, Heron, two different men named Apollonius, Ammonius, and other Alexandrians as key contributors, as well as Tryphon, Euelpistus, and Meges in Rome. See Marganne 1998 for the development of surgery as attested in papyrological sources.

147. Bliquez (2015: 4 n6) notes that, as the collections from the Naples Museum reveal, physicians were interred with their tools, which did not happen as much with other professions.

148. For Basel Relief and grave and votive monuments, see Berger 1970; Tabanelli 1958; Jackson 1988; and Hillert 1990. For the cupping vessel as the emblem of medicine, both on coinage and gravestones, see Lambros 1895: 15–19; Penn 1994: 141–43; and Berger 1970: 19–23; cf. Bliquez 2015: 50. See also note 152 on p. 59, which mentions the "Telemachus Relief," Founding Stele of the Sanctuary of Asklepios, Athens, ca. 420/419 BCE, Acropolis Museum, EAM 2490, which features a cupping glass next to forceps and an indiscernible implement above the head of the goddess Hygeia.

149. The sanctuary of Asclepius in Athens also provided a statue base, potentially for use under a votive offering, with two large cupping glasses flanking a box of six surgical tools, Sanctuary of Asklepios, Athens, ca. 320 BCE, Acropolis Museum, EAM 1378. We should also consider the temple of Kom Ombo, built by Ptolemy VI Philometor during the period of 180–47 BCE, which includes an interior rear wall cover in hieroglyphs of medical implements including cupping glasses, scalpels, forceps, scales, shears, spoons, and other such material technologies.

150. Along with these two examples, there is an early 4th century CE sarcophagus of a Greek physician with book rolls and a similar box of surgical tools as

FIGURE 18. Grave or votive for a physician, 1st century BCE, 1st century CE, Roman, carved from stone and likely manufactured in the Peloponnese, Greece, Antikensammlung der Staatlichen Museen zu Berlin—Preußischer Kulturbesitz, Berlin, no. SK 804.

appeared from medical reliefs,[151] by Galen's lifetime specialized cutting implements had become a more common emblem of medical practice.

Certain competitions even suggest that people celebrated and encouraged novel technologies and the development of new tools. An inscription from Ephesus from the reign of Antoninus Pius (138–161 CE) records competitions in medical instrument design, alongside those in pharmacology

---

the icons of his profession (Ostia, Rome, Metropolitan Museum of Art, New York, no. 48.76.1). This shift in medical iconography toward cutting implements occurs despite the fact that cupping glasses are still among the most common finds of the Roman imperial period and the most referred to medical implement (Bliquez 2015). For scalpels on monuments as medical emblems, see Krug 2008. Cases for such medical instruments seem to appear quite early, which might suggest physicians carried small sets of metal tools with them, but evidence is hard to parse (see Bliquez 2015: 17–18; Hipp. *Decent.* 8 = 9.236L mentions one).

151. See the second-century CE marble tombstone of Jason the physician from Athens, at the British Museum in London, GR 1865,0103.3.

FIGURES 19A AND 19B. Funerary plaque for a surgeon and a plaster reproduction of the same plaque, ca. 160 BCE; tomb of Marcus Ulpius Amerimnus, Ostia, Italy, tomb 100.

and surgery.[152] Galen himself mentions a (hypothetical) four-sided ax that an interlocutor designed (but did not have made) to excise a section of artery with a single swipe, such that its contents (air or blood) would remain inside.[153] Recent grave finds have unearthed a number of highly special-

152. Engelman, Knibbe, and Merkelbach 1980: 108–19 (nos. 1161–69); cf. Bliquez 2015: 16 for reference.

153. Gal. *AA* 7.16 Singer = 2.643–644K.

ized tools, including those for knives for the removal of bladder stones, female catheters, a *poulkos* syringe, tools for bone surgery, and more.[154] The details evince a culture of innovation surrounding the material implements of medicine. As such, Galen's natural faculties were enacted with a set of dissection tools that had developed in the centuries leading up to his experimental vivisections. The vitalist body was produced by tools developed in the first and second centuries CE. As a consequence, we should understand that Galen's assertion that the living tissues possess biotic natural faculties was not a step away from tools but a step into articulating the body with a growing set of specialized implements and the techniques that they facilitated.

Rebecca Flemming has emphasized how Galen's demiurge acts as a guarantor of knowledge, marshaling nature and its components in a way that mirrors both Galen's position as authoritative organizer of knowledge and the Roman emperor himself, who sustains the empire, its integrity, and its systematicity.[155] We might also acknowledge that Galen mirrors the divine demiurge insofar as both are technicians who manufacture tools perfectly designed for specific tasks. Indeed, seeing the precision engineering of the body, a feat so integral to his project, requires his precision-engineered tools. As a consequence, in dismantling and reverse engineering the body, Galen becomes a reflection of Nature. Moreover, since these material tools enact the phenomena of his vitalist organism, his theory of corporeality rests on a set of technologies that resemble the object that they articulate. This mutually generative and mutually reinforcing conceptual harmony reveals the degree to which the technological changes of the first and second centuries CE in which Galen himself participated were crucial for the organism to exist as he structured it. This explanatory object was a technological product of its age, emergent from his technological program.

6.7 CONCLUSION

The Roman Empire of the first and second centuries CE saw the status of anatomical investigations once again grow. This practice had lingered on as part of medical instruction in Alexandria throughout the Hellenistic period, at least in some form, but now medical theorists, performing physicians, and their patient-filled audiences paid greater attention to dissection as a valuable epistemic practice. With the return of anatomy came the

154. Bliquez 2015: 5.
155. Flemming 2009.

return of thinking about the body as a functional object, best understood by examining the tool-like operations of the corporeal interior. Galen took up this project as his own, insisting that Nature was not just goal oriented but had designed the human body in the best possible way, such that the components of this biotic artifact could not be improved. Such a philosophical position produced a physiology that liberally incorporated tool heuristics to conceptualize both the function and operation of interior parts. At the same time, Galen insisted that biotic tissues possess certain natural capacities that nonliving matter does not. Different tissues possessed different, specialized faculties, whether for expansion and contraction, attraction of a certain type, blood production, or any of the other biotic necessities. Rather than representing a step away from tool heuristics, as the vitalist/mechanist debate would suggest, Galen saw these capacities as the logical extension of the claim that interior parts had activities and functions. For him, the latter implied the former, such that agreeing that the body "worked" according to a structural teleology necessitated that its parts possessed the intrinsic powers to complete these operations. Nevertheless, these vitalist natural powers were not simply theoretical forces. Galen argued that he could demonstrate these capacities and make their properties manifest through vivisection experiments that lemmatized the body and induced malfunction. Insofar as these vitalist powers were enacted in technical interventions, the specific behaviors and properties of the natural faculties are inseparable from the material circumstances that made them visible. Rather than representing a step away from tools, vitalism represents a step toward further integration of the body and technologies.

Overall, then, this chapter illustrates how the overlap of biotic and technological entities cannot be addressed solely through the question of originating motions or unique living capacities, since the influence of tools on theories of corporeality stretch far beyond this individual question. This particular point of analysis is valuable and revealing, but it cannot provide a full picture. Instead of essentializing, this chapter has attempted to show the various registers at which technologies are crucial for understanding Galen's theories of corporeality, and to highlight moments of recursion where the tools that he uses to articulate the body and enact its properties blend with the objects and behaviors they disclose. His organized body was inseparable from a host of technologies that enabled its emergence as the proper object of medical science.

Conclusion

The previous six chapters have explored a single question at the heart of this book: how did ancient technological realities affect theories of corporeality in Greek and Greco-Roman medicines? To do so, it has followed three interwoven vectors, which I have called the structural, investigative, and analogical use of technologies to interrogate and conceptualize corporeality. Chapter 1 illustrated that pre-Hellenistic Hippocratic treatises did not typically use function to consolidate and arrange the body as a whole, even if they discussed certain roles that individual parts play. More than asserting some common Hippocratic notion of what type of object the body is, this chapter emphasized that multiple theories—sometimes overlapping, sometimes competing—ran through different texts. It highlighted how different authors privileged different therapeutic technologies to make the nature of the body manifest, and how these different technologies built different bodies. Chapter 2 returned to the pre-Socratic physikoi to examine how Empedocles used the clepsydra to explain the act of breathing, even as he himself did not focus on assigning this corporeal behavior to a particular organ or describing its function. Instead, the chapter argued that his use of a technological analogy to account for respiration facilitated later thinking in terms of individuated corporeal tools. This technological analogy was not imported to explain a part's function. It helped make functional parts. Indeed, Plato picked up the vocabulary of tools as he started discussing corporeal organa as part of the teleological framework developed in the *Timaeus*. Although the body articulated in this text was ultimately oriented toward the maintenance of rationality, not the production of life. Chapter 3 illustrated that it was Aristotle who fully organized the body, establishing it as a hierarchical arrangement of tool-like functional parts that were oriented toward the maintenance of the living whole. It also illustrated the complexities of the systems that then needed to be described within such an object, as Aristotle used philosophical argumentation, anatomical investigation, and tool analogies to describe various corporeal systems that sometimes meshed poorly. Lastly, it emphasized how contemporary projectile technologies that used animal tissues to generate explosion force shaped his assumptions about how the body generated motion. Chapter 4 examined the expansion of the organism as an epistemic object in the Hellenistic period and why this expansion was paired with a new medical interest in anatomical investigations, placing these new tool-like bodies within the development of pneumatics during this period. Chapter 5 examined the pushback against the organism as a model of corporeality in the late Hellenistic period but illustrated how Roman imperial infrastructure started to impact the theories of Asclepiades of Bithynia, who drew from aqueducts and artificial water systems when describing the flow of corpuscles through the pores of the body. Lastly, chapter 6 examined Galen of Pergamon, who championed the organism as a concept and established a set of investigative techniques and dissection tools to demonstrate new somatic capacities that he made essential to the functioning of his tool-like body.

Although multiple lines of inquiry culminated in Galen, the interwoven histories of bodies and tools did not end in the second century CE, and there is no era in which notions about bodies do not reflect the technologies that helped construct them. In late antique, Arabic and Islamic, medieval, and early modern medicines, the tools used to investigate, treat, or imitate bodies likewise colored or structured somatic theories or individual explanations in various ways. Nevertheless, Galen offers a way to consolidate several lines of discourse that consolidated the human body as an organism. These other temporal eras will need to be explored in another book.

Although we still live with the organism as an inheritance from Greek and Greco-Roman medicine, there is no undisturbed line between Galen's views of the body and twentieth-century medical science. Multiple centuries and medical theories intervened, and these ideas moved into and through multiple medical traditions. Nevertheless, the strength of the organism as a concept was such that it survived through competing theoretical frames, including spiritual, mechanistic, vitalist, and genetic ones, such that many medical theories of the twentieth and twenty-first centuries have adopted it as a nonmetaphorical description of the body. Conceptualizing each part as a functional object seems to be the baseline framework for both medical and evolutionary interpretations of corporeality. Indeed, imagining the body as machine fit well with the increasing number of technological interventions into corporeal processes or mechanical replacements for functional parts. Some of these machines were external prosthetics, which did the "work" of the organs on their behalf, including iron lungs and kidney dialysis machines. Others were internal, including titanium hips or pacemakers. Others still, such as colostomy bags or battery-driven artificial hearts, establish lifelong prosthetic links. For all their success, the mechanical replacement of inner organs to this day remains relatively limited, and we have had far more success with replacing damaged parts with other living tissue, whether from another area of our body, as in the case of skin grafts or soft-tissue reconstruction, or from other humans (and potentially animals). These technical and mechanical failures can help illustrate the limits of the organism as an explanatory concept, where biotic "functions" are multi-vectored and multidirectional.

Over the last few decades, investigations into our microbiomes have revealed that human bodies are made up of many different living beings, perhaps trillions of microbes, which could outnumber human cells by a factor of ten to one.[1] Bacteria live on our skin. They colonize our intestinal tract. They thrive inside our nose. We are—by cellular count—mostly comprised of biotic life-forms that are not "us" in genetic terms. Similarly, as viral pathogens become ever more elevated in our social consciousness and conceptions of our long-term health, there is great likelihood that we will discover that viruses, too, play a considerable part in our embodied lives and cannot be so easily separated out from our normative ideas about how human corporeality operates in the physical world. Although we might be inclined to separate these living things from what we consider to be "our" body, we cannot live without them any more than we can live without the kidneys or a liver. This reality shows the limitations of using the organism as a totalizing description of the human whole. Instead, we can also productively conceptualize the body as a type of ecology that needs to be nurtured, balanced, and maintained, wherein different life-forms and cellular colonies thrive within integrated environments. Such an ecological approach loosens the grip that organs have as the agents of action, and they instead return to being locations or sites where processes and interactions take place. Understood this way, our viscera act in a way quite similar to the metaphorical scheme presented by *On Fleshes*, rather than operating as functional objects. Cells and microbes can be seen as independent entities providing benefits to the body, but only insofar as each acts in its own best interests. These ecological approaches to corporeality have generated useful shifts in healthcare, such that medical practitioners and patients have envisioned new ways to maintain health be-

1. National Institutes of Health 2012.

yond simply shoring up the integrity of functional parts. Bacteria are now viewed as essential to corporeal systems, encouraged in the gut, monitored in the mouth, and preserved across generations through birthing practices. Rather than viewing external agents as productive of dysfunction, we now recognize the complex multi-organism interactions that are required even to maintain homeostasis within living entities.

If ecological metaphors revive top-down models of corporeality, consolidating the body as a network of environments rather than an assemblage of parts, another dominant strain of biology conceptualizes corporeality from the bottom up, seeing the body as the product manufactured from DNA. In this vision, DNA not only provides the blueprints for producing proteins but also embodies the self-driven mechanisms of assembly. Human wholes are the by-product of genetic replication. This approach, too, has its strengths. For example, Raghuveer Parthasarathy argues that four simple principles underlie the complex assembly of living things.[2] He notes that DNA does not generally encode for the structure of the parts it builds, but for certain proteins whose basic physical structures self-assemble according to essential physical forces. DNA does not provide instructions for building the body as a product, as it were. It provides instructions for building factories that will build other factories. Indeed, these two models of corporeality could be understood as updated versions of the mechanistic vs. vitalist positions. Yet the point is that mechanical self-assembly is not incompatible with either tool-like functions or ecological interpretations. In fact, no single metaphor or conceptual scheme is likely to capture all the dynamics at work in systems as complex as those that underpin life. Recognizing the strengths each metaphorical scheme provides can help reveal how the tool-like logic of corporeal functions is a powerful heuristic but does not exhaust or capture bodies in their entirety.

Throughout these chapters, this book has argued that recovering ancient theories of the body requires acknowledging that they were not pure intellectual exercises or doctrinal commitments weighed against each other in a vacuum. Instead, medical theories were sustained by certain material practices. Some medical theorists used their therapeutic technologies as the essential investigative tools, while others developed new epistemic techniques or simply drew from the technological milieu that surrounded them. Assessing how technologies impact the body generated in this way has highlighted the reciprocal, recursive relationships that consolidated and articulated the human form. Nevertheless, as much as

2. Parthasarathy 2022.

this book has emphasized the importance of technologies to ancient medical theories from Greek and Greco-Roman sources, authors in antiquity assembled their epistemic object from multiple exemplars and multiple species, examined and interrogated in different ways, and mixed with theoretical commitments, hearsay observations, and basic inferential logic. Technologies may be essential to the very formation of corporeality as a concept, but they are not totalizing. Nevertheless, as I hope the previous pages have illustrated, tools were indispensable to biomedical conceptions of the body, such that any history of medicine should acknowledge the powerful role they played for constructing the object physicians sought to soothe, cure, and understand.

Acknowledgments

With a project such as this one, which has grown in multiple directions over the past decade, recognizing the many contributions and kindness that helped it grow feels all but impossible. This is especially true since a project never truly starts, but one day becomes visible long after its life, roots, and limbs have started to coalesce. My hope is that this project also never really ends, even though it is far past time for this instantiation to be complete.

My first thanks go to those who have read iterations of these arguments over the years, especially Katharina Volk, whose practical advice structured, but never diminished, my always overambitious plans. Nancy Worman, Matthew Jones, and Katja Vogt also provided comments that have stuck with me for years, as did Brooke Holmes, who continually supplies an inspiring body of scholarship and has been a generous interlocutor and mentor. Calloway Scott has slogged through many chapter fragments, letters, and grant applications, and all should know that he is a human of great character and strength. Anna Bonnell Freidin helped me keep focus on a single project at a time and has always been an immensely helpful intellectual partner and encouraging friend. Henry Cowles likewise pulled me more out into the history of science beyond Classics, and conversations with him would be enough to sustain a full life of the mind.

At UC Davis, I have been blessed with truly wonderful colleagues, and I thank all of them for creating an environment of mutual respect, support, and friendship. Thanks to Emily Albu, Tim Brelinski, Mike Chin, Katie Cruz, Ralph Hexter, Shennan Hutton, Valentina Popescu, John Rundin, Carey Seal, Alexandra Sofroniew, David Traill, Melissa Stem, and Rex Stem, our much-missed colleague. Special thanks to Anna Uhlig, who has read multiple chapters and side projects, and each time manages to offer insightful, clarifying comments. Beyond my department, thanks to Daniel Stolzenberg, John Slater, Claire Waters, Claire Goldberg, Fran Dolan, and all the others who have helped organize and participate in the Early Science Workshop. Joan Cadden has been especially generous with her time and offered valuable comments, as did Daryn Lehoux, Sylvia Berryman, and Maria Gerolemou, all of whom participated in a manuscript workshop and made the arguments of this book better and more refined. Other interlocutors include Claire Bubb and Kassandra Miller, who have been great sources of solidarity as we have moved through the publishing process, and I have learned so much from their scholarship. Sean Coughlin was kind enough to offer an advanced look at his work on the fragments of the Pneumatists, while Jennifer Stager provided some valuable last-minute references. The *Society for Ancient Medicine* executives, including Aileen Das and Jessica Wright, have also kept me excited about the future of the discipline and the possibilities it can present.

Paul Keyser, Susan Mattern, Courtney Roby, Rebecca Flemming, Julie Laskaris, Tony Long, and Karen ní Mheallaigh have all shown me great kindness, and their small tokens of encouragement kept me motivated and hopeful. I am particularly thankful to Heinrich von Staden, whose work not only provided the spark for this project, but who changed my scholarly practice for the better with a few soft demonstrations of intellectual generosity. In many ways, this book feels like a long footnote and tribute to his scholarship.

During my days at Columbia, Caleb Dance and Mathias Hanses both read early chapters, and they joined numerous others in making my years there incredibly rewarding and incredibly fun. Thanks to Julia Perrin, Zoe Tan, James Tan, Kate Stanley, Royden Kadyschuk, Anne Diebel, Anjuli Raza Kolb, Alicia DeSantis, Kate Brassel, Charles McNamara, Joe Sheppard, and Kimberley Regler. Each deserves a poem of praise, but I can here only offer brief gratitude. Before New York, I also spent many wonderful years in Halifax, Nova Scotia, where I lived by the harbor, trudged through the snow, and savored every spring day above freezing. Many thanks to Gary McGonagill, Peter O'Brien, Dennis House, Neil Robertson, Gordon McOuat, and my friends Ben Frenken, Ross Gower, Bethie Baxter, Andra Striowski, Dan Wilband, and Michelle Wilband. Thanks also to Chris Evans, Leslie Najgebauer Landon Tresise, Jared Aldridge, Max and Lily Cameron, Brandon and Michele Maultash, Scott and Lily Pearsall, David and Laura Costain, and our wonderful neighbor Karen, who has put up with half-finished home projects languishing in our shared driveway as I completed this book.

I have been fortunate enough to employ research assistants for parts of this project. Taylor Bell has been instrumental, reading, revising, and helping me bring many chapters to fruition, while David Gilstrap has helped revise, while also allowing me the opportunity to read Greek and chat about this material. Many thanks to all my teaching assistants who helped deliver courses on materials related to this book, including Dominique Paz, Lauren Hitt, Bex Jones, Naida Nooristani, and C. W. Wilkinson.

Several fellowships have given me the time to both expand my research program's scope and then contract it for the purpose of completing this book. These include a UC Davis Service Award and a Hellman Fellowship, as well as a year spent at the Stanford Humanities Center, where I benefited from friendships with Kayla Schuller, Liz Marcus, Aileen Robinson (who helped with my book proposal), and many others. Thanks to Caroline Winterer for making that fellowship possible, and to Giovanna Ceserani and Reviel Netz for making me feel welcome in the Classics Department while I was on campus. The UC Davis Humanities Institute has also been instrumental in helping me find the network of colleagues needed to sustain a project like this, and I've received generous support from the Dean's Office of the UC Davis College of Letters and Science.

Many thanks to my editor, Karen Darling, who has provided guidance and displayed great patience, both of which allowed me to write the best version of this project that my abilities would allow. Thanks to Fabiola Enríquez Flores, Jessica Wilson, Jonathan Farr, Adriana Smith, Anne Strother, and the editorial assistants who have brought this work to print, as well as the anonymous readers who gave excellent guidance to help improve this book. As with all projects, completing my own tasks would not have been possible without the labor of many others, some of whom took care of my children during the day, while others maintained my spaces of employment, staffed libraries, filled out paperwork, facilitated grants, completed payrolls, and kept my internet working. Thanks in particular to Adam Siegel, Mandy Bachman, and Omar Mojaddedi, among many others.

Deep thanks to my father, Cole Webster, whose brilliance found its expression in tools and techniques, and whose depths of knowledge and endless reserves of endurance are profoundly, perpetually missed. Also, to my mother, Nancy Newman, who continues to learn about the world and its fascinating offerings, and whose practice of creativity has always been inspiring. Her knowledge of textiles is woven through this work as well. Thanks also to my stepfather, John Milne, who has been a plank of constancy and kindness for decades and deserves so much more recognition than he receives, and to Neil and Carolyn Turnbull, my uncle and aunt, for giving me my first job swinging a hammer and for continuing to commit to both the care for others and need for one's own joy. Lastly, many thanks to my brother, Liam, whose friendship and intelligence have helped stabilize the foundation on which any life's work needs to be built.

My greatest thanks go to my wife, Christina. My gratitude to and for her could not be measured by the number of teaspoons in the ocean. Her partnership and generosity have taught me so much about myself, and she allowed me to figure out what this book was and how it fit within the life we continue to build with one another. During the completion of this project, we have welcomed two boys into the world, Cy and Ames, and their wondrous capacity for learning only makes me feel how great a privilege it is to construct a career around curiosity. I hope that I can help them figure out the questions that excite them and provide them the encouragement to take that excitement seriously.

Bibliography

Abel, Karlhans. 1958. "Die Lehre vom Blutkreislauf im Corpus Hippocraticum." *Hermes* 86, no. 2: 192–219.

Abou-Aly, Amal Mohamed Abdullah. 1992. "The Medical Writings of Rufus of Ephesus." PhD diss., University College, London. https://discovery.ucl.ac.uk/id/eprint/1317541.

Adair, Mark J. 1996. "Plato's View of the 'Wandering Uterus.'" *Classical Journal* 91, no. 2: 153–63.

Adamson, Peter. 2014. "Galen on Void." In *Philosophical Themes in Galen*, edited by Peter Adamson, Rotraud E. Hansberger, and James Wilberding, 197–211. London: University of London.

Année, Magali, ed. 2019. *Alcméon de Crotone: Fragments; Traité scientifique en prose ou poème médical?* Paris: Librairie Philosophique J. Vrin.

Ariew, André. 2002. "Platonic and Aristotelian Roots of Teleological Arguments." In *Functions: New Essays in the Philosophy of Psychology and Biology*, edited by André Ariew, Robert Cummins, and Mark Perlman, 7–32. Oxford: Oxford University Press.

Artelt, Walter. 1968. *Studien zur Geschichte der Begriffe 'Heilmittel' und 'Gift.'* Darmstadt: Wissenschaftliche Buchgesellschaft.

Asmis, Elizabeth. 1993. "Asclepiades of Bithynia Rediscovered?" Review of *The Lost Theory of Asclepiades of Bithynia*, by J. T. Vallance. *Classical Philology* 88, no. 2: 145–56.

Balme, David M. 2002. *Aristotle: Historia Animalium: Volume 1. Books I-X: Text.* Cambridge: Cambridge University Press.

———. 1987. "Teleology and Necessity." In *Philosophical Issues in Aristotle's Biology*, edited by Allan Gotthelf and James G. Lennox, 275–85. Cambridge: Cambridge University Press.

Barnes, Jonathan. 1991. "Galen on Logic and Therapy." In *Galen's Method of Healing*, edited by F. Kudlien and Richard J. Durling, 50–102. Leiden: Brill.

———. 1993. "Galen and the Utility of Logic." In *Galen und das hellenistische Erbe*: Verhandlungen des IV. Internationalen Galen-Symposiums, Sudhoffs Archiv, Beiheft XXXII, edited by Jutta Kollesch and Diethard Nickel, 33–52. Stuttgart: Franz Steiner Verlag.

———. 2003. "Proofs and Syllogisms in Galen." In *Galien et la philosophie: huit exposés suivis de discussions*. Entretiens sur l'Antiquité classique 49, edited by Jonathan Barnes and Jacques Jouanna, 1–30. Geneva: Fondation Hardt.

Barnes, Jonathan, and Jacques Jouanna, eds. 2003. *Galien et la philosophie*. Entretiens sur l'Antiquité classique 49. Geneva: Fondation Hardt.

Barras, Vincent, Terpsichore Birchler, and Anne-France Morand, eds. and trans. 1995. *Galien: L'Âme et ses passions*. Paris: Les Belles Lettres.

Bartoš, Hynek. 2015. *Philosophy and Dietetics in the Hippocratic "On Regimen": A Delicate Balance of Health*. Leiden: Brill.

Beardslee, John Walter. 1918. *The Use of ΦΥΣΙΣ in Fifth-Century Greek Literature*. Chicago: University of Chicago Press.

Bensaude-Vincent, Bernadette, and William R. Newman, eds. 2007. *The Artificial and the Natural: An Evolving Polarity*. Cambridge, MA: MIT Press.

Berger, Ernst. 1970. *Das Basler Arztrelief: Studien zum griechischen Grabund Votivrelief um 500 v. Chr. und zur vorhippokratischen Medizin*. Mainz: Philip von Zabern.

Berrey, Marquis. 2017a. *Hellenistic Science at Court*. Berlin: De Gruyter.

———. 2017b. "Methodists." *Oxford Classical Dictionary*. https://doi.org/10.1093/acrefore/9780199381135.013.8165.

Berryman, Sylvia. 1997. "*Horror Vacui* in the Third Century BC: When Is a Theory Not a Theory?" In "Aristotle and After," edited by Richard Sorabji, *Bulletin of the Institute of Classical Studies* supplement 68, 147–57.

———. 2002a. "Aristotle on *Pneuma* and Animal Self-Motion." *Oxford Studies in Ancient Philosophy* 23: 85–97.

———. 2002b. "Galen and the Mechanical Philosophy." *Apeiron* 35, no. 3: 235–53.

———. 2003. "Ancient Automata and Mechanical Explanation." *Phronesis* 48, no. 4: 344–69.

———. 2009. *The Mechanical Hypothesis in Ancient Greek Natural Philosophy*. Cambridge: Cambridge University Press.

Betegh, Gábor. 2010. "The Transmission of Ancient Wisdom: Texts, Doxographies, Libraries." In *The Cambridge History of Philosophy in Late Antiquity*, vol. 1, edited by Lloyd P. Gerson, 25–38. Cambridge: Cambridge University Press.

———. 2021. "Plato on Illness in the *Phaedo*, the *Republic*, and the *Timaeus*." In *Plato's Timaeus: Proceedings of the Tenth Symposium Platonicum Pra-*

gense, edited by Chad Jorgenson, Filip Karfík, and Štěpán Špinka, 228–58. Leiden: Brill.

Bianchi, Emanuela. 2017. "Aristotle's Organism, and Ours." In *Contemporary Encounters with Ancient Metaphysics*, edited by Abraham J. Greenstine and Ryan J. Johnson, 138–57. Edinburgh: Edinburgh University Press.

Black, Max. 1962. *Models and Metaphors: Studies in Language and Philosophy*. Ithaca, NY: Cornell University Press.

Bliquez, Lawrence. 2015. *The Tools of Asclepius: Surgical Instruments in Greek and Roman Times*. Leiden: Brill.

Booth, N. B. 1960. "Empedocles' Account of Breathing." *Journal of Hellenic Studies* 80: 10–15.

Bos, Abraham P. 2003. *The Soul and Its Instrumental Body*. Leiden: Brill.

Boudon-Millot, Véronique. 2012. *Galen de Pergame: Un médecin grec à Rome*. Paris: Les Belles Lettres.

Boudon-Millot, Véronique, and Antoine Pietrobelli. 2005. "Galien ressuscité: Édition *princeps* du texte grec du *De propriis placitis*." *Revue des études grecques* 118, no. 1: 168–213.

Bourgey, Louis. 1980. "Hippocrate et Aristote: L'origine, chez le philosophe, de la doctrine concernant la nature." In *Hippocratica: Actes du colloque hippocratique de Paris (4–9 septembre 1978)*, edited by M. D. Grmek, 59–64. Paris: Éditions du CNRS.

Bowker, Geoffrey C., and Susan Leigh Star. 1999. *Sorting Things Out: Classification and Its Consequences*. Boston, MA: MIT Press.

Boylan, Michael. 1982. "The Digestive and "Circulatory" Systems in Aristotle's Biology." *Journal of the History of Biology* 15, no. 1: 89–118.

Brunt, Peter A. 1994. "The Bubble of the Second Sophistic." *Bulletin of the Institute of Classical Studies* 39: 25–52.

Bruun, Christer. 2000. "Water Legislation in the Ancient World (c. 2200 B.C.–c. A.D. 500): The Greek World." In *Handbook of Ancient Water Technology*, edited by Örjan Wikander, 557–74. Leiden: Brill.

Bubb, Claire. 2019. "Hollows in the Heart: A Lexical Approach to Cardiac Structure in Aristotle." *Sudhoffs Archiv* 103, no. 2: 128–40.

———. 2020. "Blood Flow in Aristotle." *Classical Quarterly* 70, no. 1: 137–53.

———. 2022. *Dissection in Classical Antiquity: A Social and Medical History*. New York: University of Cambridge Press.

Buddensiek, Friedemann. 2009. "Aristoteles' Zirbeldrüse? Zum Verhaltnis von Seele und pneuma in Aristoteles' Theorie der Ortsbewegung der Lebewesen." In *Body and Soul in Ancient Philosophy*, edited by Dorothea Frede and Burkhard Reis, 309–29. Berlin: De Gruyter.

Burgière, Paul, and Danielle Gourevitch, eds. and trans. 2003. *Soranos d'Éphèse, Maladies des Femmes, Vol. 1–IV*. Paris: Les Belles Lettres.

Byl, Simon. 1971. "Note sur la polysémie d'ΟΡΓΑΝΟΝ et les origines du finalisme." *L'Antiquité classique* 40, no. 1: 121–33.

———. 1980. *Recherches sur les grandes traités biologiques d'Aristote: Sources écrites et préjugés*. Brussels: Palais des Académies.

———. 1989. "L'odeur végétale dans la thérapeutique gynécologique de Corpus hippocratique." *Revue belge de philologie et d'histoire* 67, no. 1: 53–64.

Camardo, Domenico, Elly Heirbaut, Gemma C. M. Jansen, Anne-Marie Jouquand-Thomas, Antonella Merletto, Jacques Seigne, and Jeroen van Vaerenberg. 2011. "Toilets in the Urban and Domestic Water Infrastructure." In *Roman Toilets: Their Archaeology and Cultural History*, edited by Gemma C. M. Jansen, Ann Olga Koloski-Ostrow, and Eric Moorman, 71–94. Leuven: Peeters.

Canguilhem, Georges. 2008. *Knowledge of Life*. Translated by Stefanos Geroulanos and Daniela Ginsberg. Edited by Paola Marrati and Todd Meyers. New York: Fordham University Press. First published in 1952 as *La connaissance de la vie*. Paris: Hachette.

Carra de Vaux, Bernard. 1902. *Le livre des appareils pneumatiques et des machines hydrauliques par Philon de Byzance, edité d'après les versions arabes d' Oxford et de Constantinople et traduit en français par le Baron Carra de Vaux*. Paris: C. Klincksieck.

Casadei, Elena. 1997. "La dottrina corpuscolare di Asclepiade e i suoi rapporti con la tradizione atomista." *Elenchos* 18: 77–106.

Charles, David. 1991. "Teleological Causation in the *Physics*." In *Aristotle's* Physics*: A Collection of Essays*, edited by Lindsay Judson, 101–28. Oxford: Oxford University Press.

Ciszuk, Martin, and Lena Hammarlund. 2008. "Roman Looms—A Study of Craftsmanship and Technology in the Mons Claudianus Textile Project." In *Purpureae Vestes II: Vestidos, textiles y tintes. Estudios sobre la producción de bienes de consume en la Antigüedad*, edited by Carmen Alfaro and L. Karali, 119–33. Valencia: University of Valencia.

Cocchi, Antonio. 1758. *Discorso primo sopra Asclepiade*. Florence: G. Albizzini. Translated in 1955 by Robert M. Green in *Asclepiades, His Life and Writings: A Translation of Cocchi's "Life of Asclepiades" and Gumpert's "Fragments of Asclepiades."* New Haven, CT: E. Licht.

Connell, Sophia M. 2016. *Aristotle on Female Animals: A Study of the "Generation of Animals."* Cambridge: Cambridge University Press.

———, ed. 2021. *The Cambridge Companion to Aristotle's Biology*. Cambridge: Cambridge University Press.

Cooper, John M. 1982. "Aristotle on Natural Teleology." In *Language and Logos: Studies in Ancient Greek Philosophy Presented to G. E. L. Owen*, edited by Malcolm Schofield and Martha C. Nussbaum, 197–222. Cambridge: Cambridge University Press.

———. 1987. "Hypothetical Necessity and Natural Teleology." In *Philosophical Issues in Aristotle's Biology*, edited by Allan Gotthelf and James G. Lennox, 243–74. Cambridge: Cambridge University Press.

———. 2020. "The Role of Thought in Animal Voluntary Self-Locomotion, *MA* 7 (through 701b1)." In *Aristotle's De Motu Animalium: Symposium Aristotelicum*, edited by Christof Rapp and Oliver Primavesi, 345–86. Oxford: Oxford University Press.

Corcilius, Klaus. 2008. *Streben und Bewegen: Aristoteles' Theorie der animalischen Ortsbewegung*. Berlin: De Gruyter.

———. 2020. "Resuming Discussion of the Common Cause of Animal Self-Motion: How Does the Soul Move the Body? *MA* 6." In *Aristotle's De Motu Animalium: Symposium Aristotelicum*, edited by Christof Rapp and Oliver Primavesi, 299–344. Oxford: Oxford University Press.

Corcilius, Klaus, and Pavel Gregoric. 2013. "Aristotle's Model of Animal Motion." *Phronesis* 58, no. 1: 52–97.

Cornford, Francis M. (1937) 1997. *Plato's Cosmology: The "Timaeus" of Plato*. Indianapolis, IN: Hackett.

Craik, Elizabeth, ed. and trans. 2006. *Two Hippocratic Treatises: "On Sight" and "On Anatomy."* Leiden: Brill.

———. 2015. *The 'Hippocratic' Corpus: Content and Context*. New York: Routledge.

———. 2017. "Teleology in Hippocratic Texts: Clues to the Future?" In *Teleology in the Ancient World: Philosophical and Medical Approaches*, edited by Julius Rocca, 203–15. Cambridge: Cambridge University Press.

———. 2018. "The 'Hippocratic Question' and the Nature of the Hippocratic Corpus." In *The Cambridge Companion to Hippocrates*, edited by Peter E. Pormann, 25–37. Cambridge: Cambridge University Press.

Crouch, Dora P. 1993. *Water Management in Ancient Greek Cities*. Oxford: Oxford University Press.

Dalby, Andrew. 2003. *Food in the Ancient World from A to Z*. London: Routledge.

Davidson, James. 1997. *Courtesans & Fishcakes: The Consuming Passions of Classical Athens*. New York: St. Martin's Press.

Dean-Jones, Lesley. 1994a. "Medicine: The 'Proof' of Anatomy." In *Women in the Classical World: Image and Text*, edited by Elaine Fantham, Helene P. Foley, Natalie B. Kampen, Sarah B. Pomeroy, and H. A. Shapiro, 183–205. Oxford: Oxford University Press.

———. 1994b. *Women's Bodies in Classical Greek Science*. Oxford: Clarendon Press.

———. 2013. "Too Much of a Good Thing: The Health of Olympic Athletes in Ancient Greece." In *From Athens to Beijing: West Meets East in the Olympic Games*, vol. 1, *Sport, the Body and Humanism in Ancient Greece and China*, edited by Susan Brownell, 49–65. New York: Greekworks.

———. 2017. "Aristotle's Heart and the Heartless Man." *The Comparable Body: Analogy and Metaphor in Ancient Mesopotamian, Egyptian, and Greco-Roman Medicine*, edited by John Wee, 122–41. Leiden: Brill.

Debru, Armelle. 1996. *Le Corps Respirant: La pensée physiologique chez Galien*. Leiden: Brill.

———. 2008. "Physiology." In *The Cambridge Companion to Galen*, edited by R. J. Hankinson, 263–82. Cambridge: Cambridge University Press.

De Groot, Jean. 2008. "Dunamis and the Science of Mechanics: Aristotle on Animal Motion." *Journal of the History of Philosophy* 46, no. 1: 43–67.

———. 2014. *Aristotle's Empiricism: Experience and Mechanics in the Fourth Century BC*. Las Vegas, NV: Parmenides.

Demont, Paul. 2014. "The Tongue and the Reed: Instruments and Organs in the Philosophical Part of Hippocratic *Regimen*." *Journal of Hellenic Studies* 134: 12–22.

Derrida, Jacques. 1974. "White Mythology: Metaphor in the Text of Philosophy." Translated by F. C. T. Moore. *New Literary History* 6, no. 1: 5–74.

Devereux, Daniel, and Pierre Pellegrin, eds. 1990. *Biologie, logique et métaphysique chez Aristote*. Paris: Éditions du CNRS.

Diels, Hermann. 1893. "Über das physikalische System des Straton." *Sitzungsberichte der Prüssichen Akademie der Wissenschaften zu Berlin* 1: 101–27.

Diller, Hans. 1952. "Hippokratische Medizin und attische Philosophie." *Hermes* 80, no. 4: 385–409. Reprinted in *Kleine Schriften zur antiken Literatur*, edited by Hans-Joachim Newiger and Hans Seyffert. Munich: Beck, 1973.

Draaisma, Douwe. 2000. *Metaphors of Memory: A History of Ideas about the Mind*. Cambridge: Cambridge University Press.

Drabkin, Israel E., ed. and trans. 1950. *Caelius Aurelianus: On Acute Diseases and on Chronic Diseases*. Chicago: University of Chicago Press.

Drachmann, Aage G. 1948. *Ktesibios, Philo and Heron. A Study in Ancient Pneumatics*. Copenhagen: E. Munksgaard.

DuBois, Page. 1991. *Torture and Truth*. London: Routledge.

Duhem, Pierre. 1906. *La Théorie physique: Son objet et sa structure*. Paris: Chevalier & Rivière. Translated into English in 1954 by Philip P. Wiener as *The Aim and Structure of Physical Theory*. Princeton, NJ: Princeton University Press.

Duminil, Marie-Paule. 1983. *Le Sang, les vaisseaux, le coeur dans la Collection hippocratique: Anatomie et physiologie*. Paris: Les Belles Lettres.

Ebrey, David, ed. 2015. *Theory and Practice in Aristotle's Natural Science*. Cambridge: Cambridge University Press.

Edelstein, Ludwig. 1935. "The Development of Greek Anatomy." *Bulletin of the History of Medicine* 3, no. 4: 235–48.

———. 1967. *Ancient Medicine: Selected Papers of Ludwig Edelstein*. Edited by Owsei Temkin and Clarice L. Temkin. Baltimore, MD: Johns Hopkins University Press.

Engelman, Helmut, Dieter Knibbe, and Reinhold Merkelbach, eds. 1980. *Die Inschriften von Ephesos*, part 4. Vol. 14 of *Inschriften griechischer Städte aus Kleinasien*. Bonn: R. Habelt.

Ermerins, Franz Zacharias, ed. and trans. 1864. *Hippocratis et aliorum veterum reliquiae*. Vol. 3. Utrecht: Kemink.

Evans, Harry B. 1994. *Water Distribution in Ancient Rome: The Evidence of Frontinus*. Ann Arbor: University of Michigan Press.

Fagan, Garrett. 2001. *Bathing in Public in the Roman World*. Ann Arbor: University of Michigan Press.

Fahlbusch, Heinrich. 1994. "Die Abschätzung der Leistungsfähigkeit der archaischen Wasserleitung Athens." In *Das archaische Wasserleitungsnetz für Athen und seine späteren Bauphasen*, edited by Renate Tölle-Kastenbein, 109–10. Mainz: Zabern.

Flashar, Hellmut. 2016. *Hippokrates: Meister der Heilkunst; Leben und Werk*. Munich: C. H. Beck.

Flemming, Rebecca. 2000. *Medicine and the Making of Roman Women: Gender, Nature, and Authority from Celsus to Galen*. Oxford: Oxford University Press.

———. 2009. "Demiurge and Emperor in Galen's World of Knowledge." In *Galen and the World of Knowledge*, edited by Christopher Gill, Tim Whitmarsh, and John Wilkins, 59–84. Cambridge: Cambridge University Press.

———. 2013. "Baths and Bathing in Greek Medicine." In *Greek Baths and Bathing Culture: New Discoveries and Approaches*, edited by Sandra K. Lucore and Monika Trümper, 23–32. Leuven: Peeters.

Foucault, Michel. 1970. *The Order of Things*. New York: Random House.

Franssen, Maarten, Gert-Jan Lokhorst, and Ibo Van de Poel. 2018. "Philosophy of Technology." *Stanford Encyclopedia of Philosophy*, revised September 6. https://plato.stanford.edu/entries/technology/.

Fraser, Peter M. 1969. "The Career of Erasistratus of Ceos." *Rendiconti Instituto Lombardo* 103: 518–37.

———. 1972. *Ptolemaic Alexandria*. 3 vols. Oxford: Clarendon Press.

Frede, Michael. 1982. "The Method of the So-Called Methodical School of Medicine." In *Science and Speculation: Studies in Hellenistic Theory and*

Practice, edited by Jonathan Barnes, Jacques Brunschwig, Myles Burnyeat, and Malcolm Schofield, 1–23. Cambridge: Cambridge University Press. Reprinted in 1987 in Michael Frede, *Essays in Ancient Philosophy*, 261–78. Minneapolis: University of Minnesota Press.

———. 2003. "Galen's Theology." In *Galen et la philosophie*, Entretiens sur l'Antiquité classique 49, edited by Jonathan Barnes and Jacques Jouanna, 73–129. Geneva: Fondation Hardt.

———. 2004. "Aristotle's Account of the Origins of Philosophy." *Rhizai* 1, no. 1: 9–44.

Freeland, Cynthia. 1990. "Scientific Explanation and Empirical Data in Aristotle's *Meteorology*." *Oxford Studies in Ancient Philosophy* 8: 67–102.

Freidin, Anna Bonnell. Forthcoming. *Birthing Romans: Childbearing and Risk in Imperial Rome*. Princeton, NJ: Princeton University Press.

Freudenthal, Gad. 1995. *Aristotle's Theory of Material Substance: Heat Pneuma, Form and Soul*. New York: Oxford University Press.

Frigg, Roman, and Stephan Hartmann. 2020. "Models in Science." *Stanford Encyclopedia of Philosophy*, revised February 4. https://plato.stanford.edu/entries/models-science/.

Furley, David J. 1957. "Empedocles and the Clepsydra." *Journal of Hellenic Studies* 77, no. 1: 31–34.

———. 1985. "The Rainfall Example in *Physics* II.8." In *Aristotle on Nature and Living Things: Philosophical and Historical Studies Presented to David M. Balme on his Seventieth Birthday*, edited by Allan Gotthelf, 177–82. Pittsburgh, PA: Mathesis. Reprinted in 1989 in *Cosmic Problems: Essays on Greek and Roman Philosophy of Nature*, edited by David J. Furley, 115–20. Cambridge: Cambridge University Press.

Furley, David J., and James S. Wilkie. 1984. *Galen on Respiration and the Arteries*. Princeton, NJ: Princeton University Press.

Gagarin, Michael. 1996. "The Torture of Slaves in Athenian Law." *Classical Philology* 91, no. 1: 1–18.

Galison, Peter. 1997. *Image and Logic: A Material Culture in Microphysics*. Chicago: University of Chicago Press.

Gallego Pérez, M. T. 1996. "*Physis* dans la Collection hippocratique." In *Hippokratische Medizin und antike Philosophie*, vol. 2, edited by Renate Wittern and Pierre Pellegrin, 419–36. Hildesheim: Olms.

Garnsey, Peter. 1999. *Food and Society in Classical Antiquity*. Cambridge: Cambridge University Press.

Garofalo, Ivan, ed. 1988. *Erasistrati fragmenta: Collegit et digessit*. Pisa: Giardini.

———, ed. 1997. *Anonymi Medici, De Morbis Acutis et Chroniis*. Translated by Brian Fuchs. Leiden: Brill.

Gelber, Jessica. 2021. "Teleological Perspectives in Aristotle's Biology." In *The Cambridge Companion to Aristotle's Biology*, edited by Sophia M. Connell, 97–113. Cambridge: Cambridge University Press.

Geller, Markham J. 2010. *Ancient Babylonian Medicine: Theory and Practice*. Oxford: Wiley-Blackwell.

Gentner, Dedre. 1982. "Are Scientific Analogies Metaphors?" In *Metaphor: Problems and Perspectives*, edited by David S. Miall, 106–32. Brighton: Harvester Press.

Gentner, Dedre, and Catherine Clement. 1988. "Evidence for Relational Selectivity in the Interpretation of Analogy and Metaphor." In *The Psychology of Learning and Motivation: Advances in Research and Theory*, edited by Gordon H. Bower, 307–58. New York: Academic Press.

Gentner, Dedre, and Michael Jeziorski. 1989. "Historical Shifts of the Use of Analogy in Science." In *The Psychology of Science*, edited by Barry Gholson, William R. Shadish Jr., Robert A. Niemeyer, and Arthur C. Houts, 296–325. Cambridge: Cambridge University Press.

Gerolemou, Maria. 2023. *Technical Automation in Classical Antiquity*. London: Bloomsbury.

Gerolemou, Maria, and George Kazantzidis. 2023. *Iatromechanics: Body and Machine*. Cambridge: Cambridge University Press.

Gerolemou, M., Isabel Ruffell, and T. Bur. Forthcoming. *Technical Animation in Classical Antiquity*. Oxford: Oxford University Press.

Gersh, Carolyn. 2012. "Naming the Body: A Translation with Commentary and Interpretive Essays of Three Anatomical Works Attributed to Rufus of Ephesus." PhD diss., University of Michigan. https://deepblue.lib.umich.edu/handle/2027.42/95946.

Gill, Christopher. 2007. "Galen and the Stoics: Mortal Enemies or Blood Brothers?" *Phronesis* 52, no. 1: 88–120.

Ginouvès, René. 1962. *Balaneutike: recherches sur le bain dans l'Antiquite grecque*. Paris: E. de Boccard.

Gleason, Maud W. 2009. "Shock and Awe: The Performance Dimension of Galen's Anatomy and Demonstrations." In *Galen and the World of Knowledge*, edited by Christopher Gill, Tim Whitmarsh, and John Wilkins, 85–114. Cambridge: Cambridge University Press.

Goldin, Owen. 1996. *Explaining an Eclipse: Aristotle's* Posterior Analytics *2.1–10*. Ann Arbor: University of Michigan Press.

Goltz, Dietlinde. 1974. *Studien zur altorientalischen und grieschischen Heilkunde: Therapie–Arzneibereitung–Rezeptstruktur*. Wiesbaden: Franz Steiner.

Gotthelf, Allan. 1976. "Aristotle's Conception of Final Causality." *Review of Metaphysics* 30, no. 2: 226–54. Revised and reprinted in 1987 in *Philosoph-*

ical Issues in Aristotle's Biology, edited by Allan Gotthelf and James G. Lennox, 204–42. Cambridge: Cambridge University Press.

———, ed. 1985. *Aristotle on Nature and Living Things: Philosophical and Historical Studies Presented to David M. Balme on His Seventieth Birthday*. Pittsburgh, PA: Mathesis.

———. 2012. *Teleology, First Principles, and Scientific Method in Aristotle's Biology*. Oxford: Oxford University Press.

Gotthelf, Allan, and James G. Lennox, eds. 1987. *Philosophical Issues in Aristotle's Biology*. Cambridge: Cambridge University Press.

Gottschalk, Hans B. 1964. "Strato of Lampsacus: Some Texts." *Proceedings of Leeds Philosophical and Literary Society* 11: 95–182.

———. 1980. *Heraclides of Pontus*. Oxford: Clarendon Press.

Gourevitch, Danielle. 1984. *Le mal d'être femme: La femme et la médicine dans la Rome antique*. Paris: Les Belles Lettres.

———. 1991. "Le pratique méthodique." In *Les écoles médicales à Rome*, edited by Philippe Mudry and Jackie Pigeaud, 57–82. Geneva: Droz.

Grant, Mark. 2018. "Dietetics: Regimen for Life and Health." In *The Oxford Handbook of Science and Medicine in the Classical World*, edited by Paul T. Keyser and John Scarborough, 543–54. Oxford: Oxford University Press.

Gregoric, Pavel. 2020. "The Origin and the Instrument of Animal Motion." In *Aristotle's De Motu Animalium: Symposium Aristotelicum*, edited by Christof Rapp and Oliver Primavesi, 416–44. Oxford: Oxford University Press.

Gregoric, Pavel, and Martin Kuhar. 2014. "Aristotle's Physiology of Animal Motion: On *Neura* and Muscles." *Apeiron* 41, no. 1: 95–115.

Grensemann, Hermann. 1975. *Knidische Medizin*. Part 1, *Die Testimonien zur ältesten knidischen Lehre und Analysen knidischer Schriften im Corpus Hippocraticum*. Berlin: De Gruyter.

Grillo, Francesco. 2019. "Hero of Alexandria's Automata: A Critical Edition and Translation, Including a Commentary on Book One." PhD diss., University of Glasgow. https://theses.gla.ac.uk/76774/.

Grmek, Mirko. 1997. *Le chaudron de Médée: L'expérimentation sur le vivant dans l'Antiquité*. Paris: Institut Synthélabo.

Grmek, Mirko, and Danielle Gourevitch. 1994. "Aux sources de la doctrine médicale de Galien: L'enseignement de Marinus, Quintus et Numisianus." *Aufstieg und Niedergang der römischen Welt II* 37, no. 2: 1491–528.

Gundert, Beate. 1992. "Parts and their Roles in Hippocratic Medicine." *Isis* 83, no. 3: 453–65.

Hacking, Ian. 1983. *Representing and Intervening: Introductory Topics in the Philosophy of Natural Science*. New York: Cambridge University Press.

———. 1998. "Canguilhem amid the Cyborgs." *Economy and Society* 27, no. 2–3: 202–16.

Hankinson, R. J. 1988. "Galen Explains the Elephant." In "Philosophy and Biology," edited by Mohan Matthen and Bernard Linsky. *Canadian Journal of Philosophy*, supplement 14: 135–57.

———. 1989. "Galen and the Best of All Possible Worlds." *Classical Quarterly* 39, no. 1: 206–27.

———. 1991. *Galen. On the Therapeutic Method. Books I and II*. New York: Oxford University Press.

———. 1994. "Galen's Anatomical Procedures." In *Aufstieg und Niedergang der römischen Welt II* 37, no. 2: 1834–55.

———. 2008a. "Epistemology." In *The Cambridge Companion to Galen*, edited by R. J. Hankinson, 157–83. Cambridge: Cambridge University Press.

———. 2008b. "The Man and his Work." In *The Cambridge Companion to Galen*, edited by R. J. Hankinson, 1–33. Cambridge: Cambridge University Press.

———. 2014. "Galen and the Ontology of Powers." *British Journal for the History of Philosophy* 22, no. 5: 951–73.

Hankinson, R. J., and Matyáš Havrda, eds. 2020. *Galen's Epistemology: Experience, Reason, and Method in Ancient Medicine*. New York: Cambridge University Press.

Hanson, Ann. 1991. "Continuity and Change: Three Case Studies in Hippocratic Gynecological Therapy and Theory." In *Women's History and Ancient History*, edited by Sarah B. Pomeroy, 73–110. Chapel Hill: University of North Carolina Press.

———. 1992. "Conception, Gestation, and the Origin of Female Nature in the *Corpus Hippocraticum*." *Helios* 19, no. 1–2: 31–71.

———. 1998. "Talking Recipes in the Gynaecological Texts of the *Hippocratic Corpus*." In *Parchments of Gender: Deciphering the Bodies of Antiquity*, edited by Maria Wyke, 71–94. Oxford: Oxford University Press.

———. 1999. "A Hair on Her Liver Has Been Lacerated . . ." In *Aspetti della terapia nel Corpus Hippocraticum*, edited by Ivan Garofalo, Alessandro Lami, Daniela Manetti, and Amneris Roselli, 235–54. Florence: Leo S. Olschki.

Haraway, Donna. 1991. "A Cyborg Manifesto." In Donna Haraway, *Simians, Cyborgs, and Women: The Reinvention of Nature*, 149–82. New York: Routledge.

———. 2016. *Staying with the Trouble: Making Kin in the Chthulucene*. Durham, NC: Duke University Press.

Harig, Georg. 1983. "Die philosophischen Grundlagen des medizinischen Systems des Asklepiades von Bithynien." *Philologus* 127, no. 1–2: 43–60.

Harris, Charles R. S. 1973. *The Heart and the Vascular System in Ancient Greek Medicine*. Oxford: Clarendon Press.

Harte, Verity. 2002. *Plato on Parts and Wholes: The Metaphysics of Structure*. Oxford: Oxford University Press.

Hershbell, Jackson P. 1974. "Empedoclean Influences on the *Timaeus*." *Phoenix* 28, no. 2: 145–66.

Hesse, Mary. 1966. *Models and Analogies in Science*. Notre Dame, IN: University of Notre Dame Press.

Hillert, Andreas. 1990. *Antike Ärztedarstellungen*. Marburger Schriften zur Medizingeschichte 25. Frankfurt am Main: P. Lang.

Hodge, A. Trevor. (1992) 2002. *Roman Aqueducts and Water Supply*. Reprint ed. London: Duckworth.

———. 2000. "Aqueducts." In *Handbook of Ancient Water Technology*, edited by Örjan Wikander, 39–66. Leiden: Brill.

Hofstadter, Douglas, and Emmanuel Sander. 2013. *Surfaces and Essences: Analogy as the Fuel and Fire of Thinking*. New York: Basic Books.

Holmes, Brooke. 2010a. "Medical Knowledge and Technology." In *A Cultural History of the Human Body*, vol. 1, *In Antiquity*, edited by Daniel H. Garrison, 83–105. Oxford: Berg.

———. 2010b. *The Symptom and the Subject: The Emergence of the Physical Body in Ancient Greece*. Princeton, NJ: Princeton University Press.

———. 2012. *Gender: Antiquity and Its Legacy*. London: I. B. Tauris.

———. 2014. "Proto-Sympathy in the Hippocratic Corpus." In *Hippocrate et les hippocratismes: Médecine, religion, société; Actes du XIVe Colloque International Hippocratique*, edited by Jacques Jouanna and Michel Zink, 123–38. Paris: Académie des Inscriptions et Belles Lettres.

———. 2017. "Pure Life: The Limits of the Vegetal Analogy in the Hippocratics and Galen." In *The Comparable Body: Imagination and Analogy in Ancient Anatomy and Physiology*, edited by John Wee, 358–86. Leiden: Brill.

———. 2018. "Body." In *The Cambridge Companion to Hippocrates*, edited by Peter E. Pormann, 63–88. Cambridge: Cambridge University Press.

Humphrey, John W. 2006. *Ancient Technology*. Westport, CT: Greenwood Press.

Huxley, Thomas H. 1879. "On Certain Errors Respecting the Structure of the Heart Attributed to Aristotle." *Nature* 21, no. 523: 1–5.

Ioannidi, Hélène. 1981. "Les notions de partie du corps et d'organe." In *Formes de pensée dans la collection hippocratique: Actes du IVe colloque international hippocratique; Lausanne, 21–26 septembre 1981*, edited by François Lasserre and Philippe Mudry, 327–30. Geneva: Droz.

Jackson, Ralph. 1988. *Doctors and Diseases in the Roman Empire*. London: British Museum Publications.

———. 2010. "Cutting for Stone: Roman Lithotomy Instruments in the Museo Nazionale Romano." *Medicina nei Secoli* 22, no. 1–3: 393–418.

Jaeger, Werner. 1938. *Diokles von Karystos: Die griechische Medizin und die Schule des Aristoteles*. Berlin: De Gruyter.

———. 1940. "Diocles of Carystus: A New Pupil of Aristotle." *Philosophical Review* 49, no. 4: 393–414.

Jansen, Gemma C. M. 2000. "Urban Water Transport and Distribution." In *Handbook of Ancient Water Technology*, edited by Örjan Wikander, 103–26. Leiden: Brill.

Johansen, Thomas Kjeller. 2004. *Plato's Natural Philosophy: A Study of the Timaeus-Critias*. Cambridge: Cambridge University Press.

Johnson, Monte R. 2005. *Aristotle on Teleology*. Oxford: Oxford University Press.

———. 2017. "Aristotelian Mechanistic Explanation." In *Teleology in the Ancient World: Philosophical and Medical Approaches*, edited by Julius Rocca, 125–50. Cambridge: Cambridge University Press.

Joly, Robert. 1966. *Le niveau de la science hippocratique*. Paris: Les Belles Lettres.

Jouanna, Jacques. 1974. *Hippocrate: Pour une archéologie de l'École de Cnide*. Paris: Les Belles Lettres.

———. 1999. *Hippocrates*. Translated by M. B. DeBevoise. Baltimore: Johns Hopkins University Press. Originally published in 1992 as *Hippocrate*. Paris: Arthème Fayard.

———. 2003. "La notion de la nature chez Galien." In *Galen et la philosophie*, Entretiens sur l'Antiquité classique 49, edited by Jonathan Barnes and Jacques Jouanna, 229–68. Geneva: Fondation Hardt.

———. 2012. "Rhetoric and Medicine in the Hippocratic Corpus: A Contribution to the History of Rhetoric in the Fifth Century." In *Greek Medicine from Hippocrates to Galen: Selected Papers*, edited by Philip van der Eijk, 39–53. Leiden: Brill.

———. 2018. "Textual History." In *The Cambridge Companion to Hippocrates*, edited by Peter E. Pormann, 38–62. Cambridge: Cambridge University Press.

Kefalidou, E. 2003. "'What Is This Then? Not a Clepsydra?' (Aristophanes, *Vespai*, 858): A Group of Peculiar Archaic Vases." *Egnatia* 7: 61–107.

Kieffer, John Spangler, ed. (1964) 2020. *Galen's Institutio Logica: English Translation, Introduction, and Commentary*. Baltimore, MD: Johns Hopkins University Press.

King, Helen. 1993. "Once upon a Text: The Hippocratic Origins of Hysteria." In *Hysteria Beyond Freud*, edited by Sander L. Gilman, Helen King, Roy Porter, G. S. Rousseau, and Elaine Showalter, 3–90. Berkeley: University of California Press.

———. 1995a. "Medical Texts as a Source for Women's History." In *The Greek World*, edited by Anton Powell, 199–218. London: Routledge.

———. 1995b. "Self-Help, Self-Knowledge: In Search of the Patient in Hippocratic Gynaecology." In *Women in Antiquity: New Assessments*, edited by Richard Hawley and Barbara Levick, 135–48. London: Routledge.

———. 1998. *Hippocrates' Woman: Reading the Female Body in Ancient Greece*. London: Routledge.

———. 2020. *Hippocrates Now: The "Father of Medicine" in the Internet Age*. New York: Bloomsbury Academic.

Kleijn, Gerda de. 2001. *The Water Supply of Ancient Rome: City Area, Water, and Population*. Amsterdam: J. C. Gieben.

Kollesch, Jutta. 1973. *Untersuchungen zu den pseudogalenischen Definitiones medicae*. Berlin: Akademie-Verlag,

———. 1981. "Galen und die zweite Sophistik." In *Galen: Problems and Prospects*, edited by Vivian Nutton, 1–11. London: Wellcome Institute for the History of Medicine.

Kovačić, Franjo. 2001. *Der Begriff der Physis bei Galen vor dem Hintergrund seiner Vorgänger*. Stuttgart: Franz Steiner.

Kövecses, Zoltan. 2010. *Metaphor: A Practical Introduction*. New York: Oxford University Press.

Krug, Antje. 2008. *Das Berliner Arztrelief*. Winckelmannsprogramm der Archäologischen Gesellschaft zu Berlin 142. Berlin: De Gruyter.

———. 2012. "Doktorspiele? Der Aryballos Peytel." *Boreas, Münstersche Beiträge zur Archäologie* 35: 11–13.

Kuhn, Thomas. 1979. "Metaphor in Science." In *Metaphor and Thought*, edited by Andrew Ortony, 409–19. Cambridge: Cambridge University Press.

Künzl, Ernst. 1983. *Medizinische Instrumente aus Sepulkrafinden der römischen Kaiserzeit*. Bonn: R. Habelt.

Kuriyama, Shigehisa. 1999. *The Expressiveness of the Body and the Divergence of Greek and Chinese Medicine*. Princeton, NJ: Zone Books.

Lakoff, George, and Mark Johnson. 1980. *Metaphors We Live By*. Chicago: University of Chicago Press.

Lambros, K. P. I. 1985. Περὶ σικύων και σικυάσεως παρὰ τοῖς ἀρχαίοις. Athens: J. Angelopolou.

Lane Fox, Robin. 2020. *The Invention of Medicine: From Homer to Hippocrates*. New York: Basic Books.

Laskaris, Julie. 1999. "Archaic Healing Cults as a Source for Hippocratic

Pharmacology." In *Aspetti della terapia nel Corpus Hippocraticum*, edited by Ivan Garofalo, Alessandro Lami, Daniela Manetti, and Amneris Roselli, 1–12. Florence: Leo S. Olschki.

Latour, Bruno. 2004. "How to Talk About the Body? The Normative Dimension of Science Studies." *Body and Society* 10, no. 2–3: 205–29.

Le Blay, Frédéric. 2005. "Microcosm and Macrocosm: The Dual Direction of Analogy in Hippocratic Thought and the Meteorological Tradition." In *Hippocrates in Context: Papers Read at the XIth International Hippocrates Colloquium (University of Newcastle upon Tyne, 27–31 August 2002)*, edited by Philip van der Eijk, 251–69. Leiden: Brill.

Lee, Keel Yong, Sung-Jin Park, David G. Matthews, Sean L. Kim, Carlos Antonio Marquez, John F. Zimmerman, Herdeline Ann M. Ardoña, Andre G. Kleber, George V. Lauder, and Kevin Kit Parker. 2022. "An Autonomously Swimming Biohybrid Fish Designed with Human Cardiac Biophysics." *Science* 375, no. 6581: 639–47.

Lefèvre, Charles. 1972. *Sur l'evolution d'Aristote en psychologie*. Louvain: Éditions de l'Institut supérieur de Philosophie.

Lehoux, Daryn. 1999. "All Voids Great and Small, Being a Discussion of Place and Void in Strato of Lampsacus's Matter Theory." *Apeiron* 32, no. 1: 1–36.

———. 2012. *What Did the Romans Know? An Inquiry into Science and Worldmaking*. Chicago: University of Chicago Press.

———. 2017. *Creatures Born of Mud and Slime: The Wonder and Complexity of Spontaneous Generation*. Baltimore, MD: Johns Hopkins University Press.

Leith, David. 2008. "The *Diatritus* and Therapy in Graeco-Roman Medicine." *Classical Quarterly* 58, no. 2: 581–600.

———. 2009. "The Qualitative Status of the *Onkoi* in Asclepiades' Theory of Matter." *Oxford Studies in Ancient Philosophy* 36: 283–320.

———. 2012. "Pores and Void in Asclepiades' Physical Theory." *Phronesis* 57, no. 2: 164–91.

———. 2015. "Erasistratus' *Triplokia* of Arteries, Veins and Nerves." *Apeiron* 48, no. 3: 251–62.

———. 2016. "How Popular Were the Medical Sects?" In *Popular Medicine in Graeco-Roman Antiquity: Explorations*, edited by William Harris, 231–50. Leiden: Brill.

Lennox, James G. 1982. "Teleology, Chance, and Aristotle's Theory of Spontaneous Generation." *Journal of the History of Philosophy* 20, no. 3: 219–38.

———. 1985. "Plato's Unnatural Teleology." In *Platonic Investigations*, edited by Dominic J. O'Meara, 195–218. Washington, DC: Catholic University of America Press. Reprinted in 2001 in *Aristotle's Philosophy of Biology: Studies in the Origins of Life Science*, edited by James G. Lennox, 280–302. Cambridge: Cambridge University Press.

———. 2001. *Aristotle's Philosophy of Biology: Studies in the Origins of Life Science*. Cambridge: Cambridge University Press.

———. (2006) 2021. "Aristotle's Biology." *Stanford Encyclopedia of Philosophy*, revised July 16. https://plato.stanford.edu/entries/aristotle-biology/#Bib.

Leroi, A. M. 2014. *The Lagoon: How Aristotle Invented Science*. London: Bloomsbury.

Leunissen, Mariska. 2010. *Explanation and Teleology in Aristotle's Science of Nature*. Cambridge: Cambridge University Press.

———. 2017. "Biology and Teleology in Aristotle's Account of the City." In *Teleology in the Ancient World: Philosophical and Medical Approaches*, edited by Julius Rocca, 107–24. Cambridge: Cambridge University Press.

Lewis, Michael J .T. 2000a. "The Hellenistic Period." In *Handbook of Ancient Water Technology*, edited by Örjan Wikander, 631–48. Leiden: Brill.

———. 2000b. "Theoretical Hydraulics, Automata, and Water Clocks." In *Handbook of Ancient Water Technology*, edited by Örjan Wikander, 343–70. Leiden: Brill.

Lewis, Orly. 2017. *Praxagoras of Cos on Arteries, Pulse and Pneuma*. Leiden: Brill.

Littman, R. 2008. "Hippokratic Corpus, *Heart* (*ca* 350–250 BCE)." In *The Encyclopedia of Ancient Natural Scientists: The Greek Tradition and Its Many Heirs*, edited by Paul Keyser and Georgia L. Irby-Massie, 413. London: Routledge.

Lloyd, Geoffrey E. R. 1963. "Who Is Attacked in *On Ancient Medicine?*" *Phronesis* 8, no. 2: 108–26. Reprinted in 1991 in *Methods and Problems in Greek Science*, edited by Geoffrey E. R. Lloyd, 49–69. Cambridge: Cambridge University Press.

———. 1966. *Polarity and Analogy: Two Types of Argumentation in Early Greek Thought*. Cambridge: Cambridge University Press.

———. 1973. *Greek Science after Aristotle*. London: W. W. Norton.

———. 1975a. "Alcmaeon and the Early History of Dissection." *Sudhoffs Archiv* 59, no. 2: 113–47. Reprinted in 1991 in *Methods and Problems in Greek Science: Selected Papers*, edited by Geoffrey E. R. Lloyd, 164–93. Cambridge: Cambridge University Press.

———. 1975b. "The Hippocratic Question." *Classical Quarterly* 25, no. 2: 171–92.

———. 1975c. "A Note on Erasistratus of Ceos." *Journal of Hellenic Studies* 95: 172–75.

———, ed. 1978. *Hippocratic Writings*. Translated by John Chadwick and W. N. Mann. New York: Penguin Books.

———. 1983. *Science, Folklore and Ideology: Studies in the Life Sciences in Ancient Greece*. Cambridge: Cambridge University Press.

———. 1991. *Methods and Problems in Greek Science: Selected Papers*. Cambridge: Cambridge University Press.

———. 1996. *Aristotelian Explorations*. Cambridge: Cambridge University Press.

———. 1999. *Magic, Reason, and Experience: Studies in the Origins and Development of Greek Science*. Indianapolis, IN: Hackett.

———. 2008. "Galen and His Contemporaries." In *The Cambridge Companion to Galen*, edited by R. J. Hankinson, 34–48. Cambridge: Cambridge University Press.

———. 2015. *Analogical Investigations: Historical and Cross-Cultural Perspectives on Human Reasoning*. Cambridge: Cambridge University Press.

Long, Christopher P. 2006. "Saving *Ta Legomena*: Aristotle and the History of Philosophy." *Review of Metaphysics* 60, no. 2: 247–67.

Longrigg, James. 1963. "Philosophy and Medicine: Some Early Interactions." *Harvard Studies in Classical Philology* 67: 147–75.

———. 1983. "[Hippocrates] *Ancient Medicine* and Its Intellectual Context." In *Formes de pensée dans la collection hippocratique: Actes du IVe colloque international hippocratique; Lausanne, 21–26 septembre 1981*, edited by François Lasserre and Philippe Mudry, 249–56. Geneva: Droz.

———. 1988. "Anatomy in Alexandria in the Third Century B.C." *British Journal for the History of Science* 21, no. 4: 455–88.

———. 1993. *Greek Rational Medicine: Philosophy and Medicine from Alcmaeon to the Alexandrians*. New York: Routledge.

Lonie, Iain M. 1964. "Erasistratus, the Erasistrateans, and Aristotle." *Bulletin of the History of Medicine* 38, no. 5: 426–43.

———. 1965a. "The Cnidian Treatises of the *Corpus Hippocraticum*." *Classical Quarterly* 15, no. 1: 1–30.

———. 1965b. "Medical Theory in Heraclides of Pontus." *Mnemosyne* 18, no. 2: 126–43.

———. 1969. "On the Botanical Excursus in *De Natura Pueri* 22–27." *Hermes* 97, no. 4: 391–411.

———. 1973a. "The Paradoxical Text 'On the Heart': Part I." *Medical History* 17, no. 1: 1–15.

———. 1973b. "The Paradoxical Text 'On the Heart': Part II." *Medical History* 17, no. 2: 136–53.

———. 1977. "A Structural Pattern in Greek Dietetics and the Early History of Greek Medicine." *Medical History* 21, no. 3: 235–60.

———. 1978a. "Cos versus Cnidus and the Historians: Part 1." *History of Science* 16, no. 1: 42–75.

———. 1978b. "Cos versus Cnidus and the Historians: Part 2." *History of Science* 16, no. 2: 77–92.

———. 1981a. "Hippocrates the Iatromechanist." *Medical History* 25, no. 2: 113–50.

———. 1981b. *The Hippocratic Treatises "On Generation," "On the Nature of Child," "Diseases IV": A Commentary*. Berlin: De Gruyter.

Lopez-Salva, Mercedes. 1999. "Terapia Quirúrgica in el Corpus Hippocraticum: Estudio Lexico del Instrumental." In *Aspetti della terapia nel Corpus Hippocraticum*, edited by Ivan Garofalo, Alessandro Lami, Daniela Manetti, and Amneris Roselli, 299–312. Florence: Leo S. Olschki.

Louis, Pierre, ed. and trans. 1973/2015. *Aristote: Marche des animaux—Mouvement des animaux*, Paris: Les Belles Lettres.

Lucore, Sandra K., and Monika Trümper, eds. 2013. *Greek Baths and Bathing Culture: New Discoveries and Approaches*. Leuven: Peeters.

MacKinney, Loren. 1964. "The Concept of Isonomia in Greek Medicine." In *Isonomia: Studien zur Gleichheitsvorstellung im griechischen Denken*, edited by Jürgen Mau and Ernst Günther Schmidt, 79–88. Berlin: De Gruyter.

Maire, Brigitte, ed. 2014. *'Greek' and 'Roman' in Latin Medical Texts: Studies in Cultural Change and Exchange in Ancient Medicine*. Leiden: Brill.

Majno, Guido. 1975. *The Healing Hand: Man and Wound in the Ancient World*. Cambridge, MA: Harvard University Press.

Manetti, Daniela. 2005. "Medici contemporanei a Ippocrate: Problemi di identificazione dei medici di nome Erodico." In *Hippocrates in Context: Papers Read at the XIth International Hippocrates Colloquium (University of Newcastle upon Tyne, 27–31 August 2002)*, edited by Philip van der Eijk, 295–313. Leiden: Brill.

———. 2008a. "Diokles of Karustos (400–300 BCE)." In *The Encyclopedia of Ancient Natural Scientists. The Greek Tradition and Its Many Heirs*, edited by Paul Keyser and Georgia L. Irby-Massie, 255–57. London: Routledge.

———. 2008b. "Eudemos of Alexandria (285–235 BCE)." In *The Encyclopedia of Ancient Natural Scientists. The Greek Tradition and Its Many Heirs*, edited by Paul Keyser and Georgia L. Irby-Massie, 308. London: Routledge.

———, ed. 2011. *Anonymus Londiniensis: De Medicina*. Berlin: Bibliotheca Teubneriana.

Mansfeld, Jaap. 1975. "Alcmaeon: *Physikos* or Physician? With Some Remarks on Calcidius' *On Vision* Compared to Galen, *Plac. Hipp. Plat*. VII." In *Kephalaion: Studies in Greek Philosophy and Its Continuation Offered to Professor C. J. de Vogel*, edited by Jaap Mansfeld and Lambertus M. de Rijk, 26–38. Assen: Van Gorcum.

———. 1990. "Doxography and Dialectic: The *Sitz im Leben* of the 'Placita.'" *Aufstieg und Niedergang der römischen Welt II* 36, no. 4: 3056–229.

———. 1999. "Sources." In *The Cambridge History of Hellenistic Philosophy*, edited by Keimpe A. Algra, Jonathan Barnes, Jaap Mansfeld, and Malcolm Schofield, 3–30. Cambridge: Cambridge University Press.

———. 2013. "The Body Politic. Aëtius on Alcmaeon on Isonomia and Monarchia." In *Politeia in Greek and Roman Philosophy*, edited by Verity Harte and Melissa Lane, 78–95. Cambridge: Cambridge University Press.

———. 2017. "Ancient Philosophy and the Doxographical Tradition." In *Ancient Philosophy. Historical Paths and Explorations*, edited by Lorenzo Perilli and Daniela P. Taormina, 41–64. New York: Routledge.

Maréchal, Sadi. 2020. *Public Baths and Bathing Habits in Late Antiquity: A Study of the Evidence from Italy, North Africa and Palestine A.D. 285–700*. Leiden: Brill.

Marganne, Marie-Hélène. 1998. *La chirurgie dans l'Égypte gréco-romaine d'après les papyrus littéraires grecs*. Leiden: Brill.

Mattern, Susan. 2008. *Galen and the Rhetoric of Healing*. Baltimore, MD: Johns Hopkins University Press.

———. 2013. *The Prince of Medicine: Galen in the Roman Empire*. Oxford: Oxford University Press.

Mayhew, Robert. 2004. *The Female in Aristotle's Biology: Reason or Rationalization*. Chicago: University of Chicago Press.

Mayor, Adrienne. 2018. *Gods and Robots: Myths, Machines, and Ancient Dreams of Technology*. Princeton, NJ: Princeton University Press.

McKirahan, Richard D. 1992. *Principles and Proofs: Aristotle's Theory of Demonstrative Science*. Princeton, NJ: Princeton University Press.

———. 2010. *Philosophy Before Socrates: An Introduction with Texts and Commentary*. 2nd ed. Indianapolis, IN: Hackett.

Mejer, Jørgen. 2006. "Ancient Philosophy and the Doxographical Tradition." In *Blackwell's Companion to Ancient Philosophy*, edited by Mary Louise Gill and Pierre Pellegrin, 20–33. Malden, MA: Blackwell.

Miller, Harold W. 1952. "*Dynamis* and *Physis* in *On Ancient Medicine*." *Transactions and Proceedings of the American Philological Association* 83: 184–97.

———. 1957. "The Flux of the Body in Plato's *Timaeus*." *Transactions and Proceedings of the American Philological Association* 88: 103–13.

———. 1959. "The Concept of *Dynamis* in *De victu*." *Transactions and Proceedings of the American Philological Association* 90: 147–64.

———. 1962. "The Aetiology of Disease in Plato's *Timaeus*." *Transactions and Proceedings of the American Philological Association* 93: 175–87.

Miller, Kassandra. Forthcoming 2023. *Time and Ancient Medicine: How Sundials and Water Clocks Changed Medical Science*. New York: Oxford University Press.

Milne, John Stewart. 1907. *Surgical Instruments in Greek and Roman Times*. Oxford: Clarendon Press. Reprinted in 1970. New York: Augustus M. Kelley.

Mol, Annemarie. 2002. *The Body Multiple: Ontology in Medical Practice.* Durham, NC: Duke University Press.

Morison, Ben. 2008. "Language." In *The Cambridge Companion to Galen*, edited by R.J. Hankinson, 116–56. Cambridge: Cambridge University Press.

Mugler, Charles. 1958. "Alcméon et les cycles physiologiques de Platon." *Review des Études grecques* 71: 42–50.

National Institutes of Health. 2012. "NIH Human Microbiome Project Defines Normal Bacterial Makeup of the Body." National Institutes of Health, US Department of Health and Human Services, Wednesday, June 13. https://www.nih.gov/news-events/news-releases/nih-human-microbiome-project-defines-normal-bacterial-makeup-body.

Nunn, John F. 1996. *Ancient Egyptian Medicine.* Norman: University of Oklahoma Press.

Nussbaum, Martha. (1976) 1985. *Aristotle's De Motu Animalium: Text with Translation, Commentary, and Interpretive Essays.* Princeton, NJ: Princeton University Press

———. 1987. "Saving Aristotle's Appearances." In *Language and Logos: Studies in Ancient Greek Philosophy*, edited by Malcolm Schofield and Martha Nussbaum, 267–93. Cambridge: Cambridge University Press.

Nutton, Vivian. 1985. "The Drug Trade in Antiquity." *Journal of the Royal Society of Medicine* 78: 138–45.

———. 1990. "Therapeutic Methods and Methodist Therapeutics in the Roman Empire." In *History of Therapy*, edited by Yosio Kawakita, Shizu Sakai, and Yasuo Otsuka, 1–36. Tokyo: Ishiyaku EuroAmerica.

———. 1993. "Roman Medicine: Tradition, Confrontation, Assimilation." *Aufstieg und Niedergang der römischen Welt II* 37, no. 1: 49–78.

———. 2004 (2nd ed. 2013). *Ancient Medicine.* New York: Routledge.

———. 2020. *Galen: A Thinking Doctor in Imperial Rome.* New York: Routledge.

O'Brien, D. 1970. "The Effect of a Simile: Empedocles' Theories of Seeing and Breathing." *Journal of Hellenic Studies* 90: 140–79.

Oleson, John Peter. 1984. *Greek and Roman Mechanical Water-Lifting Devices: The History of a Technology.* Toronto: University of Toronto Press.

Oppenheimer, J. Robert. 1956. "Analogy in Science." *American Psychologist* 11: 127–35.

Ortloff, C. R. 2009. *Water Engineering in the Ancient World: Archaeological and Climate Perspectives on Societies of Ancient South America, the Middle East, and South-East Asia.* Oxford: Oxford University Press.

Osborne, Robin. 2011. *The History Written on the Classical Greek Body.* Cambridge: Cambridge University Press.

Oser-Grote, Carolin. 2004. *Aristoteles und das Corpus Hippocraticum: Die Anatomie und Physiologie des Menschen.* Stuttgart: Franz Steiner.

Owen, Gwilym E. L. 1986. "*Tithenai ta phainomena*." In *Logic, Science and Dialectic: Collected Papers in Greek Philosophy*, edited by Martha Nussbaum, 239–51. London: Duckworth.

Papakonstantinou, Zinon. 2021. *Sport and Identity in Ancient Greece*. New York: Routledge.

Parthasarathy, Raghuveer. 2022. *So Simple a Beginning: How Four Physical Principles Share Our Living World*. Princeton, NJ: Princeton University Press.

Pelavski, Andres. 2014. "Physiology in Plato's *Timaeus*." *Cambridge Classical Journal* 60: 61–74.

Pender, E. E. 2003. "Plato on Metaphors and Models." In *Metaphor, Allegory and the Classical Tradition*, edited by G. R. Boys-Stones, 55–82. New York: Oxford University Press.

Penn, Raymond George. 1994. *Medicine on Ancient Greek and Roman Coins*. London: Batsford.

Pérez, Jesus Acero, Miko Flohr, Barry Hobson, Jens Koehler, Ann Olga Koloski-Ostrow, Silvia Radbauer, and Jeroen van Vaerenbergh. 2011. "Location and Contexts of Toilets." In *Roman Toilets: Their Archaeology and Cultural History*, edited by Gemma C. M. Jansen, Ann Olga Koloski-Ostrow, and Eric Moorman, 113–30. Leuven: Peeters.

Pigeaud, Jackie. 1980. "La physiologie de Lucrèce." *Revue des études latines* 58: 176–200.

———. 1991. "Les fondements du méthodisme." In *Les écoles médicales à Rome*, edited by Philippe Mudry and Jackie Pigeaud, 9–50. Geneva: Droz.

Polito, Roberto. 1999. "On the Life of Asclepiades of Bithynia." *Journal of Hellenic Studies* 119: 48–66.

———. 2006. "Matter, Medicine, and the Mind: Asclepiades vs. Epicurus." *Oxford Studies in Ancient Philosophy* 30: 285–335.

———. 2007. "Frail or Monolithic? A Note on Asclepiades' Corpuscles." *Classical Quarterly* 57, no. 1: 314–17.

Potter, Paul, trans. 1995. *Hippocrates: Vol. VIII: Places in Man. Glands. Fleshes. Prorrhetic 1–2. Physician. Use of Liquids. Ulcers. Haemorrhoids and Fistulas*. Loeb Classical Library 482. Cambridge, MA: Harvard University Press.

Prager, Frank D. 1974. *Pneumatica: The First Treatise on Experimental Physics: Western Version and Eastern Version*. Wiesbaden: Reichert.

Preus, Anthony. 1981. *Aristotle and Michal of Ephesus: On the Movement and Progression of Animals*. Hildesheim: Georg Olms.

———. 2020. "The Techne of Nutrition in Ancient Greek Philosophy." *Archai* 29: e02904.

Prevosti, Marta. 2013. "A Textile Workshop from Roman Times: The *Villa dels Antigons*." *Datatèxtil* 28: 10–18.

Primavesi, Oliver, and Benjamin Morison. 2020. "Aristotelis *De motu animalium*: A New Critical Edition of the Greek Text with English Translation." *In Aristotle's De motu animalium: Symposium Aristotelicum*, edited by Christof Rapp and Oliver Primavesi, 159–202. New York: Oxford University Press.

Rapp, Christof. 2020. "Introduction Part I: The Argument of *De Motu Animalium*." In *Aristotle's De motu animalium: Symposium Aristotelicum*, edited by Christof Rapp and Oliver Primavesi, 1–66. New York: Oxford University Press.

Rapp, Christof, and Oliver Primavesi, eds. 2020. *In Aristotle's De motu animalium: Symposium Aristotelicum*, New York: Oxford University Press.

Rawson, Elizabeth. 1982. "On the Life and Death of Asclepiades of Bithynia." *Classical Quarterly* 32, no. 2: 358–70.

Reiche, Harald A. T. 1965. "Aristotle on Breathing in the *Timaeus*." *American Journal of Philology* 86, no. 4: 404–8.

Rheinberger, Hans-Jörg. 1997. *Towards a History of Epistemic Things: Synthesizing Proteins in the Test Tube*. Stanford, CA: Stanford University Press.

Ricoeur, Paul. 1978. *The Rule of Metaphor: Multi-Disciplinary Studies of the Creation of Meaning in Language*. Toronto: University of Toronto Press.

Rihll, Tracey. 2007. *The Catapult: A History*. Yardley, PA: Westholme Publishing.

Rivaud, Albert, ed. 1925. *Platon, Oeuvres Completes, Tome X: Timée—Critias*. Paris: Les Belles Lettres.

Rocca, Julius. 2003. *Galen on the Brain: Anatomical Knowledge and Physiological Speculation in the Second Century AD*. Leiden: Brill.

———. 2008. "Anatomy." In *The Cambridge Companion to Galen*, edited by R. J. Hankinson, 242–62. Cambridge: Cambridge University Press.

———, ed. 2017. *Teleology in the Ancient World: Philosophical and Medical Approaches*. Cambridge: Cambridge University Press.

Rochberg, Francesca. 2016. *Before Nature: Cuneiform Knowledge and the History of Science*. Chicago: University of Chicago Press.

Roosth, Sophia. 2017. *Synthetic: How Life Got Made*. Chicago: University of Chicago Press.

Rubinstein, G. L. 1985. "The Riddle of the Methodist Method: Understanding a Roman Medical Sect." PhD diss., University of Cambridge.

Rumor, Maddalena. 2016. "Babylonian Pharmacology in Graeco-Roman *Dreckapotheke*. With an Edition of Uruanna III, 1–143 (138)." PhD diss., Freie Universität Berlin.

Rüsche, Franz. 1930. *Blut, Leben und Seele: Ihr Verhaeltnis nach Auffassung der griechischen und hellenistischen Antike, der Bibel und der alten alexandrini-*

schen Theologen; Eine Vorarbeit zur Religionsgeschichte des Opfers. Paderborn: F. Schöningh.

Russo, Lucio. 2004. *The Forgotten Revolution: How Science Was Born in 300 BC and Why It Had to Be Reborn*. Translated by Silvio Levy. Berlin: Springer.

Salas, Luis A. 2020a. *Cutting Words: Polemical Dimensions of Galen's Anatomical Experiments*. Leiden: Brill.

———. 2020b. "Galen on the Definition of Disease." *American Journal of Philology* 141, no. 4: 603–34.

Sarton, George. 1959. *Hellenistic Science and Culture in the Last Three Centuries B.C.* New York: Dover.

Savage-Smith, Emilie, Simon Swain, and Geert J. van Gelder, eds. 2020. *Ibn Abī Uṣaybiʿah: A Literary History of Medicine*. Leiden: Brill.

Scarborough, John. 1968. "Roman Medicine and the Legions: A Reconsideration." *Medical History* 12, no. 3: 254–61.

———. 1969. *Roman Medicine*. London: Thames and Hudson.

———. 1983. "Theoretical Assumptions in Hippocratic Pharmacology." In *Formes de pensée dans la collection hippocratique: Actes du IVe colloque international hippocratique; Lausanne, 21–26 septembre 1981*, edited by François Lasserre and Philippe Mudry, 307–25. Geneva: Droz.

———. 1992. "The Pharmacy of Methodist Medicine: The Evidence of Soranus' Gynecology." In *Les écoles médicales à Rome*, edited by Philippe Mudry and Jackie Pigeaud, 204–16. Geneva: Droz.

———. 2008. "Rufus of Ephesos." In *The Encyclopedia of Ancient Natural Scientists*, edited by Paul T. Keyser and Georgia L. Irby-Massie, 720–21. New York: Routledge.

———. 2010. *Pharmacy and Drug Lore in Antiquity*. Farnham, UK: Ashgate.

Schatzberg, Eric. 2006. "*Technik* Comes to America: Changing Meanings of *Technology* before 1930." *Technology and Culture* 47, no. 3: 486–512.

———. 2018. *Technology: Critical History of a Concept*. Chicago: University of Chicago Press.

Schiefsky, Mark J., trans. 2005. *Hippocrates: On Ancient Medicine: Translated with Introduction and Commentary*. Leiden: Brill.

———. 2007. "Galen's Teleology and Functional Explanation." *Oxford Studies in Ancient Philosophy* 33: 369–400.

Schlange-Schöningen, Heinrich. 2003. *Die römische Gesellschaft bei Galen*. Berlin: De Gruyter.

Schomberg, Anette. 2008. "Ancient Water Technology: Between Hellenistic Innovation and Arabic Tradition." *Syria* 85: 119–28.

Schöne H. I. 1920. "Hippokrates Peri Pharmakon." *Rheinisches Museum* 73: 434–48.

Schubert, C. 1996. "Menschenbild und Normwandel in der klassischen Zeit." In *Médicine et morale dans l'Antiquité*, edited by Hellmut Flashar and Jacques Jouanna, 121–55. Geneva: Fondation Hardt.

Scolnicov, Samuel. 2017. "Atemporal Teleology in Plato." In *Teleology in the Ancient World: Philosophical and Medical Approaches*, edited by Julius Rocca, 45–57. Cambridge: Cambridge University Press.

Scurlock, Joann. 2014. *Sourcebook for Ancient Mesopotamian Medicine*. Atlanta: Society of Biblical Literature.

Sedley, David. 1982. "On Signs." In *Science and Speculation: Studies in Hellenistic Theory and Practice*, edited by Jonathan Barnes, Jacques Brunschwig, Myles Burnyeat, and Malcolm Schofield, 239–72. Cambridge: Cambridge University Press.

———. 1991. "Is Aristotle's Teleology Anthropocentric?" *Phronesis* 36, no. 2: 179–96.

———. 2007. *Creationism and Its Critics in Antiquity*. Berkeley: University of California Press.

———. 2010. "Teleology, Aristotelian and Platonic." In *Being, Nature, and Life in Aristotle: Essays in Honor of Allan Gotthelf*, edited by James G. Lennox and Robert Bolton, 5–29. Cambridge: Cambridge University Press.

———. 2017. "Socrates, Darwin, and Teleology." In *Teleology in the Ancient World: Philosophical and Medical Approaches*, edited by Julius Rocca, 25–42. Cambridge: Cambridge University Press.

Sharples, Robert W. 2017. "The Purpose of the Natural World: Aristotle's Followers and Interpreters." In *Teleology in the Ancient World: Philosophical and Medical Approaches*, edited by Julius Rocca, 151–68. Cambridge: Cambridge University Press.

Shaw, James R. 1972. "Models for Cardiac Structure and Function in Aristotle." *Journal of the History of Biology* 5, no. 2: 355–88.

Singer, P. N. 2013. "Galen and the Philosophers: Philosophical Engagement, Shadowy Contemporaries, Aristotelian Transformations." *Bulletin of the Institute of Classical Studies*, supplement 114: 7–38.

Smith, Robin. 1995. "Logic." In *The Cambridge Companion to Aristotle*, edited by Jonathan Barnes, 27–65. Cambridge: Cambridge University Press.

Smith, Wesley. 1980. "The Development of Classical Dietetic Theory." In *Hippocratica: Actes du colloque hippocratique de Paris (4–9 septembre 1978)*, edited by Mirko D. Grmek, 439–48. Paris: Éditions du CNRS.

———. 1989. "Notes on Ancient Medical Historiography." *Bulletin of the History of Medicine* 63, no. 1: 73–109.

———, ed. and trans. 1990. *Hippocrates: Pseudepigraphic Writings; Letters—Embassy—Speech from the Altar—Decree*. Leiden: Brill.

———. 1992. "Regimen, χρῆσις, and the History of Dietetics." In *Tratados*

Hipocráticos, edited by Jose Antonio López Férez, 263–71. Madrid: Universidad Nacional de Educatión a Distancia.

Solmsen, Friedrich. (1956) 1968. "On Plato's Account of Respiration." In *Kleine Schriften*, vol. 1, 583–87. Hildesheim: G. Olms. Originally published in *Studi italiani di filologia classica* 27–28: 544–48.

———. 1961. "Greek Philosophy and the Discovery of the Nerves." *Museum Helveticum* 18: 150–97. Reprinted in 1968 in *Kleine Schriften*, vol. 1, 536–82. Hildesheim: Georg Olms.

Stalley, R. F. 1996. "Punishment and the Physiology of Plato's *Timaeus*." *Classical Quarterly* 46, no. 2: 357–70.

Stannard, Jerry. 1961. "Hippocratic Pharmacology." *Bulletin of the History of Medicine* 35, no. 6: 497–518.

Steckerl, Fritz, ed. 1958. *The Fragments of Praxagoras of Cos and His School*. Leiden: Brill.

Stein, Richard. 2004. "Roman Wooden Force Pumps: A Case Study in Innovation." *Journal of Roman Archaeology* 17: 221–50.

Stewart, Keith A. 2018. *Galen's Theory of Black Bile: Hippocratic Tradition, Manipulation, Innovation*. Studies in Ancient Medicine 51. Leiden: Brill.

Strouhal, Eugen, Bretislav Vachala, and Hana Vymazalová. 2014. *The Medicine of the Ancient Egyptians*. Vol. 1, *Surgery, Gynecology, Obstetrics, and Pediatrics*. Cairo: American University in Cairo Press.

Switalski, Bronislaus W. 1902. *Des Chalcidius Kommentar zu Plato's "Timaeus": Eine historisch-kritische Untersuchung*. Münster: Aschendorff.

Tabanelli, Mario. 1958. *Chirurgia nell'antica Roma*. Turin: Minerva Medica.

Tamburrino, Aldo. 2010. "Water Technology in Ancient Mesopotamia." In *Ancient Water Technologies*, edited by Larry W. Mays, 29–51. New York: Springer.

Tarrant, Harold. 2017. "Teleology and Names in the Platonic and Anaxagorean Traditions." In *Teleology in the Ancient World: Philosophical and Medical Approaches*, edited by Julius Rocca, 58–75. Cambridge: Cambridge University Press.

Taylor, Alfred E. 1928. *A Commentary on Plato's "Timaeus."* Oxford: Clarendon Press.

Tecusan, Manuela. 2004. *The Fragments of the Methodists. Vol. 1: Text and Translation*. Leiden: Brill.

Thivel, Antoine. 1999. "Quale scoperta ha reso celebre Ippocrate." In *Aspetti della terapia nel Corpus Hippocraticum*, edited by Ivan Garofalo, Alessandro Lami, Daniela Manetti, and Amneris Roselli, 149–61. Florence: Leo S. Olschki.

———. 2005. "Air, Pneuma and Breathing from Homer to Hippocrates." In *Hippocrates in Context: Papers Read at the XIth International Hippocrates*

Colloquium (University of Newcastle upon Tyne, 27–31 August 2002), edited by Philip van der Eijk, 239–49. Leiden: Brill.

Thomson, Homer A., and R. E. Wycherley. 1972. *The Agora of Athens: The History, Shape and Uses of an Ancient City Center*. Princeton, NJ: American School of Classical Studies at Athens.

Tieleman, Teun. 2008. "Methodology." In *The Cambridge Companion to Galen*, edited by R. J. Hankinson, 49–65. Cambridge: Cambridge University Press.

Tölle-Kastenbein, R. 1991. "Entlüftung antiker Wasserleitungsrohre." *Archäologischer Anzeiger* 1: 25–30.

———. 1994. *Das archaische Wassleitungsnetz für Athen und seine späteteren Bauphasen*. Mainz: von Zabern.

———. 1996. "Das archaische Wasserleitungsnetz für Athen." In *Cura Aquarum in Campania*, Babesch Supplements 4, edited by N. de Haan and Gemma C. M. Jansen, 129–36. Leiden: Babesch.

Totelin, Laurence. 2009. *Hippocratic Recipes: Oral and Written Transmission of Pharmacological Knowledge in Fifth- and Fourth-Century Greece*. Leiden: Brill.

———. 2015. "When Foods Become Remedies in Ancient Greece: The Curious Case of Garlic and Other Substances." *Journal of Ethnopharmacology* 167: 30–37.

———. 2018. "Therapeutics." In *The Cambridge Companion to Hippocrates*, edited by Peter E. Pormann, 200–16. Cambridge: Cambridge University Press.

Touwaide, Alain. 1996. "The Aristotelian School and the Birth of Theoretical Pharmacology in Ancient Greece." In *The Pharmacy: Windows on History*, edited by Regine Pötzsch, 11–21. Basel: Roche.

Tracy, Theodore J. 1969. *Physiological Theory and the Doctrine of the Mean in Plato and Aristotle*. The Hague: Mouton.

Triller, Daniel Wilhelm. 1766. *Opuscula Medica*. Vol. 2. 2nd ed. Leipzig. Originally published 1728. Leiden.

Trümper, Monika. 2014. "Baths and Bathing, Greek." In *Encyclopedia of Global Archaeology*, edited by Claire Smith, 784–99. New York: Springer.

Tybjerg, Karin. 2008. "Ktēsibios of Alexandria (290–250 BCE)." In *The Encyclopedia of Ancient Natural Scientists. The Greek Tradition and Its Many Heirs*, edited by Paul Keyser and Georgia L. Irby-Massie, 496. London: Routledge.

Vallance, J. T. 1990. *The Lost Theory of Asclepiades of Bithynia*. Oxford: Clarendon Press.

Valleriani, Matteo. 2023. "The Mechanics of the Heart in Antiquity." In *Iatro-*

mechanics: Body and Machine, edited by George Kazantzidis and Maria Gerolemou, 245–61. Cambridge: Cambridge University Press.

Van der Eijk, Philip. 2001. *Diocles of Carystus*. Vol. 2, *Commentary*. Leiden: Brill.

———. 2004a. "Divination, Prognosis, Prophylaxis: The Hippocratic Work 'On Dreams' (*De victu* 4) and Its Near Eastern Background." In *Magic and Rationality in Ancient Near Eastern and Graeco-Roman Medicine*, edited by Herman F. J. Horstmanshoff and Marten Stol, 187–218. Leiden: Brill.

———. 2004b. "Introduction." In *Magic and Rationality in Ancient Near Eastern and Graeco-Roman Medicine*, edited by Herman F. J. Horstmanshoff and Marten Stol, 1–10. Leiden: Brill.

———. 2014. "Galen on the Nature of Human Beings." In "Philosophical Themes in Galen," edited by Peter Adamson, Rotraud Hansberger, and James Wilberding. *Bulletin of the Institute of Classical Studies*, supplement 114: 89–134.

———. 2016. "On 'Hippocratic' and 'Non-Hippocratic' Medical Writings." In *Ancient Concepts of the Hippocratic*, edited by Lesley Dean-Jones and Ralph Rosen, 17–47. Leiden: Brill.

Vegetti, Mario. 1995. "L'épistémologie d' Érasistrate et la technologie hellénistique." In *Ancient Medicine in Its Socio-Cultural Context*, vol. 2, edited by H. F. J. Horstmanshoff, Philip J. van der Eijk, and P. H. Schrijvers, 461–71. Amsterdam: Brill.

———. 1998. "Empedocle: 'medico e sofista' (*Antica Medicina* 20)." *Elenchos* 19: 347–59.

Vlastos, Gregory. 1975. "Plato's Universe." *Philosophy* 51: 483–85.

Volk, Katharina. 2010. "Literary Theft and Roman Water Rights in Manilius' Second Proem." *Materiali e discussioni per l'analisi dei testi classici* 65: 187–97.

Von Staden, Heinrich. 1975. "Experiment and Experience in Hellenistic Medicine." *Bulletin of the Institute of Classical Studies* 22: 178–99.

———. 1989. *Herophilus: The Art of Medicine in Early Alexandria*. Cambridge: Cambridge University Press.

———. 1991. "Matière et signification: Rituel, sexe et pharmacologie dans le Corpus hippocratique." *L'Antiquité classique* 60: 42–61.

———. 1992a. "The Discovery of the Body: Human Dissection and Its Cultural Contexts in Ancient Greece." *Yale Journal of Biology and Medicine* 65, no. 3: 223–41.

———. 1992b. "Jaeger's 'Skandalon der historischen Vernunft': Diocles, Aristotle, and Theophrastus." In *Werner Jaeger Reconsidered*, edited by William M. Calder III, 227–65. Atlanta: Scholars Press.

———. 1992c. "Review: *The Lost Theory of Asclepiades of Bithynia* by J. T. Vallance." *Isis* 83, no. 4: 642–43.

———. 1992d. "Women and Dirt." *Helios* 19, no. 1–2: 7–30.

———. 1995. "Teleology and Mechanism: Aristotelian Biology and Early Hellenistic Medicine." In *Aristotelische Biologie: Intentionem, Methoden, Ergebniss*, edited by Wolfgang Kullmann and Sabine Föllinger, 183–208. Stuttgart: Franz Steiner.

———. 1996. "Body and Machine: Interactions between Medicine, Mechanics, and Philosophy in Early Alexandria." In *Alexandria and Alexandrianism: Papers Delivered at a Symposium Organized by The J. Paul Getty Museum and The Getty Center for the History of Art and the Humanities and Held at the Museum, April 22–25, 1993*, 85–106. Los Angeles: J. Paul Getty Museum.

———. 1997. "Galen and the Second Sophistic." In "Aristotle and After," edited by Richard Sorabji. *Bulletin of the Institute of Classical Studies*, supplement 68: 33–54.

———. 1998. "Andréas de Caryste et Philon de Byzance: Médecine et mécanique à Alexandrie." In *Sciences exactes et sciences appliquées à Alexandrie*, edited by Gilbert Argoud and Jean-Yves Guillaumin, 147–72. Saint-Étienne: Université de Saint-Étienne.

———. 1999. "Reading the Agonal Body: The Hippocratic Corpus." In *Medicine and the History of the Body: Proceedings of the 20th, 21st and 22nd International Symposium on the Comparative History of Medicine; East and West*, edited by Y. Otsuka, S. Sakai, and S. Kuriyama, 287–94. Tokyo: Ishiyaku EuroAmerica.

———. 2007. "*Physis* and *Technē* in Greek Medicine." In *The Artificial and the Natural: An Evolving Polarity*, edited by Bernadette Bensaude-Vincent and William R. Newman, 21–51. Cambridge, MA: MIT Press.

Wardy, Robert. 1993. "Aristotelian Rainfall or the Lore of Averages." *Phronesis* 38, no. 1: 18–30.

Webster, Colin. 2015. "Heuristic Medicine: The Methodists and Metalepsis." *Isis* 106, no. 3: 657–68.

———. 2023. "Hippocrates, *Diseases* 4 and the Technological Body." In *Iatromechanics*, edited by George Kazantzidis and Maria Gerolemou, 155–78. New York: Cambridge University Press.

———. Forthcoming. "Manufacturing Motion in Aristotle's *De Motu Animalium*." In *Technological Animation in Classical Antiquity*, edited by Tatiana Bur, Maria Gerolemou, and Isabel A. Ruffell. Oxford: Oxford University Press.

Wehrli, Fritz. 1850. *Die Schule des Aristoteles: Texte und Kommentar*. Vol. 5, *Straton von Lampsakos*. Basel: B. Schwabe.

Wellmann, Max. 1901. *Die Fragmente der sikelischen Ärtze Akron, Philistion und des Diokles von Karystos*. Berlin: Weidmannsche Buchhandlung.

———. 1908. "Asklepiades aus Bithynien von einem herrschenden Vorurteil befreit." *Neue Jahrbücher für das klassische Altertum* 21: 684–703.

Wikander, Örjan, ed. 2000. *Handbook of Ancient Water Technology*. Leiden: Brill.

Wilson, Andrew I. 2000. "Drainage and Sanitation." In *Handbook of Ancient Water Technology*, edited by Örjan Wikander, 151–79. Leiden: Brill.

———. 2008. "Hydraulic Engineering and Water Supply." In *The Oxford Handbook of Engineering and Technology in the Classical World*, edited by John Peter Oleson, 285–318. Oxford: Oxford University Press.

Wolfe, Charles T. 2021. "Vitalism in Early Modern Medical and Philosophical Thought." In *Encyclopedia of Early Modern Philosophy and the Sciences*, edited by Dana Jalobeanu and Charles T. Wolfe. Cham: Springer. https://doi.org/10.1007/978-3-319-20791-9_314-1.

Worthen, Thomas. 1970. "Pneumatic Action in the Klepsydra and Empedocles' Account." *Isis* 61, no. 4: 520–30.

Yegül, Fikret. 1992. *Baths and Bathing in Classical Antiquity*. Cambridge, MA: MIT Press.

———. 2010. *Bathing in the Roman World*. Cambridge: Cambridge University Press.

Zhmud, Leonid. 2008. "Alkmaion of Kroton." In *The Encyclopedia of Ancient Natural Scientists. The Greek Tradition and Its Many Heirs*, edited by Paul Keyser and Georgia L. Irby-Massie, 61. London: Routledge.

———. 2012. *Pythagoras and the Early Pythagoreans*. Translated by Kevin Windle and Rosh Ireland. Oxford: Oxford University Press.

Zierlein, Stephan. 2005. "Aristoteles' anatomische Vorstellung vom menschlichen Herzen." *Antike Naturwissenschaft und ihre Rezeption* 15: 43–71.

Zucconi, Laura M. 2019. *Ancient Medicine: From Mesopotamia to Rome*. Grand Rapids, MI: William B. Eerdmans.

Index

Page numbers in italics refer to figures.